Stochastic Methods for Estimation and Problem Solving in Engineering

Seifedine Kadry
Beirut Arab University, Lebanon

A volume in the Advances in Mechatronics and Mechanical Engineering (AMME) Book Series

IGI Global
DISSEMINATOR OF KNOWLEDGE

Published in the United States of America by
IGI Global
Engineering Science Reference (an imprint of IGI Global)
701 E. Chocolate Avenue
Hershey PA, USA 17033
Tel: 717-533-8845
Fax: 717-533-8661
E-mail: cust@igi-global.com
Web site: http://www.igi-global.com

Library of Congress Cataloging-in-Publication Data

Names: Kadry, Seifedine, 1977- editor.
Title: Stochastic methods for estimation and problem-solving in engineering / Seifedine Kadry, editor.
Description: Hershey, PA : Engineering Science Reference, [2018]
Identifiers: LCCN 2017034172| ISBN 9781522550457 (h/c) | ISBN 9781522550464 (eISBN)
Subjects: LCSH: Engineering--Statistical methods. | Estimation theory. | Industrial engineering--Mathematics.
Classification: LCC TA340 .S76 2018 | DDC 620.001/51922--dc23 LC record available at https://lccn.loc.gov/2017034172

This book is published in the IGI Global book series Advances in Mechatronics and Mechanical Engineering (AMME) (ISSN: 2328-8205; eISSN: 2328-823X)

British Cataloguing in Publication Data
A Cataloguing in Publication record for this book is available from the British Library.

For electronic access to this publication, please contact: eresources@igi-global.com.

Advances in Mechatronics and Mechanical Engineering (AMME) Book Series

J. Paulo Davim
University of Aveiro, Portugal

ISSN:2328-8205
EISSN:2328-823X

Mission

With its aid in the creation of smartphones, cars, medical imaging devices, and manufacturing tools, the mechatronics engineering field is in high demand. Mechatronics aims to combine the principles of mechanical, computer, and electrical engineering together to bridge the gap of communication between the different disciplines.

The **Advances in Mechatronics and Mechanical Engineering (AMME) Book Series** provides innovative research and practical developments in the field of mechatronics and mechanical engineering. This series covers a wide variety of application areas in electrical engineering, mechanical engineering, computer and software engineering; essential for academics, practitioners, researchers, and industry leaders.

Coverage

- Mechanisms and machines
- Intelligent Navigation
- Control Systems Modelling and Analysis
- Vibration and acoustics
- Tribology and surface engineering
- Manufacturing Methodologies
- Control Methodologies
- Autonomous Systems
- Nanomaterials and nanomanufacturing
- Computer-Based Manufacturing

IGI Global is currently accepting manuscripts for publication within this series. To submit a proposal for a volume in this series, please contact our Acquisition Editors at Acquisitions@igi-global.com or visit: http://www.igi-global.com/publish/.

The Advances in Mechatronics and Mechanical Engineering (AMME) Book Series (ISSN 2328-8205) is published by IGI Global, 701 E. Chocolate Avenue, Hershey, PA 17033-1240, USA, www.igi-global.com. This series is composed of titles available for purchase individually; each title is edited to be contextually exclusive from any other title within the series. For pricing and ordering information please visit http://www.igi-global.com/book-series/advances-mechatronics-mechanical-engineering/73808. Postmaster: Send all address changes to above address.

Titles in this Series

For a list of additional titles in this series, please visit: www.igi-global.com/book-series

Design and Optimization of Mechanical Engineering Products
K. Kumar (Birla Institute of Technology, India) and J. Paulo Davim (University of Aveiro, Portugal)
Engineering Science Reference • copyright 2018 • 347pp • H/C (ISBN: 9781522534013) • US $235.00 (our price)

Managerial Approaches Toward Queuing Systems and Simulations
Salvador Hernandez-Gonzalez (Tecnológico Nacional de México en Celaya, Mexico) and Manuel Dario Hernandez Ripalda (Tecnológico Nacional de México en Celaya, Mexico)
Engineering Science Reference • copyright 2018 • 311pp • H/C (ISBN: 9781522552642) • US $195.00 (our price)

Socio-Technical Decision Support in Air Navigation Systems Emerging Research and Opportunities
Tetiana Shmelova (National Aviation University, Ukraine) Yuliya Sikirda (National Aviation University, Ukraine) Nina Rizun (Gdansk University of Technology, Poland) Abdel-Badeeh M. Salem (Ain Shams University, Egypt) and Yury N. Kovalyov (National Aviation University, Ukraine)
Engineering Science Reference • copyright 2018 • 305pp • H/C (ISBN: 9781522531081) • US $175.00 (our price)

Soft Computing Techniques and Applications in Mechanical Engineering
Mangey Ram (Graphic Era University, India) and J. Paulo Davim (University of Aveiro, Portugal)
Engineering Science Reference • copyright 2018 • 336pp • H/C (ISBN: 9781522530350) • US $215.00 (our price)

Advanced Numerical Simulations in Mechanical Engineering
Ashwani Kumar (Government of Uttar Pradesh, India) Pravin P. Patil (Graphic Era University, India) and Yogesh Kr. Prajapati (National Institute of Technology Uttarakhand, India)
Engineering Science Reference • copyright 2018 • 242pp • H/C (ISBN: 9781522537229) • US $165.00 (our price)

Numerical and Analytical Solutions for Solving Nonlinear Equations in Heat Transfer
Davood Domiri Ganji (Babol Noshirvani University of Technology, Iran) and Roghayeh Abbasi Talarposhti (University of Mazandaran, Iran)
Engineering Science Reference • copyright 2018 • 275pp • H/C (ISBN: 9781522527138) • US $195.00 (our price)

Nanofluid Technologies and Thermal Convection Techniques
Ramesh Chand (Government Arya Degree College Nurpur (HP), India)
Engineering Science Reference • copyright 2017 • 229pp • H/C (ISBN: 9781683180067) • US $185.00 (our price)

Mathematical Concepts and Applications in Mechanical Engineering and Mechatronics
Mangey Ram (Graphic Era University, India) and J. Paulo Davim (University of Aveiro, Portugal)
Engineering Science Reference • copyright 2017 • 495pp • H/C (ISBN: 9781522516392) • US $215.00 (our price)

IGI Global
DISSEMINATOR OF KNOWLEDGE

701 East Chocolate Avenue, Hershey, PA 17033, USA
Tel: 717-533-8845 x100 • Fax: 717-533-8661
E-Mail: cust@igi-global.com • www.igi-global.com

Editorial Advisory Board

Table of Contents

Detailed Table of Contents

ramp rate limit, and prohibited operating zone. The output is also noted when transmission loss is taken into consideration. The results are found better than those of many other hybrid methods. Considering the quality of the solution obtained and nature of convergence, PSO-ACO may be accepted as a good alternative for solving ELD problems of varying complexity. Though PSO has been extensively used in ELD problems for its flexibility, robustness, and fast convergence, it often produces suboptimal solution due to its premature convergence. ACO, on the other hand, known for its good global exploration feature, imparts better balance between local and global search when combined with PSO.

Soraia Oueida, American University of Middle East, Kuwait
Seifedine Kadry, Beirut Arab University, Lebanon
Sorin Ionescu, Polytechnic University of Bucharest, Romania

In complex systems, such as healthcare, patient satisfaction is identified as the patient experience of care and has been referred to as the "indispensable outcome." The main goals of ED practitioners are the patient satisfaction along with maintaining optimal outcomes. Patient satisfaction has become a very important outcome measure when assessing healthcare systems performance. Nevertheless, it is a complex confusing concept. Some providers suggest several activities in order to enhance the satisfaction without being sure if these actions really improve satisfaction or no. Also, patient satisfaction enhancement activities should not conflict with cost efficacy. Several factors fall under patient satisfaction. Interviews with physicians prove that patients have specific expectations during a clinical encounter; being aware of these expectations, physicians can fulfill patient satisfaction. The goal of this chapter is to determine the contribution and effect of these factors in influencing patient satisfaction.

G. V. Alferov, Saint Petersburg State University, Russia
G. G. Ivanov, Saint Petersburg State University, Russia
P. A. Efimova, Saint Petersburg State University, Russia
A. S. Sharlay, Saint Petersburg State University, Russia

To study the dynamics of mechanical systems and to define the construction parameters and control laws, it is necessary to have computational models accurately describing properties of real mechanisms. From a mathematical point of view, the computational models of mechanical systems are actually the systems of differential equations. These models can contain equations that also describe non-mechanical phenomena. In this chapter, the problems of stability and asymptotic stability conditions for the motion of mechanical systems with holonomic and non-holonomic constraints are under consideration. Stability analysis for the systems of differential equations is given in term of the second Lyapunov's method. With the use of the set-theoretic approach, the necessary and sufficient conditions for stability and asymptotic stability of zero solution of the considered system are formulated. The proposed approaches can be used to study the stability of the motion for robot manipulators, transport, space, and socio-economic systems.

High-performance networks featuring advance bandwidth reservation have been developed and deployed to support big data transfer in extreme-scale scientific applications. The performance of such big data transfer largely depends on the transport protocols being used. For a given protocol in a given network environment, different parameter settings may lead to different performance, and oftentimes the default settings do not yield the best performance. It is, however, impractical to conduct an exhaustive search in the large parameter space of transport protocols for a set of suitable parameter values. This chapter proposes a stochastic approximation-based transport profiler, namely FastProf, to quickly determine the optimal operational zone of a protocol over dedicated connections. The proposed method is evaluated using both emulations based on real-life measurements and experiments over physical connections. The results show that FastProf significantly reduces the profiling overhead while achieving a comparable level of transport performance with the exhaustive search-based approach.

In this chapter, the author presents stochastic methods in aviation. The stochastic methods are presented as methods of decision making (DM) of operators of air navigation systems (ANS) in risk and uncertainly. The ANS is presented as a socio-technical system (STS). Analysis influences the factors of professional and non-professional activities on DM of STS's operators. The author made an analysis of the International Civil Aviation Organization (ICAO) documents on risk assessment. To determine the quantitative characteristics of risk levels, models for DM by the operator of the aviation system under risk and uncertainty have been developed. The author demonstrates some interesting advantages offered by the new methodology of DM in STS and forecasting the behavior of the operator in an emergency situation (ES) for using models of DM in risk and uncertainty.

The main objective of this chapter is to present a stochastic modeling and simulation methodology for estimating the variation of functional parameters of MEMS devices as a result of fabrication tolerances. The approach used for achieving this objective is to implement in the simulation process the variations of the geometrical parameters caused by the fabrication dispersion as random variables and to model the influence of these variations on the performance of the simulated device. The proposed method is demonstrated using two test structures: a micro-com resonator and a passive micromixer. In each example there are underlined important characteristics of the proposed simulation method: the ability to predict variation ranges of functional parameters, the ability to improve the design in function of the fabrication process, and the possibility of obtaining more precise results than the traditional deterministic methods.

Social media is revolutionizing the way people connect and share information. LinkedIn, Facebook, YouTube, Twitter, and other social media are changing the way we interact, and many organizations are struggling to respond. Given the potential risks and benefits of social media in the workplace, it is critical for managers to develop policies and procedures governing its appropriate use. This chapter identifies key issues and poses strategic questions to help guide managers in making more informed decisions when navigating social media issues in their organizations. After a brief introduction, current most popular social websites and tools are described concisely. Relationships between social media and human resources are then discussed. Utilizing social media in organizations are not without risks, which are thoroughly talked about further along with the benefits of such websites for recruitment. Finally, recommendations are made for companies that are considering utilizing social media and for companies that have already benefitted from such networks to improve their strategies.

Uncertainty modelling with random variables motivates the adoption of advanced PTM for reliability analysis to solve problems of mechanical systems. Probabilistic transformation method (PTM) is readily applicable when the function between the input and the output of the system is explicit. When these functions are implicit, a technique is proposed that combines finite element analysis (FEA) and probabilistic transformation method (PTM) that is based on the numerical simulations of the finite element analysis (FEA) and the probabilistic transformation method (PTM) using an interface between finite element software and Matlab. Structure problems are treated with the proposed technique, and the obtained results are compared to those obtained by the reference Monte Carlo method. A second aim of this work is to develop an algorithm of global optimization using the local method SQP. The proposed approach MSQP is tested on test functions comparing with other methods, and it is used to resolve a structural problem under reliability constraints.

Preface

Stochastic methods are used in a vast number of areas, including aerospace, medicine, transportation, and finance, to name but a few. Whether the goal is refining the design of a missile or aircraft, determining the effectiveness of a new drug, developing the most efficient timing strategies for traffic signals, or making investment decisions in order to increase profits, stochastic algorithms can help researchers and practitioners devise optimal solutions to countless real-world problems.

Individuals and organizations are often faced with making tradeoffs in order to achieve desirable outcomes. Choosing these tradeoffs in the "best" way is the essence of the stochastic problem. Mathematical algorithms for search and optimization play a large role in finding the best options in many problems in engineering, business, medicine, and the physical and social sciences. Stochastic methods pertain to problems where there is random noise in the measurements provided to the algorithm and/or there is injected (Monte Carlo Simulation) randomness in the algorithm itself. A countless number of stochastic problems arise in industry, academia, and government.

Researchers refer to physical systems in which they are uncertain about the values of parameters, measurements, expected input and disturbances as "stochastic systems". In probability theory, a purely stochastic system is one whose state is randomly determined, having a random probability distribution or pattern that may be analyzed statistically but may not be predicted precisely. In this regard, it can be classified as non-deterministic (i.e., "random") so that the subsequent state of the system is determined probabilistically. From one hand, any system or process that must be analyzed using probability theory is stochastic at least in part. Stochastic systems and processes play a fundamental role in mathematical models of phenomena in many fields of science, engineering, finance and economics. From the other hand, stochastic methods are for the purpose to estimate, understand and solve the behavior of such systems.

This book is a collection of nine chapters that could completely cover the huge coverage that Stochastic Methods for Estimation and Problem-Solving in Engineering encompasses. The chapters draw from different science fields like mechanical and industrial engineering, manufacturing and electrical engineering, big data, decision making, management, system reliability and optimization.

Chapter 1, "A New Technique for Optimization of Product Acceptance Process in Terms of Misclassification Probability", promotes the novel ideas for effective optimization (constrained or unconstrained) of product acceptance processes in terms of misclassification probability under parametric uncertainty of underlying models with real practical applications. It is expected that these ideas will give interesting and novel contributions to statistical theory and its applications at a good mathematical level, where the theoretical results are obtained through the frequentist (non-Bayesian) statistical approach. Frequentist probability interpretations of the methods proposed are clear. Bayesian methods are not considered here. It will be noted, however, that although subjective Bayesian approach has a clear personal probability interpretation, it is not generally clear how this should be applied to non-personal prediction or decisions. Objective Bayesian methods, on the other hand, do not have clear probability interpretations in finite samples. The methodology described here can be extended in several different directions to handle various problems that arise in practice. It is based on the pivotal quantity averaging (PQA) technique which allows one to eliminate the unknown parameters from the problem and to use available statistical information as completely as possible. While the details of problems considered in the chapter can change significantly from one industry to the next, the focus is always on making better statistical decisions and predictive inferences.

Chapter 2, "Combination of Stochastic Methods for Solving ELD Problem of Thermal Power Generation System", reports a hybrid optimization technique, a combination of stochastic methods - particle swarm optimization (PSO) and ant colony optimization (ACO), which is applied to find economic dispatch schedule and minimum generation cost for convex and non-convex power generation system simulated in MATLAB. 40-generator system is considered here with combinations of valve point loading, ramp rate limit and prohibited operating zone. The output is also noted when transmission loss is taken into consideration. The results are found better than those of many other hybrid methods. Considering the quality of the solution obtained and nature of convergence PSO-ACO may be accepted as a good alternative for solving ELD problems of varying complexity. Though PSO has been extensively used in ELD problems for its flexibility, robustness and fast convergence, it often produces sub optimal solution due to its premature convergence. ACO, on the other hand, known for its good global exploration feature, imparts better balance between local and global search when combined with PSO.

Chapter 3, "Factors Influencing Patients' Satisfaction", determines the contribution and effect of some factors in influencing patient satisfaction. In complex systems, such as healthcare, patient satisfaction is identified as the patient experience of care and has been referred to as the "indispensable Outcome". The main goals of ED practitioners are the patient satisfaction along with maintaining optimal outcomes. Patient satisfaction has become a very important outcome measure when assessing health care systems performance. Nevertheless, it is a complex confusing concept. Some providers suggest several activities in order to enhance the satisfaction without being sure if these actions really improve satisfaction or no. Also, patient satisfaction enhancement activities should not conflict with cost efficacy. Interviews with physicians prove that patients have specific expectations during a clinical encounter; being aware of these expectations, physician can then fulfill patient satisfaction. Sometimes physicians may misperceive patient expectations and thus leading to poor medical practice where unnecessary care is provided.

Chapter 4, "Stability of Linear Systems With Multitask Right-Hand Member", the stability problem of parametrically perturbed linear systems is considered. In it is shown that the problem of the zero solution stability of a selector linear differential inclusion reduces to the problem of solvability of a algebraic inequalities system whose coefficients are uniquely defined by coefficients of the initial differential inclusion. In section 2 and 3, the problem of stability of linear differential systems is considered in terms of the second Lyapunov's method, i.e. this reduces to the problem of a function with certain properties existence. The set-theoretical approach to the solution of the considered problem is the new in this paper. Such an approach is not only of theoretical interest, but it also helps to formalize the necessary and sufficient conditions for the stability and asymptotic stability of zero solution of the considered system with the use of Lyapunov's functions and to simplify the proof of the introduced assertions validity.

Chapter 5, "Stochastic Approximation-Based Transport Profiling for Big Data Movement Over Dedicated Connections", proposed a stochastic approximation-based transport profiler, FastProf, to accelerate the profiling process for improving big data transfer performance in high-performance networks. FastProf was implemented based on the existing Transport Profile Generator (TPG) toolkit. Extensive profiling emulations in comparison with other search algorithms, and profiling experiments over physical network connections with various delays are both conducted to evaluate the performance superiority of FastProf. The emulation and experimental results showed that FastProf significantly reduces the profiling time while achieving a comparable level of end-to-end throughput performance, which enables "on-line" profiling to achieve optimal big data transfer performance in high-performance networks.

Chapter 6 is "Stochastic Methods of Decision Making in Aviation". In Safety Management Manual was written safety requirements in aviation for estimation of "Risk". Risks are often expressed as probabilities; however, the concept of risk involves more than probabilities. To illustrate this with a hypothetical example, let us assume that the probability of the supporting cable of a 100-passenger cable car failing and allowing the cable car to fall was assessed as being the same as the probability of a 12-passenger elevator failing and allowing the elevator to fall. While the probabilities of the events occurring may be the same, the potential consequences of the cable car accident are much more severe. The risk is therefore two-dimensional. Evaluation of the acceptability of a given risk associated with a particular hazard must always take into account both the likelihood of occurrence of the hazard and the severity of its potential consequences. Therefore, it is proposed to assess the Risk taking into account the probabilities, outcomes, risk levels at each DM's stage.

Chapter 7 is "Stochastic Methods in Microsystems Engineering". In the recent years, fault tolerance analyses and design for manufacturability methodologies for MEMS have become a major concern in the scientific community as well as in the industry. However, from this perspective, the research has focused mainly on the development of experimental techniques for reliability testing of MEMS devices. The fabrication tolerances lead to variations of the geometrical parameters of the fabricated device, which leads to variations of the performance parameters. These variations in performance need to be quantified and controlled. Most of the techniques developed up to date for monitoring the effects of the deviations of geometrical parameters on the behaviour of microsystems are based on experimental methods. The experimental testing of the device is not always convenient, since many applications like bio-medicine, pharmaceutics or environmental monitoring require the development of sterile or disposable devices and physical testing would imply the device contamination. In other applications like biophysics or biochemistry, functionalised surfaces are needed and the experimental testing would imply the damaging of these surfaces. Another approach has been to integrate testing mechanisms in the device (like optical sensors, capacitance measurement electrodes, impedance sensors, etc). This approach brings along two main disadvantages: the design complexity and implicitly the cost of the device are increased and on the other hand, there are needed supplementary fabrication steps, which are not always compatible with the materials used for manufacturing the MEMS device. Given the unique challenges presented by the experimental testing of MEMS devices, in this chapter, it is proposed a stochastic simulation based methodology for estimating the influence of geometrical parametric deviations generated by the microfabrication tolerances on the general performance of the device. This methodology is developed by following the general flow of uncertainty modelling methods for studying the influence of random factors on the behaviour of engineering systems, but by adapting it to the design specifics of microsystems. This simulation method brings important benefits such as the possibility of determining values of certain parameters that cannot be measured experimentally and the development of precise and detailed behavioural models of MEMS devices without contaminating or damaging the device.

Chapter 8 is "Utilization of Social Networks for Human Resources Purposes". Social media is revolutionizing the way people connect and share information. LinkedIn, Facebook, YouTube, Twitter and other social media are changing the way we interact, and many organizations are struggling to respond. Given the potential risks and benefits of social media in the workplace, it is critical for managers to develop policies and procedures governing its appropriate use. This chapter will identify key issues and pose strategic questions to help guide managers in making more informed decisions when navigating social media issues in their organizations. After a brief introduction, current most popular social websites and tools are described concisely. Relationships between social media and human resources are then discussed. Utilizing social media in organizations are not without risks which are thoroughly talked about further along with the benefits of such websites for recruitment. Finally, recommendations are made for companies that are considering utilizing social media and for companies that have already benefitted from such networks to improve their strategies.

Chapter 9 is "Stochastic Methods Applied to Structural Mechanics: Reliability and Optimization Methods". Uncertainty modelling with random variables motivates the adoption of advanced PTM for reliability analysis to solve problems of mechanical systems. Probabilistic Transformation Method (PTM) is readily applicable when the function between the input and the output of the system is explicit. When these function is implicit, a technique is proposed that combines Finite Element Analysis (FEA) and Probabilistic Transformation Method (PTM) that is based on the numerical simulations of the Finite Element Analysis (FEA) and the Probabilistic Transformation Method (PTM) using an interface between Finite Element software and Matlab. Structures problems are treated with the proposed technique and the obtained results are compared to those obtained by the reference Monte Carlo method. A second aim of this work is to develop an algorithm of global optimization using the local method SQP. The proposed approach MSQP is tested on test functions comparing with other methods, and it is used to resolve, a structural problem under reliability constraints.

WHO AND HOW TO READ THIS BOOK

This book has three groups of people as its potential audience: (1) undergraduate students and postgraduate students conducting research in the areas of stochastic methods, (2) researchers at universities and other institutions working in these fields, and (3) practitioners in the Research and Development departments of industrial settings. This book differs from other books that have comprehensive case study and real data from industrial settings. The book can be used as an advanced reference for a course taught at the postgraduate level in industrial engineering, electrical engineering, mechanical engineering, manufacturing intelligence, management, and social sciences.

Seifedine Kadry
Beirut Arab University, Lebanon

Acknowledgment

I would like to express my gratitude to the many people who saw me through this book; to all those who provided support, talked things over, read, wrote, offered comments, allowed me to quote their remarks and assisted in the editing, proofreading, and design.

I would like to thank IGI publisher and Colleen Moore for enabling me to publish this book. Above all I want to thank my wife, Diana and my son Hadi, who supported and encouraged me in spite of all the time it took me away from them. It was a long and difficult journey for them.

Thanks to all contributors and reviewers, without them this book would never find its way to the Web and to printed version.

Last and not least: I beg forgiveness of all those who have been with me over the course of the years and whose names I have failed to mention.

Chapter 1
A New Technique for Optimization of Product Acceptance Process in Terms of Misclassification Probability

Nicholas A. Nechval
University of Latvia, Latvia

Konstantin N. Nechval
Transport and Telecommunication Institute, Latvia

ABSTRACT

A product acceptance process is an inspecting one in statistical quality control or reliability tests, which are used to make decisions about accepting or rejecting lots of products to be submitted. This process is important for industrial and business purposes of quality management. To determine the optimal parameters of the product acceptance process under parametric uncertainty of underlying lifetime models (in terms of misclassification probability), a new optimization technique is proposed. The most popular lifetime distribution used in the field of product acceptance is a two-parameter Weibull distribution, with the assumption that the shape parameter is known. Such oversimplified assumptions can facilitate the follow-up analyses, but may overlook the fact that the lifetime distribution can significantly affect the estimation of the failure rate of a product. Therefore, the situations are also considered when both Weibull distribution parameters are unknown. An illustrative numerical example is given.

INTRODUCTION

The product acceptance process is used to make decisions of accepting or rejecting a lot of products based on a random sample collected from the lot. Optimization of an acceptance process specifies the minimum sample size required to be used along with the acceptance and non-acceptance criteria for the lot. So the optimized acceptance process specifies the most effective number of units, say n, to be used for testing, and the acceptance criterion for the specified producer and consumer risks. For a given product acceptance process, the consumer's and producer's risks are the probabilities that a bad lot is

DOI: 10.4018/978-1-5225-5045-7.ch001

accepted and a good lot is rejected, respectively. Usually, with every product acceptance process, the associated consumer's and producer's risks are also provided.

A typical application of product acceptance process is as follows: a company receives a shipment of product from a vendor. This product is often a component or raw material used in the company's manufacturing process. A sample is taken from the lot and the relevant quality characteristic of the units in the sample is inspected. On the basis of the information in this sample, a decision is made regarding lot disposition. Traditionally, when the life test indicates that the mean life of products exceeds the specified one, the lot of products is accepted, otherwise it is rejected. Accepted lots are put into production, while rejected lots may be returned to the vendor or may be subjected to some other lot disposition action. While it is customary to think of acceptance sampling as a receiving inspection activity, there are also other uses. For example, frequently a manufacturer samples and inspects its own product at various stages of production. Lots that are accepted are sent forward for further processing, while rejected lots may be reworked or scrapped. For the purpose of reducing the test time and cost, a truncated life test may be conducted to determine the smallest sample size to ensure a certain mean life of products when the life test is terminated at a preassigned time t_0 and the number of failures observed does not exceed a given number.

A product acceptance process in the case that the sample observations are lifetimes of products put to test aims at verifying that the actual population mean exceeds a required minimum. The population mean stands for the mean lifetime of the product, say μ. If μ_0 is a specified minimum value, then one would like to verify that $\mu \geq \mu_0$, this means that the true unknown population mean lifetime of the product exceeds the specified value. On the basis of a random sample of size n, the lot is accepted, if by means of a suitable decision criterion, the product acceptance model decides in favour of $\mu \geq \mu_0$. Otherwise the lot is rejected.

Extensive work has been done on product acceptance processes since their inception. Several text books and papers are available which provide different acceptance processes of product for different probability distribution functions, see, for example, Epstein (1954), Sobel and Tischendrof (1959), Goode and Kao (1961), Gupta and Groll (1961), Gupta (1962), Fertig and Mann (1980), Kantam and Rosaiah (1998), Kantam et al. (2001), Baklizi (2003), Wu and Tsai (2005), Rosaiah and Kantam (2005), Rosaiah et al. (2006), Tsai and Wu (2006), Balakrishnan et al. (2007), Srinivasa Rao and Kantam (2010), Stephens (2001), Squeglia (2008), and the references cited therein. All these authors considered the design of product acceptance processes based on the population mean or median.

In this chapter, a new technique for optimization of product acceptance process under parametric uncertainty (in terms of misclassification probability) is proposed. The mathematical solution to the optimization problem of product acceptance process does not represent any difficulties for the case when the parametric values of the underlying lifetime distributions are known with certainty. In actual practice, such is simply not the case.

When the model is applied to solve real-world problems, the parameters are estimated and then treated as if they were the true values. The risk associated with using estimates rather than the true parameters is called estimation risk and is often ignored. For efficient optimization of statistical decisions under parametric uncertainty, the classical theory of statistical estimation has little to offer in this type of situation. The bulk of the classical theory has been developed about the assumption of a quadratic, or at least symmetric and analytically simple loss structure. In some cases this assumption is made explicit, although in most it is implicit in the search for estimating procedures that have the "nice" statistical properties of unbiasedness and minimum variance. Such procedures are usually satisfactory if the estimates so generated are to be used solely for the purpose of reporting information to another party for an unknown purpose, when the loss structure is not easily discernible, or when the number of observations is large enough to support Normal approximations and asymptotic results. Unfortunately, we seldom are fortunate enough to be in asymptotic situations. Small sample sizes are generally the rule when the small sample properties of estimators do not appear to have been thoroughly investigated. They have long been recognized as deficient, however, when the purpose of estimation is the making of a specific decision (or sequence of decisions) on the basis of a limited amount of information in a situation where the losses are clearly asymmetric. For this case, a pivotal quantity averaging (PQA) approach is suggested. It allows one to eliminate unknown parameters from the problem through the PQA and to find the better decision rules, which have smaller risk than any of the well-known decision rules (Nechval & Vasermanis, 2004; Nechval et al., 2008; Nechval et al., 2010).

WEIBULL LIFETIME MODEL

In this chapter, the two-parameter Weibull lifetime distribution with the probability density function (pdf)

$$f_{\theta=(\beta,\delta)}(x) = \frac{\delta}{\beta}\left(\frac{x}{\beta}\right)^{\delta-1} \exp\left[-\left(\frac{x}{\beta}\right)^{\delta}\right], \quad x > 0, \quad \beta > 0, \quad \delta > 0, \tag{1}$$

and cumulative distribution function (cdf)

$$F_\theta(x) = 1 - \exp\left[-\left(\frac{x}{\beta}\right)^{\delta}\right], \quad x > 0, \quad \beta > 0, \quad \delta > 0, \tag{2}$$

indexed by scale parameter β and shape parameter δ, is used as the underlying distribution of a random variable X in a sample of the lifetime data.

The Weibull distribution is widely used in reliability and survival analysis due to its flexible shape and ability to model a wide range of failure rates. It can be derived theoretically as a form of extreme value distribution, governing the time to occurrence of the "weakest link" of many competing failure processes. Its special case with shape parameter δ=2 is the Rayleigh distribution which is commonly used for modelling the magnitude of radial error when x and y coordinate errors are independent normal variables with zero mean and the same standard deviation while the case δ=1 corresponds to the widely used exponential distribution.

Cumulative Distribution Function of $\widehat{\beta}^{\delta}$ When the Shape Parameter δ Is Known

Consider the case when the shape parameter δ is known. Let $(X_1 \le \ldots \le X_n)$ be a random sample of n observations from the two-parameter Weibull distribution (1). Then a maximum likelihood estimator (MLE) of β^{δ}, based on the random sample of size n from (1), is given by

$$\widehat{\beta}^{\delta} = \sum_{i=1}^{n} X_i^{\delta} \Big/ n, \tag{3}$$

It can be shown that

$$\widehat{\beta}^{\delta} \sim f_{\theta}(\widehat{\beta}^{\delta} \mid n) = \frac{1}{\Gamma(n)(\beta^{\delta}/n)^{n}} \left(\widehat{\beta}^{\delta}\right)^{n-1} \exp\left(-\frac{\widehat{\beta}^{\delta}}{\beta^{\delta}/n}\right), \quad \widehat{\beta}^{\delta} \geq 0, \tag{4}$$

where a function of the data $\widehat{\beta}$ and parameters (β, δ),

$$\frac{\widehat{\beta}^{\delta}}{\beta^{\delta}} = V, \tag{5}$$

is said to be a pivot or a pivotal quantity since its probability density function,

$$f(v \mid n) = \frac{1}{\Gamma(n)(1/n)^{n}} v^{n-1} \exp\left(-\frac{v}{1/n}\right), \quad v \geq 0, \tag{6}$$

does not depend on the parameters (β, δ).

It follows from (6) that the cdf of $\widehat{\beta}^{\delta}$ is given by

$$\Pr\left(\widehat{\beta}^{\delta} \leq h\right) = \Pr\left(\frac{\widehat{\beta}^{\delta}}{\beta^{\delta}} \leq \frac{h}{\beta^{\delta}}\right) = \Pr\left(V \leq \frac{h}{\beta^{\delta}}\right)$$

$$= \int_{0}^{h/\beta^{\delta}} f(v \mid n) dv = \int_{0}^{h/\beta^{\delta}} \frac{1}{\Gamma(n)(1/n)^{n}} v^{n-1} \exp\left(-\frac{v}{1/n}\right) dv. \tag{7}$$

Cumulative Distribution Function of $\widehat{\beta}^{\widehat{\delta}}$ When the Shape Parameter δ Is Unknown

Consider the case when the shape parameter δ is unknown. Let $(X_1, \ldots, X_n)$ be a random sample from the two-parameter Weibull distribution (1), and let $\widehat{\beta}$, $\widehat{\delta}$ be the maximum likelihood estimates of β, δ, respectively, computed on the basis of $(X_1, \ldots, X_n)$:

$$\widehat{\beta} = \left(\sum_{i=1}^{n} x_i^{\widehat{\delta}} \Big/ n\right)^{1/\widehat{\delta}}, \tag{8}$$

and

$$\widehat{\delta} = \left[\left(\sum_{i=1}^{n} x_i^{\widehat{\delta}} \ln x_i\right)\left(\sum_{i=1}^{n} x_i^{\widehat{\delta}}\right)^{-1} - \frac{1}{n}\sum_{i=1}^{n} \ln x_i\right]^{-1}, \tag{9}$$

In terms of the Weibull variates, we have that

$$V_1 = \left(\frac{\widehat{\beta}}{\beta}\right)^{\delta},\ V_2 = \frac{\delta}{\widehat{\delta}},\ V_3 = \left(\frac{\widehat{\beta}}{\beta}\right)^{\widehat{\delta}} \tag{10}$$

are pivotal quantities. Furthermore, let

$$Z_i = (X_i / \widehat{\beta})^{\widehat{\delta}}, i=1, \ldots, n. \tag{11}$$

It is readily verified that any $n-2$ of the Z_i's, say $Z_1, \ldots, Z_{n-2}$ form a set of $n-2$ functionally independent ancillary statistics. The appropriate conditional approach is to consider the distributions of V_1, V_2, V_3 conditional on the observed value of $\mathbf{Z}^{(n)} = (Z_1, \ldots, Z_n)$. (For purposes of symmetry of notation we include all of $Z_1, \ldots, Z_n$ in expressions stated here; it can be shown that Z_n, Z_{n-1}, can be determined as functions of $Z_1, \ldots, Z_{n-2}$ only.)

- **Theorem 1:** (Joint pdf of the pivotal quantities V_1,V_2 from the two-parameter Weibull distribution) Let $(X_1, \ldots, X_n)$ be a random sample of n observations from the two-parameter Weibull distribution (1). Then the joint pdf of the pivotal quantities

$$V_1 = \left(\frac{\widehat{\beta}}{\beta}\right)^{\delta},\ V_2 = \frac{\delta}{\widehat{\delta}}, \tag{12}$$

conditional on fixed

$$\mathbf{z}^{(n)} = (z_i, \ldots, z_n), \tag{13}$$

where

$$Z_i = \left(\frac{X_i}{\widehat{\beta}}\right)^{\widehat{\delta}}, \quad i = 1, \ldots, n, \tag{14}$$

are ancillary statistics, any $n-2$ of which form a functionally independent set, $\widehat{\beta}$ and $\widehat{\delta}$ are the maximum likelihood estimates for β and δ, respectively, based on a random sample of n observations $(X_1, \ldots, X_n)$ from the two-parameter Weibull distribution (1), is given by

$$f_n(v_1, v_2 \mid \mathbf{z}^{(n)}) = \left(\frac{1}{\Gamma(n)} \left[\sum_{i=1}^{n} z_i^{v_2} \right]^n v_1^{n-1} \exp\left(-v_1 \sum_{i=1}^{n} z_i^{v_2} \right) \right) \left(\frac{1}{\vartheta(\mathbf{z}^{(n)})} v_2^{n-2} \prod_{i=1}^{n} z_i^{v_2} \left[\sum_{i=1}^{n} z_i^{v_2} \right]^{-n} \right)$$

$$= f_n(v_1 \mid \mathbf{z}^{(n)}, v_2) f_n(v_2 \mid \mathbf{z}^{(n)}), v_1 \in (0, \infty), v_2 \in (0, \infty), \tag{15}$$

where

$$f_n(v_1 \mid \mathbf{z}^{(n)}, v_2) = \frac{1}{\Gamma(n)} \left[\sum_{i=1}^{n} z_i^{v_2} \right]^n v_1^{n-1} \exp\left(-v_1 \sum_{i=1}^{n} z_i^{v_2} \right), v_1 \in (0, \infty). \tag{16}$$

$$f_n(v_2 \mid \mathbf{z}^{(n)}) = \frac{1}{\vartheta(\mathbf{z}^{(n)})} v_2^{n-2} \prod_{i=1}^{n} z_i^{v_2} \left(\sum_{i=1}^{n} z_i^{v_2} \right)^{-n}, \ v_2 \in (0, \infty), \tag{17}$$

$$\vartheta(\mathbf{z}^{(n)}) = \int_0^\infty v_2^{n-2} \prod_{i=1}^{n} z_i^{v_2} \left(\sum_{i=1}^{n} z_i^{v_2} \right)^{-n} dv_2 \tag{18}$$

is the normalizing constant.

Proof. The joint density of $X_1, ..., X_n$ is given by

$$f_{(\beta,\delta)}(x_1, \ ..., \ x_n) = \prod_{i=1}^{n} \frac{\delta}{\beta} \left(\frac{x_i}{\beta} \right)^{\delta-1} \exp\left(-\left(\frac{x_i}{\beta} \right)^{\delta} \right). \tag{19}$$

Using the invariant embedding technique (Nechval & Vasermanis, 2004; Nechval et al., 2008; Nechval et al., 2010), we transform (19) to

$$f_{(\beta,\delta)}(x_1, \ ..., \ x_n) \ d\widehat{\beta} \ d\widehat{\delta} = \prod_{i=1}^{n} \frac{\delta}{x_i} \left(\frac{x_i}{\beta} \right)^{\delta} \exp\left(-\sum_{i=1}^{n} \left(\frac{x_i}{\beta} \right)^{\delta} \right) d\widehat{\beta} \ d\widehat{\delta}$$

$$= -\widehat{\beta}\widehat{\delta}^{\,n} \prod_{i=1}^{n} \frac{1}{x_i} \left(\frac{\delta}{\widehat{\delta}} \right)^{n-2} \prod_{i=1}^{n} \left(\frac{x_i}{\widehat{\beta}} \right)^{\delta} \left(\frac{\widehat{\beta}}{\beta} \right)^{\delta(n-1)} \exp\left(-\left(\frac{\widehat{\beta}}{\beta} \right)^{\delta} \sum_{i=1}^{n} \left(\frac{x_i}{\widehat{\beta}} \right)^{\delta} \right) \left(\frac{\delta}{\beta} \left(\frac{\widehat{\beta}}{\beta} \right)^{\delta-1} d\widehat{\beta} \right) \left(-\frac{\delta}{\widehat{\delta}^2} d\widehat{\delta} \right)$$

$$= -\widehat{\beta}\widehat{\delta}^{\,n}\prod_{i=1}^{n}\frac{1}{x_i}\left(\frac{\delta}{\widehat{\delta}}\right)^{n-2}\prod_{i=1}^{n}\left(\frac{x_i}{\widehat{\beta}}\right)^{\widehat{\delta}\left(\frac{\delta}{\widehat{\delta}}\right)}\left(\frac{\widehat{\beta}}{\beta}\right)^{\delta(n-1)}\exp\left(-\left(\frac{\widehat{\beta}}{\beta}\right)^{\delta}\sum_{i=1}^{n}\left(\frac{x_i}{\widehat{\beta}}\right)^{\widehat{\delta}\left(\frac{\delta}{\widehat{\delta}}\right)}\right)\left(\frac{\delta}{\beta}\left(\frac{\widehat{\beta}}{\beta}\right)^{\delta-1}d\widehat{\beta}\right)\left(-\frac{\delta}{\widehat{\delta}^{2}}d\widehat{\delta}\right)$$

$$= -\widehat{\beta}\widehat{\delta}^{\,n}\prod_{i=1}^{n}\frac{1}{x_i}\left(\frac{\delta}{\widehat{\delta}}\right)^{n-2}\prod_{i=1}^{n}\left(\frac{x_i}{\widehat{\beta}}\right)^{\widehat{\delta}\left(\frac{\delta}{\widehat{\delta}}\right)}\left(\frac{\widehat{\beta}}{\beta}\right)^{\delta(n-1)}\exp\left(-\left(\frac{\widehat{\beta}}{\beta}\right)^{\delta}\sum_{i=1}^{n}\left(\frac{x_i}{\widehat{\beta}}\right)^{\widehat{\delta}\left(\frac{\delta}{\widehat{\delta}}\right)}\right)d\left(\frac{\widehat{\beta}}{\beta}\right)^{\delta}d\left(\frac{\delta}{\widehat{\delta}}\right)$$

$$= -\widehat{\beta}\widehat{\delta}^{\,n}\prod_{i=1}^{n}\frac{1}{x_i}v_2^{n-2}\prod_{i=1}^{n}z_i^{v_2}v_1^{n-1}\exp\left(-v_1\sum_{i=1}^{n}z_i^{v_2}\right)dv_1dv_2$$

$$= -\widehat{\beta}\widehat{\delta}^{\,n}\prod_{i=1}^{n}\frac{1}{x_i}\Gamma(n)v_2^{n-2}\prod_{i=1}^{n}z_i^{v_2}\left[\sum_{i=1}^{n}z_i^{v_2}\right]^{-n}\frac{1}{\Gamma(n)}\left[\sum_{i=1}^{n}z_i^{v_2}\right]^{n}v_1^{n-1}\exp\left(-v_1\sum_{i=1}^{n}z_i^{v_2}\right)dv_1dv_2. \tag{20}$$

Normalizing (20),

$$\frac{-\widehat{\beta}\widehat{\delta}^{\,n}\prod_{i=1}^{n}\frac{1}{x_i}\Gamma(n)v_2^{n-2}\prod_{i=1}^{n}z_i^{v_2}\left[\sum_{i=1}^{n}z_i^{v_2}\right]^{-n}\frac{1}{\Gamma(n)}\left[\sum_{i=1}^{n}z_i^{v_2}\right]^{n}v_1^{n-1}\exp\left(-v_1\sum_{i=1}^{n}z_i^{v_2}\right)}{\int_0^\infty\int_0^\infty -\widehat{\beta}\widehat{\delta}^{\,n}\prod_{i=1}^{n}\frac{1}{x_i}\Gamma(n)v_2^{n-2}\prod_{i=1}^{n}z_i^{v_2}\left[\sum_{i=1}^{n}z_i^{v_2}\right]^{-n}\frac{1}{\Gamma(n)}\left[\sum_{i=1}^{n}z_i^{v_2}\right]^{n}v_1^{n-1}\exp\left(-v_1\sum_{i=1}^{n}z_i^{v_2}\right)dv_1dv_2}$$

$$= \frac{v_2^{n-2}\prod_{i=1}^{n}z_i^{v_2}\left[\sum_{i=1}^{n}z_i^{v_2}\right]^{-n}\frac{1}{\Gamma(n)}\left[\sum_{i=1}^{n}z_i^{v_2}\right]^{n}v_1^{n-1}\exp\left(-v_1\sum_{i=1}^{n}z_i^{v_2}\right)}{\int_0^\infty v_2^{n-2}\prod_{i=1}^{n}z_i^{v_2}\left[\sum_{i=1}^{n}z_i^{v_2}\right]^{-n}dv_2}$$

$$= \frac{1}{\Gamma(n)}\left[\sum_{i=1}^{n}z_i^{v_2}\right]^{n}v_1^{n-1}\exp\left(-v_1\sum_{i=1}^{n}z_i^{v_2}\right)\frac{v_2^{n-2}\prod_{i=1}^{n}z_i^{v_2}\left[\sum_{i=1}^{n}z_i^{v_2}\right]^{-n}}{\vartheta(\mathbf{z}^{(n)})} = f_n(v_1, v_2 \mid \mathbf{z}^{(n)}), \tag{21}$$

we obtain (15). This ends the proof.

Corollary 1.1 If a pivotal quantity is given by

$$W = V_1\sum_{i=1}^{n}Z_i^{V_2} = \left(\frac{\widehat{\beta}}{\beta}\right)^{\widehat{\delta}\frac{\delta}{\widehat{\delta}}}\sum_{i=1}^{n}Z_i^{V_2} = \left(\frac{\widehat{\beta}^{\widehat{\delta}}}{\beta^{\widehat{\delta}}}\right)^{V_2}\sum_{i=1}^{n}Z_i^{V_2} = V_3^{V_2}\sum_{i=1}^{n}Z_i^{V_2}, \tag{22}$$

it follows from (16) that

$$W \sim g_n(w) = \frac{1}{\Gamma(n)} w^{n-1} \exp(-w), w \in (0, \infty), \tag{23}$$

and

$$\Pr\left(W\left[=\left(\frac{\widehat{\beta}^{\delta}}{\beta^{\delta}}\right)^{V_2} \sum_{i=1}^{n} Z_i^{V_2}\right] \leq w\left[=\left(\frac{h}{\beta^{\delta}}\right)^{v_2} \sum_{i=1}^{n} z_i^{v_2}\right]\right) = \int_0^{\left(\frac{h}{\beta^{\delta}}\right)^{v_2} \sum_{i=1}^{n} z_i^{v_2}} \frac{1}{\Gamma(n)} w^{n-1} \exp(-w) dw$$

$$= G_n\left(\left(\frac{h}{\beta^{\delta}}\right)^{v_2} \sum_{i=1}^{n} z_i^{v_2}\right) = \Pr(\widehat{\beta}^{\widehat{\delta}} \leq h \mid \beta^{\widehat{\delta}}, n, \mathbf{z}^{(n)}, V_2) \tag{24}$$

Corollary 1.2 (Cumulative distribution function of $\widehat{\beta}^{\widehat{\delta}}$). It follows from (17) and (24) that the cdf of $\widehat{\beta}^{\widehat{\delta}}$ is given by

$$\Pr(\widehat{\beta}^{\widehat{\delta}} \leq h \mid \beta^{\widehat{\delta}}, n, \mathbf{z}^{(n)}) = \int_0^{\infty} \Pr(\widehat{\beta}^{\widehat{\delta}} \leq h \mid \beta^{\widehat{\delta}}, n, \mathbf{z}^{(n)}, V_2) f_n(v_2 \mid \mathbf{z}^{(n)}) dv_2$$

$$= \int_0^{\infty} G_n\left(\left(\frac{h}{\beta^{\widehat{\delta}}}\right)^{v_2} \sum_{i=1}^{n} z_i^{v_2}\right) f_n(v_2 \mid \mathbf{z}^{(n)}) dv_2$$

$$= \frac{1}{\vartheta(\mathbf{z}^{(n)})} \int_0^{\infty} \frac{v_2^{n-2} \prod_{i=1}^{n} z_i^{v_2} G_n\left(\left(\frac{h}{\beta^{\widehat{\delta}}}\right)^{v_2} \sum_{i=1}^{n} z_i^{v_2}\right)}{\left(\sum_{i=1}^{n} z_i^{v_2}\right)^{n}} dv_2. \tag{25}$$

OPTIMIZATION OF PRODUCT ACCEPTANCE PROCESS WHEN THE SCALE PARAMETER OF THE WEIBULL LIFETIME MODEL IS UNKNOWN

The Weibull distribution indexed by scale and shape parameters is generally used as a distribution of lifetime in reliability analysis because it includes increasing, decreasing or constant failure rates; for more details about the use of the Weibull distribution in reliability analysis, the reader can refer to Fertig and Mann (1980).

Suppose that product lots are submitted for inspection and the lifetimes of individual products have a Weibull distribution with the probability density function (1) and known shape parameter δ. In this case, we consider some situations of optimization of the design parameters of product acceptance process, such as the sample size "n" to test the Weibull MTTF (mean time to failure) "μ" and the separation threshold "h" to satisfy the producer's and consumer's risks. The probability of rejecting a good lot is called the producer's risk, which is denoted by α_p. The probability of accepting a bad lot is called the consumer's risk, which is denoted by α_c. In determining whether or not a product lot is accepted, we have to use the test procedure which satisfies the following conditions:

$$\Pr(\text{reject a product lot} \mid \text{Weibull MTTF} = \mu_a) \le \alpha_p \text{ (Producer's risk)}, \tag{26}$$

$$\Pr(\text{accept a product lot} \mid \text{Weibull MTTF} = \mu_r) \le \alpha_c \text{ (Consumer's risk)}, \tag{27}$$

where

$$\mu_a = E_{\theta_a=(\beta_a,\delta)}\{X\} = \int_0^\infty x f_{\theta_a}(x)dx = \int_0^\infty x \frac{\delta}{\beta_a}\left(\frac{x}{\beta_a}\right)^{\delta-1} \exp\left[-\left(\frac{x}{\beta_a}\right)^{\delta}\right]dx = \beta_a \Gamma\left(1+\frac{1}{\delta}\right) \tag{28}$$

is an acceptable Weibull MTTF,

$$\mu_r = E_{\theta_r=(\beta_r,\delta)}\{X\} = \int_0^\infty x f_{\theta_r}(x)dx = \int_0^\infty x \frac{\delta}{\beta_r}\left(\frac{x}{\beta_r}\right)^{\delta-1} \exp\left[-\left(\frac{x}{\beta_r}\right)^{\delta}\right]dx = \beta_r \Gamma\left(1+\frac{1}{\delta}\right) \tag{29}$$

is a rejectable Weibull MTTF, $\mu_a > \mu_r$. It follows from (28) and (29), respectively, that

$$\beta_a^\delta = \left[\mu_a \Big/ \Gamma\left(1+\frac{1}{\delta}\right)\right]^\delta \tag{30}$$

and

$$\beta_r^\delta = \left[\mu_r \Big/ \Gamma\left(1+\frac{1}{\delta}\right)\right]^\delta. \tag{31}$$

Situation 1

Consider the situation when the size "n" of a random sample ($X_1 \le \ldots \le X_n$) of product items to test the Weibull MTTF (mean time to failure) "μ" is preassigned, but a product lot misclassification probability,

$$\alpha(h \mid n) = \alpha_p(h \mid n) + \alpha_c(h \mid n), \tag{32}$$

is not preassigned. The problem is to find the separation threshold "h" which minimized the product lot misclassification probability (32).

Let n items be drawn at random from an acceptable two-parameter Weibull lifetime distribution with the probability density function (pdf)

$$f_{\theta_a}(x) = \frac{\delta}{\beta_a}\left(\frac{x}{\beta_a}\right)^{\delta-1} \exp\left[-\left(\frac{x}{\beta_a}\right)^{\delta}\right], \quad x > 0, \;\; \beta_a > 0, \;\; \delta > 0, \tag{33}$$

and cumulative distribution function (cdf)

$$F_{\theta_a}(x) = 1 - \exp\left[-\left(\frac{x}{\beta_a}\right)^{\delta}\right], \quad x > 0, \;\; \beta_a > 0, \;\; \delta > 0, \tag{34}$$

Then a maximum likelihood estimator (MLE) of β_a^δ, based on the random sample ($X_1 \leq ... \leq X_n$) of size n from (33), is given by

$$\widehat{\beta}^{\delta} = \sum_{i=1}^{n} X_i^{\delta} \Big/ n, \tag{35}$$

It can be shown that

$$\widehat{\beta}^{\delta} \sim g_{\theta_a}(\widehat{\beta}^{\delta} \mid n) = \frac{1}{\Gamma(n)(\beta_a^{\delta}/n)^{n}} \left(\widehat{\beta}^{\delta}\right)^{n-1} \exp\left(-\frac{\widehat{\beta}^{\delta}}{\beta_a^{\delta}/n}\right), \quad \widehat{\beta}^{\delta} \geq 0, \tag{36}$$

where a function of the data $\widehat{\beta}$ and parameters (β_a, δ),

$$\frac{\widehat{\beta}^{\delta}}{\beta_a^{\delta}} = V_a, \tag{37}$$

is said to be a pivot or a pivotal quantity since its probability density function,

$$g(v_a \mid n) = \frac{1}{\Gamma(n)(1/n)^{n}} v_a^{n-1} \exp\left(-\frac{v_a}{1/n}\right), \quad v_a \geq 0, \tag{38}$$

does not depend on the parameters (β_a, δ).

It follows from (36) and (38) that the cdf of $\widehat{\beta}^{\delta}$ is given by

$$G_{\theta_a}(h \mid n) = \int_0^h g_{\theta_a}(\widehat{\beta}^{\delta} \mid n) d\widehat{\beta}^{\delta} = \int_0^h \frac{1}{\Gamma(n)(\beta_a^{\delta}/n)^n} \left(\widehat{\beta}^{\delta}\right)^{n-1} \exp\left(-\frac{\widehat{\beta}^{\delta}}{\beta_a^{\delta}/n}\right) d\widehat{\beta}^{\delta}$$

$$= \int_0^{h/\beta_a^{\delta}} \frac{1}{\Gamma(n)(1/n)^n} v_a^{n-1} \exp\left(-\frac{v_a}{1/n}\right) dv_a = \int_0^{h/\beta_a^{\delta}} g(v_a \mid n) dv_a = \Pr\left(V_a \le \frac{h}{\beta_a^{\delta}}\right). \tag{39}$$

Let n items be drawn at random from a rejectable two-parameter Weibull lifetime distribution with the probability density function (pdf)

$$f_{\theta_r}(x) = \frac{\delta}{\beta_r}\left(\frac{x}{\beta_r}\right)^{\delta-1} \exp\left[-\left(\frac{x}{\beta_r}\right)^{\delta}\right], \quad x > 0, \;\; \beta_r > 0, \;\; \delta > 0, \tag{40}$$

and cumulative distribution function (cdf)

$$F_{\theta_r}(x) = 1 - \exp\left[-\left(\frac{x}{\beta_r}\right)^{\delta}\right], \quad x > 0, \;\; \beta_r > 0, \;\; \delta > 0, \tag{41}$$

Then a maximum likelihood estimator (MLE) of β_r^{δ}, based on the random sample $(X_1 \le ... \le X_n)$ of size n from (40), is given by

$$\widehat{\beta}^{\delta} = \sum_{i=1}^{n} X_i^{\delta} \Big/ n. \tag{42}$$

It can be shown that

$$\widehat{\beta}^{\delta} \sim g_{\theta_r}(\widehat{\beta}^{\delta} \mid n) = \frac{1}{\Gamma(n)(\beta_r^{\delta}/n)^n} \left(\widehat{\beta}^{\delta}\right)^{n-1} \exp\left(-\frac{\widehat{\beta}^{\delta}}{\beta_r^{\delta}/n}\right), \quad \widehat{\beta}^{\delta} \ge 0, \tag{43}$$

where a function of the data $\widehat{\beta}$ and parameters (β_r, δ),

$$\frac{\widehat{\beta}^{\delta}}{\beta_r^{\delta}} = V_r, \tag{44}$$

is said to be a pivot or a pivotal quantity since its probability density function,

$$g(v_r \mid n) = \frac{1}{\Gamma(n)(1/n)^n} v_r^{n-1} \exp\left(-\frac{v_r}{1/n}\right), \quad v_r \geq 0, \tag{45}$$

does not depend on the parameters (β_r, δ).

It follows from (43) and (45) that the cdf of $\widehat{\beta}^\delta$ is given by

$$G_{\theta_r}(h \mid n) = \int_0^h g_{\theta_r}(\widehat{\beta}^\delta \mid n) d\widehat{\beta}^\delta = \int_0^h \frac{1}{\Gamma(n)(\beta_r^\delta/n)^n} \left(\widehat{\beta}^\delta\right)^{n-1} \exp\left(-\frac{\widehat{\beta}^\delta}{\beta_r^\delta/n}\right) d\widehat{\beta}^\delta$$

$$= \int_0^{h/\beta_r^\delta} \frac{1}{\Gamma(n)(1/n)^n} v_r^{n-1} \exp\left(-\frac{v_r}{1/n}\right) dv_r = \int_0^{h/\beta_r^\delta} g(v_r \mid n) dv_r = \Pr\left(V_r \leq \frac{h}{\beta_r^\delta}\right). \tag{46}$$

It follows from (26) and (39) that a good product lot is rejected if

$$\widehat{\beta}^\delta \leq h, \tag{47}$$

where the separation threshold "h" satisfies the following condition:

$$G_{\theta_a}(h \mid n) = \Pr\left(\widehat{\beta}^\delta \leq h\right) = \Pr\left(\frac{\widehat{\beta}^\delta}{\beta_a^\delta} \leq \frac{h}{\beta_a^\delta}\right) = \Pr\left(V_a \leq \frac{h}{\beta_a^\delta}\right) = \alpha_p. \tag{48}$$

It follows from (27) and (46) that a bad product lot is accepted if

$$\widehat{\beta}^\delta > h, \tag{49}$$

where the separation threshold "h" satisfies the following condition:

$$\bar{G}_{\theta_r}(h \mid n) = \Pr\left(\widehat{\beta}^\delta > h \mid\right) = \Pr\left(\frac{\widehat{\beta}^\delta}{\beta_r^\delta} > \frac{h}{\beta_r^\delta}\right) = \Pr\left(V_r > \frac{h}{\beta_r^\delta}\right) = \alpha_c, \tag{50}$$

$$\bar{G}_{\theta_r}(h \mid n) = 1 - G_{\theta_r}(h \mid n). \tag{51}$$

It follows from (32), (48) and (50) that the product lot misclassification probability (see Figure 1) is given by

$$\alpha(h \mid n) = \alpha_p(h \mid n) + \alpha_c(h \mid n) = G_{\theta_a}(h \mid n) + \bar{G}_{\theta_r}(h \mid n)$$

$$= \int_0^h g_{\theta_a}(\widehat{\beta}^\delta \mid n) d\widehat{\beta}^\delta + \int_h^\infty g_{\theta_r}(\widehat{\beta}^\delta \mid n) d\widehat{\beta}^\delta$$

$$= \Pr\left(V_a \le \frac{h}{\beta_a^\delta}\right) + \Pr\left(V_r > \frac{h}{\beta_r^\delta}\right)$$

$$= \int_0^{h/\beta_a^\delta} \frac{1}{\Gamma(n)(1/n)^n} v_a^{n-1} \exp\left(-\frac{v_a}{1/n}\right) dv_a + \int_{h/\beta_r^\delta}^\infty \frac{1}{\Gamma(n)(1/n)^n} v_r^{n-1} \exp\left(-\frac{v_r}{1/n}\right) dv_r. \quad (52)$$

To minimize the probability of product lot misclassification, the separation threshold "h" is determined via (52) as follows:

$$h = \arg\min_h \alpha(h \mid n)$$

Figure 1. Product lot misclassification probability $\alpha(h \mid n)$

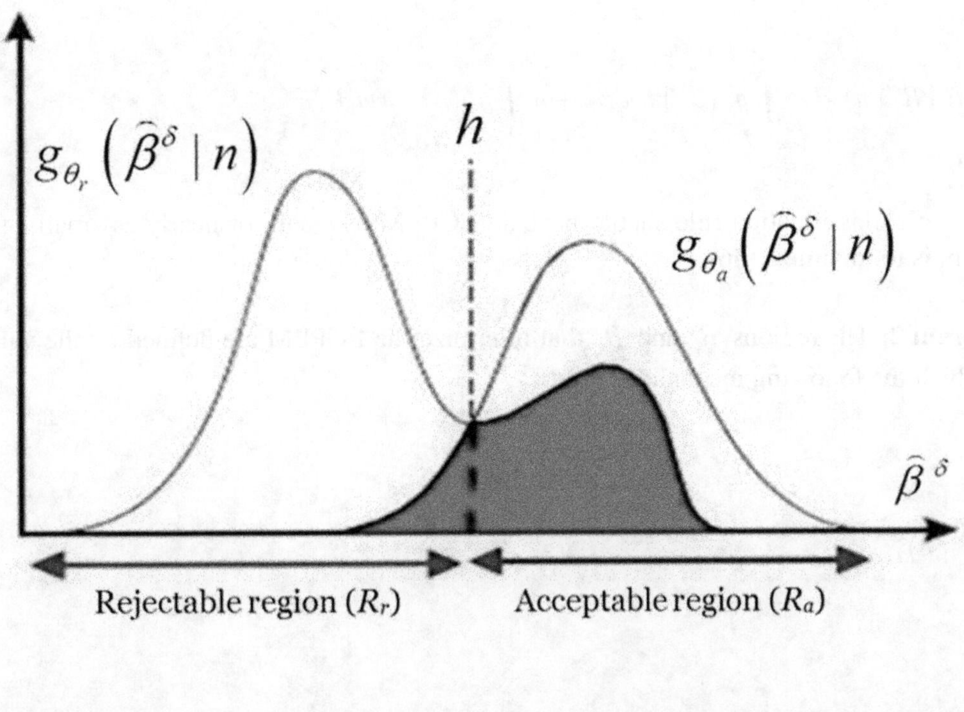

$$= \arg\min_h \left(\int_0^h g_{\theta_a}(\widehat{\beta}^\delta \mid n) d\widehat{\beta}^\delta + \int_h^\infty g_{\theta_r}(\widehat{\beta}^\delta \mid n) d\widehat{\beta}^\delta \right). \quad (53)$$

Thus, the separation threshold h, which is used to obtain the rule of determining whether or not a product lot is accepted, should be "optimal" in the sense of minimizing, on average, the number of incorrect assignments of the product lot into two different regions R_a and R_r.

It follows from the above (see Figure 1) that classification of the product lot into one of two different regions R_a and R_r (i.e., determining whether or not the product lot is accepted) is carried out through $\widehat{\beta}^\delta$ and h as follows:

$$\widehat{\beta}^\delta \in \begin{cases} R_r = \left\{ \widehat{\beta}^\delta : \widehat{\beta}^\delta \le h \right\} & \rightarrow \text{reject the product lot,} \\ R_a = \left\{ \widehat{\beta}^\delta : \widehat{\beta}^\delta > h \right\} & \rightarrow \text{accept the product lot.} \end{cases} \quad (54)$$

Remark 1. In general, the classification rule (54) can be rewritten as follows:

Assign the product lot to region R_a (or R_r) for which $g_{\theta_a}(\widehat{\beta}^\delta \mid n)$ (or $g_{\theta_r}(\widehat{\beta}^\delta \mid n)$) is largest. (55)

Situation 2

In the case of Situation 1, a classification scheme is evaluated in terms of its product lot misclassification probability, but this ignores misclassification costs. A rule that ignores costs may cause problems.

The costs of product lot misclassification can be defined as follows. The costs are: (1) zero for correct classification, (2) c_p when an estimator $\widehat{\beta}^\delta$ from R_a is incorrectly classified as R_r, and (3) c_c when an estimator $\widehat{\beta}^\delta$ from R_r is incorrectly classified as R_a.

For any classification rule, the average, or *expected cost of product lot misclassification* (ECPLM) is given by

$$\mathrm{ECPLM}(R_r, R_a \mid n) = c_p \int_{R_r} g_{\theta_a}(\widehat{\beta}^\delta \mid n) d\widehat{\beta}^\delta + c_c \int_{R_a} g_{\theta_r}(\widehat{\beta}^\delta \mid n) d\widehat{\beta}^\delta. \quad (56)$$

A reasonable classification rule should have an ECPLM as small, or nearly as small, as possible. The problem is to minimize (56).

- **Theorem 2:** The regions R_r and R_a that minimize the ECPLM are defined by the values of $\widehat{\beta}^\delta$ for which the following inequalities hold:

$$R_r : \frac{g_{\theta_r}(\widehat{\beta}^\delta \mid n)}{g_{\theta_a}(\widehat{\beta}^\delta \mid n)} \ge \frac{c_p}{c_c} \quad (57)$$

$$\begin{pmatrix}\text{density}\\ \text{ratio}\end{pmatrix} \geq \begin{pmatrix}\text{cost}\\ \text{ratio}\end{pmatrix}$$

$$R_a : \frac{g_{\theta_r}(\widehat{\beta}^\delta \mid n)}{g_{\theta_a}(\widehat{\beta}^\delta \mid n)} < \frac{c_p}{c_c} \tag{58}$$

$$\begin{pmatrix}\text{density}\\ \text{ratio}\end{pmatrix} < \begin{pmatrix}\text{cost}\\ \text{ratio}\end{pmatrix}$$

Proof. Noting that

$$\Omega = R_r \cup R_a, \tag{59}$$

so that the total probability

$$1 = \int_{\Omega} g_{\theta_a}(\widehat{\beta}^\delta \mid n) d\widehat{\beta}^\delta = \int_{R_r} g_{\theta_a}(\widehat{\beta}^\delta \mid n) d\widehat{\beta}^\delta + \int_{R_a} g_{\theta_a}(\widehat{\beta}^\delta \mid n) d\widehat{\beta}^\delta, \tag{60}$$

we can write

$$\text{ECPLM}(R_a \mid n) = c_p \left(1 - \int_{R_a} g_{\theta_a}(\widehat{\beta}^\delta \mid n) d\widehat{\beta}^\delta\right) + c_c \int_{R_a} g_{\theta_r}(\widehat{\beta}^\delta \mid n) d\widehat{\beta}^\delta. \tag{61}$$

By the additive property of integrals,

$$\text{ECPLM}(R_a \mid n) = \int_{R_a} [c_c g_{\theta_r}(\widehat{\beta}^\delta \mid n) - c_p g_{\theta_a}(\widehat{\beta}^\delta \mid n)] d\widehat{\beta}^\delta + c_p. \tag{62}$$

Now, c_c and c_p are nonnegative. In addition, $g_{\theta_r}(\widehat{\beta}^\delta \mid n)$ and $g_{\theta_a}(\widehat{\beta}^\delta \mid n)$ are nonnegative for all values of $\widehat{\beta}^\delta$ and are the only quantities in ECPLM that depend on $\widehat{\beta}^\delta$. Thus, ECPLM is minimized if R_a includes those values of $\widehat{\beta}^\delta$ for which the integrand

$$[c_c g_{\theta_r}(\widehat{\beta}^\delta \mid n) - c_p g_{\theta_a}(\widehat{\beta}^\delta \mid n)] < 0 \tag{63}$$

and excludes those values of $\widehat{\beta}^\delta$ for which this quantity is nonnegative. This completes the proof.

Corollary 2.1 If

$$c_p / c_c = 1 \text{ (Equal product lot misclassification costs)} \tag{64}$$

then

$$R_r : \frac{g_{\theta_r}(\widehat{\beta}^\delta \mid n)}{g_{\theta_a}(\widehat{\beta}^\delta \mid n)} \geq 1,\ R_a : \frac{g_{\theta_r}(\widehat{\beta}^\delta \mid n)}{g_{\theta_a}(\widehat{\beta}^\delta \mid n)} < 1. \tag{65}$$

It follows from (57) and (58) that classification of the product lot into one of two different regions R_a and R_r (i.e., determining whether or not the product lot is accepted) is carried out through $\widehat{\beta}^\delta$ and h as follows:

$$\widehat{\beta}^\delta \in \begin{cases} R_r = \{\widehat{\beta}^\delta : \widehat{\beta}^\delta \leq h\} & \to \text{reject the product lot,} \\ R_a = \{\widehat{\beta}^\delta : \widehat{\beta}^\delta > h\} & \to \text{accept the product lot,} \end{cases} \tag{66}$$

where

$$h = \arg\min_h \left(c_p \int_0^h g_{\theta_a}(\widehat{\beta}^\delta \mid n) d\widehat{\beta}^\delta + c_c \int_h^\infty g_{\theta_r}(\widehat{\beta}^\delta \mid n) d\widehat{\beta}^\delta \right). \tag{67}$$

Remark 1. In general, the classification rule (66) can be rewritten as follows:

Assign the product lot to region R_a (or R_r) for which $c_p g_{\theta_a}(\widehat{\beta}^\delta \mid n)$ (or $c_c g_{\theta_r}(\widehat{\beta}^\delta \mid n)$) is largest. 68)

Situation 3

Consider Situation 2 when there is the cost c of sampling and putting a product item on test, but the size "n" of a random sample ($X_1 \leq \ldots \leq X_n$) of product items to test the Weibull MTTF "μ", the producer's risk, which is denoted by α_p, and the consumer's risk, which is denoted by α_c, are not preassigned.

The problem is to find the optimal separation threshold "h" and integer "n", which minimize the performance index:

$$cn + c_p \alpha_p(h,n) + c_c \alpha_c(h,n)$$

$$= cn + c_p \int_0^h g_{\theta_a}(\widehat{\beta}^\delta \mid n) d\widehat{\beta}^\delta + c_c \int_h^\infty g_{\theta_r}(\widehat{\beta}^\delta \mid n) d\widehat{\beta}^\delta, \tag{69}$$

where the optimal separation threshold "h" for $\widehat{\beta}^{\delta}$ and integer "n" are given by

$$(h,n) = \arg\min_{n}\left[cn + \left(\min_{h}[c_p\alpha_p(h \mid n) + c_c\alpha_c(h \mid n)]\right)\right]$$

$$= \arg\min_{n}\left[cn + \min_{h}\left(c_p\int_{0}^{h} g_{\theta_a}(\widehat{\beta}^{\delta} \mid n)d\widehat{\beta}^{\delta} + c_c\int_{h}^{\infty} g_{\theta_r}(\widehat{\beta}^{\delta} \mid n)d\widehat{\beta}^{\delta}\right)\right]. \tag{70}$$

The classification rule of the product lot is given by

$$\widehat{\beta}^{\delta} \in \begin{cases} R_r = \left\{\widehat{\beta}^{\delta} : \widehat{\beta}^{\delta} \le h\right\} & \rightarrow \text{reject the product lot,} \\ R_a = \left\{\widehat{\beta}^{\delta} : \widehat{\beta}^{\delta} > h\right\} & \rightarrow \text{accept the product lot.} \end{cases} \tag{71}$$

Situation 4

Consider the situation when the size "n" of a random sample ($X_1 \le \ldots \le X_n$) of product items to test the Weibull MTTF "μ" is not preassigned, but a product lot misclassification probability is preassigned:

$$\alpha(h,n) = \alpha. \tag{72}$$

The problem is to find the optimal separation threshold "h" and the required smallest integer "n", which minimize the performance index:

$$[\alpha(h,n) - \alpha]^2$$

$$= \left[\int_{0}^{h} g_{\theta_a}(\widehat{\beta}^{\delta} \mid n)d\widehat{\beta}^{\delta} + \int_{h}^{\infty} g_{\theta_r}(\widehat{\beta}^{\delta} \mid n)d\widehat{\beta}^{\delta} - \alpha\right]^2$$

$$= \left[\int_{0}^{h/\beta_a^{\delta}} \frac{1}{\Gamma(n)(1/n)^n} v_a^{n-1}\exp\left(-\frac{v_a}{1/n}\right)dv_a + \int_{h/\beta_r^{\delta}}^{\infty} \frac{1}{\Gamma(n)(1/n)^n} v_r^{n-1}\exp\left(-\frac{v_r}{1/n}\right)dv_r - \alpha\right]^2, \tag{73}$$

where the optimal separation threshold "h" for $\widehat{\beta}^{\delta}$ and the required smallest integer "n" are given by

$$(h,n) = \arg\min_{n}\left(\min_{h}[\alpha(h \mid n) - \alpha]^2\right)$$

$$= \arg\min_{n}\left[\min_{h}\left(\int_{0}^{h} g_{\theta_a}(\widehat{\beta}^{\delta} \mid n) d\widehat{\beta}^{\delta} + \int_{h}^{\infty} g_{\theta_r}(\widehat{\beta}^{\delta} \mid n) d\widehat{\beta}^{\delta} - \alpha\right)^2\right]. \quad (74)$$

The classification rule of the product lot is given by

$$\widehat{\beta}^{\delta} \in \begin{cases} R_r = \left\{\widehat{\beta}^{\delta} : \widehat{\beta}^{\delta} \leq h\right\} & \rightarrow \text{reject the product lot}, \\ R_a = \left\{\widehat{\beta}^{\delta} : \widehat{\beta}^{\delta} > h\right\} & \rightarrow \text{accept the product lot}. \end{cases} \quad (75)$$

Situation 5

Consider the situation when the size "n" of a random sample ($X_1 \leq \ldots \leq X_n$) of product items to test the Weibull MTTF "μ" is not preassigned, but the producer's risk, which is denoted by α_p, and the consumer's risk, which is denoted by α_c, are preassigned.

The problem is to find the optimal separation threshold "h" and the required smallest integer "n", which minimize the performance index:

$$(\alpha_p(h,n) - \alpha_p)^2 + (\alpha_c(h,n) - \alpha_c)^2$$

$$= \left(\int_{0}^{h} g_{\theta_a}(\widehat{\beta}^{\delta} \mid n) d\widehat{\beta}^{\delta} - \alpha_p\right)^2 + \left(\int_{h}^{\infty} g_{\theta_r}(\widehat{\beta}^{\delta} \mid n) d\widehat{\beta}^{\delta} - \alpha_c\right)^2, \quad (76)$$

where the optimal separation threshold "h" for $\widehat{\beta}^{\delta}$ and the required smallest integer "n" are given by

$$(h,n) = \arg\min_{n}\left(\min_{h}[(\alpha_p(h \mid n) - \alpha_p)^2 + (\alpha_c(h \mid n) - \alpha_c)^2]\right)$$

$$= \arg\min_{n}\left(\min_{h}\left[\left(\int_{0}^{h} g_{\theta_a}(\widehat{\beta}^{\delta} \mid n) d\widehat{\beta}^{\delta} - \alpha_p\right)^2 + \left(\int_{h}^{\infty} g_{\theta_r}(\widehat{\beta}^{\delta} \mid n) d\widehat{\beta}^{\delta} - \alpha_c\right)^2\right]\right). \quad (77)$$

The classification rule of the product lot is given by

$$\widehat{\beta}^{\delta} \in \begin{cases} R_r = \left\{\widehat{\beta}^{\delta} : \widehat{\beta}^{\delta} \leq h\right\} & \rightarrow \text{reject the product lot}, \\ R_a = \left\{\widehat{\beta}^{\delta} : \widehat{\beta}^{\delta} > h\right\} & \rightarrow \text{accept the product lot}. \end{cases} \quad (78)$$

OPTIMIZATION OF PRODUCT ACCEPTANCE PROCESS WHEN THE SCALE AND SHAPE PARAMETERS OF THE WEIBULL LIFETIME MODEL ARE UNKNOWN

Suppose that product lots are submitted for inspection and the lifetimes of individual products have a Weibull distribution with the probability density function (1) and unknown parameters (β, δ).

In this case, the unknown shape parameter δ of the two-parameter Weibull lifetime distribution is replaced by the maximum likelihood estimate $\widehat{\delta}$ given by (9). It follows from (30) and (31) that

$$\beta_a^{\widehat{\delta}} = \left[\mu_a \Big/ \Gamma\left(1+\frac{1}{\widehat{\delta}}\right)\right]^{\widehat{\delta}} \tag{79}$$

and

$$\beta_r^{\widehat{\delta}} = \left[\mu_r \Big/ \Gamma\left(1+\frac{1}{\widehat{\delta}}\right)\right]^{\widehat{\delta}}, \tag{80}$$

respectively. It follows from (25) that

$$\Pr(\widehat{\beta}^{\widehat{\delta}} \le h \mid \beta_a^{\widehat{\delta}}, n, \mathbf{z}^{(n)}) = \frac{1}{\vartheta(\mathbf{z}^{(n)})} \int_0^{\infty} \frac{v_2^{n-2} \prod_{i=1}^{n} z_i^{v_2} G_n\left(\left(\frac{h}{\beta_a^{\widehat{\delta}}}\right)^{v_2} \sum_{i=1}^{n} z_i^{v_2}\right)}{\left(\sum_{i=1}^{n} z_i^{v_2}\right)^n} dv_2, \tag{81}$$

where $\vartheta(\mathbf{z}^{(n)})$ is given by (18),

$$G_n\left(\left(\frac{h}{\beta_a^{\widehat{\delta}}}\right)^{v_2} \sum_{i=1}^{n} z_i^{v_2}\right) = \int_0^{\left(\frac{h}{\beta_a^{\widehat{\delta}}}\right)^{v_2} \sum_{i=1}^{n} z_i^{v_2}} \frac{1}{\Gamma(n)} w^{n-1} \exp(-w) dw; \tag{82}$$

$$\Pr(\widehat{\beta}^{\widehat{\delta}} \le h \mid \beta_r^{\widehat{\delta}}, n, \mathbf{z}^{(n)}) = \frac{1}{\vartheta(\mathbf{z}^{(n)})} \int_0^{\infty} \frac{v_2^{n-2} \prod_{i=1}^{n} z_i^{v_2} G_n\left(\left(\frac{h}{\beta_r^{\widehat{\delta}}}\right)^{v_2} \sum_{i=1}^{n} z_i^{v_2}\right)}{\left(\sum_{i=1}^{n} z_i^{v_2}\right)^n} dv_2, \tag{83}$$

where

$$G_n\left(\left(\frac{h}{\beta_r^{\delta}}\right)^{v_2}\sum_{i=1}^{n} z_i^{v_2}\right)=\int_0^{\left(\frac{h}{\beta_r^{\delta}}\right)^{v_2}\sum_{i=1}^{n} z_i^{v_2}}\frac{1}{\Gamma(n)}w^{n-1}\exp(-w)dw. \tag{84}$$

Now, using (81) and (83), we consider the various situations of optimization of the product acceptance process that were described in the previous section.

Situation 1

Consider the situation when the size "n" of a random sample ($X_1 \le ... \le X_n$) of product items to test the Weibull MTTF (mean time to failure) "μ" is preassigned, but a product lot misclassification probability,

$$\alpha(h\,|\,n)=\alpha_p(h\,|\,n)+\alpha_c(h\,|\,n), \tag{85}$$

is not preassigned. The problem is to find the separation threshold "h" which minimized the product lot misclassification probability (85).

To minimize the probability of product lot misclassification, the separation threshold "h" is determined via (81) and (83) as follows:

$$h=\arg\min_h \alpha(h\,|\,n)=\arg\min_h[\alpha_p(h\,|\,n)+\alpha_c(h\,|\,n)]$$

$$=\arg\min_h\left(\Pr(\widehat{\beta}^{\delta}\le h\,|\,\beta_a^{\delta},n,\mathbf{z}^{(n)})+\Pr(\widehat{\beta}^{\delta}>h\,|\,\beta_r^{\delta},n,\mathbf{z}^{(n)})\right), \tag{86}$$

where

$$\Pr\left(\widehat{\beta}^{\delta}>h\,|\,\beta_r^{\delta},n,\mathbf{z}^{(n)}\right)=1-\Pr\left(\widehat{\beta}^{\delta}\le h\,|\,\beta_r^{\delta},n,\mathbf{z}^{(n)}\right). \tag{87}$$

Determining whether or not the product lot is accepted is carried out through $\widehat{\beta}^{\delta}$ and h as follows:

$$\widehat{\beta}^{\delta}\in\begin{cases}R_r=\left\{\widehat{\beta}^{\delta}:\widehat{\beta}^{\delta}\le h\right\} & \rightarrow \text{reject the product lot,}\\ R_a=\left\{\widehat{\beta}^{\delta}:\widehat{\beta}^{\delta}> h\right\} & \rightarrow \text{accept the product lot.}\end{cases} \tag{88}$$

Situation 2

In the case of Situation 1, a classification scheme is evaluated in terms of its product lot misclassification probability, but this ignores misclassification costs. A rule that ignores costs may cause problems.

The costs of product lot misclassification can be defined as follows. The costs are: (1) zero for correct classification, (2) c_p when an estimator $\widehat{\beta}^{\widehat{\delta}}$ from R_a is incorrectly classified as R_r, and (3) c_c when an estimator $\widehat{\beta}^{\widehat{\delta}}$ from R_r is incorrectly classified as R_a.

For any classification rule, the average, or *expected cost of product lot misclassification* (ECPLM) is given by

$$\mathrm{ECPLM}(h \mid n) = c_p \Pr\left(\widehat{\beta}^{\widehat{\delta}} \le h \mid \beta_a^{\delta}, n, \mathbf{z}^{(n)}\right) + c_c \Pr\left(\widehat{\beta}^{\widehat{\delta}} > h \mid \beta_r^{\delta}, n, \mathbf{z}^{(n)}\right). \tag{89}$$

A reasonable classification rule should have an ECPLM as small, or nearly as small, as possible. The problem is to minimize (89).

To minimize (89), the separation threshold "h" is determined as follows:

$$h = \arg\min_h \mathrm{ECPLM}(h \mid n)$$

$$= \arg\min_h \left(c_p \Pr\left(\widehat{\beta}^{\widehat{\delta}} \le h \mid \beta_a^{\delta}, n, \mathbf{z}^{(n)}\right) + c_c \Pr\left(\widehat{\beta}^{\widehat{\delta}} > h \mid \beta_r^{\delta}, n, \mathbf{z}^{(n)}\right)\right). \tag{90}$$

Determining whether or not the product lot is accepted is carried out through $\widehat{\beta}^{\widehat{\delta}}$ and h as follows:

$$\widehat{\beta}^{\widehat{\delta}} \in \begin{cases} R_r = \left\{\widehat{\beta}^{\widehat{\delta}} : \widehat{\beta}^{\widehat{\delta}} \le h\right\} & \rightarrow \text{reject the product lot,} \\ R_a = \left\{\widehat{\beta}^{\widehat{\delta}} : \widehat{\beta}^{\widehat{\delta}} > h\right\} & \rightarrow \text{accept the product lot.} \end{cases} \tag{91}$$

Situation 3

Consider Situation 2 when there is the cost c of sampling and putting a product item on test, but the size "n" of a random sample ($X_1 \le \ldots \le X_n$) of product items to test the Weibull MTTF "μ", the producer's risk, which is denoted by α_p, and the consumer's risk, which is denoted by α_c, are not preassigned.

The problem is to find the optimal separation threshold "h" and integer "n", which minimize the performance index:

$$cn + c_p\alpha_p(h,n) + c_c\alpha_c(h,n)$$

$$= cn + c_p \Pr\left(\widehat{\beta}^{\widehat{\delta}} \le h \mid \beta_a^{\delta}, n, \mathbf{z}^{(n)}\right) + c_c \Pr\left(\widehat{\beta}^{\widehat{\delta}} > h \mid \beta_r^{\delta}, n, \mathbf{z}^{(n)}\right), \tag{92}$$

where the optimal separation threshold "h" for $\widehat{\beta}^{\widehat{\delta}}$ and integer "n" are given by

$$(h,n) = \arg\min_n \left[cn + \left(\min_h [c_p \alpha_p (h \mid n) + c_c \alpha_c (h \mid n)] \right) \right]$$

$$= \arg\min_n \left[cn + \min_h \left(c_p \Pr\left(\widehat{\beta}^{\delta} \le h \mid \beta_a^{\delta}, n, \mathbf{z}^{(n)}\right) + c_c \Pr\left(\widehat{\beta}^{\delta} > h \mid \beta_r^{\delta}, n, \mathbf{z}^{(n)}\right) \right) \right]. \quad (93)$$

The classification rule of the product lot is given by

$$\widehat{\beta}^{\delta} \in \begin{cases} R_r = \left\{ \widehat{\beta}^{\delta} : \widehat{\beta}^{\delta} \le h \right\} & \rightarrow \text{reject the product lot,} \\ R_a = \left\{ \widehat{\beta}^{\delta} : \widehat{\beta}^{\delta} > h \right\} & \rightarrow \text{accept the product lot.} \end{cases} \quad (94)$$

Situation 4

Consider the situation when the size "n" of a random sample ($X_1 \le ... \le X_n$) of product items to test the Weibull MTTF "μ" is not preassigned, but a product lot misclassification probability is preassigned:

$$\alpha(h,n) = \alpha. \quad (95)$$

The problem is to find the optimal separation threshold "h" and the required smallest integer "n", which minimize the performance index:

$$[\alpha(h,n) - \alpha]^2$$

$$= \left[\Pr\left(\widehat{\beta}^{\delta} \le h \mid \beta_a^{\delta}, n, \mathbf{z}^{(n)}\right) + \Pr\left(\widehat{\beta}^{\delta} > h \mid \beta_r^{\delta}, n, \mathbf{z}^{(n)}\right) - \alpha \right]^2 \quad (96)$$

where the optimal separation threshold "h" for $\widehat{\beta}^{\delta}$ and the required smallest integer "n" are given by

$$(h,n) = \arg\min_n \left(\min_h [\alpha(h \mid n) - \alpha]^2 \right)$$

$$= \arg\min_n \left[\min_h \left(\Pr\left(\widehat{\beta}^{\delta} \le h \mid \beta_a^{\delta}, n, \mathbf{z}^{(n)}\right) + \Pr\left(\widehat{\beta}^{\delta} > h \mid \beta_r^{\delta}, n, \mathbf{z}^{(n)}\right) - \alpha \right)^2 \right]. \quad (97)$$

The classification rule of the product lot is given by

$$\widehat{\beta}^{\delta} \in \begin{cases} R_r = \left\{ \widehat{\beta}^{\delta} : \widehat{\beta}^{\delta} \le h \right\} & \rightarrow \text{reject the product lot,} \\ R_a = \left\{ \widehat{\beta}^{\delta} : \widehat{\beta}^{\delta} > h \right\} & \rightarrow \text{accept the product lot.} \end{cases} \quad (98)$$

Situation 5

Consider the situation when the size "n" of a random sample ($X_1 \le ... \le X_n$) of product items to test the Weibull MTTF "μ" is not preassigned, but the producer's risk, which is denoted by α_p, and the consumer's risk, which is denoted by α_c, are preassigned.

The problem is to find the optimal separation threshold "h" and the required smallest integer "n", which minimize the performance index:

$$(\alpha_p(h,n)-\alpha_p)^2+(\alpha_c(h,n)-\alpha_c)^2$$

$$=\left(\Pr\left(\widehat{\beta}^{\delta}\le h\mid\beta_a^{\delta},n,\mathbf{z}^{(n)}\right)-\alpha_p\right)^2+\left(\Pr\left(\widehat{\beta}^{\delta}>h\mid\beta_r^{\delta},n,\mathbf{z}^{(n)}\right)-\alpha_c\right)^2, \tag{99}$$

where the optimal separation threshold "h" for $\widehat{\beta}^{\delta}$ and the required smallest integer "n" are given by

$$(h,n)=\arg\min_n\left(\min_h[(\alpha_p(h\mid n)-\alpha_p)^2+(\alpha_c(h\mid n)-\alpha_c)^2]\right)$$

$$=\arg\min_n\left(\min_h\left[\left(\Pr\left(\widehat{\beta}^{\delta}\le h\mid\beta_a^{\delta},n,\mathbf{z}^{(n)}\right)-\alpha_p\right)^2+\left(\Pr\left(\widehat{\beta}^{\delta}>h\mid\beta_r^{\delta},n,\mathbf{z}^{(n)}\right)-\alpha_c\right)^2\right]\right). \tag{100}$$

The classification rule of the product lot is given by

$$\widehat{\beta}^{\delta}\in\begin{cases}R_r=\left\{\widehat{\beta}^{\delta}:\widehat{\beta}^{\delta}\le h\right\} & \to \text{reject the product lot,}\\ R_a=\left\{\widehat{\beta}^{\delta}:\widehat{\beta}^{\delta}>h\right\} & \to \text{accept the product lot.}\end{cases} \tag{101}$$

NUMERICAL EXAMPLE

A manufacturing company produces control relays, which are ideally suited for controlling everything from lights and motors to telecommunication, audio, and video signals. Some relays can be used for switching radio frequency signals. Relays come in many sizes and ratings. There are literally tens of thousands of relay varieties on the market. From the life test history, it is known that the lifetimes of the relays have a two-parameter Weibull distribution in which the shape parameter δ=3 and scale parameter varies from lot to lot. The manufacturing company uses an acceptance scheme on production items before they are shipped.

In determining whether or not a product lot is accepted, the company has to use the test procedure which satisfies the following conditions:

$$\Pr\text{ (reject a product lot I Weibull MTTF} = \mu_a) \le \alpha_p \text{ (Producer's risk),} \tag{102}$$

Pr (accept a product lot | Weibull MTTF = μ_r) ≤ α_c (Consumer's risk), (103)

where

$\mu_a = 7$ (in terms of 10^3 hours), (104)

$\mu_r = 5$ (in terms of 10^3 hours). (105)

It follows from (30), (31), (104) and (105) that

$$\beta_a^\delta = \left[\mu_a \Big/ \Gamma\left(1+\frac{1}{\delta}\right)\right]^\delta = \left[7\Big/\Gamma\left(1+\frac{1}{3}\right)\right]^3 = 481.692225 \quad (106)$$

and

$$\beta_r^\delta = \left[\mu_r \Big/ \Gamma\left(1+\frac{1}{\delta}\right)\right]^\delta = \left[5\Big/\Gamma\left(1+\frac{1}{3}\right)\right]^3 = 175.543814, \quad (107)$$

respectively.

Let $(X_1 \leq \ldots \leq X_n)$ be a random sample of n observations from the two-parameter Weibull distribution (1). Then a maximum likelihood estimator (MLE) of β^δ, based on the random sample of size n from (1), is given by (3). It follows from (39) and (46), respectively, that

$$\Pr\left(\widehat{\beta}^\delta \leq h \mid \beta_a^\delta, n\right) = \int_0^h g_{\theta_a}(\widehat{\beta}^\delta \mid n) d\widehat{\beta}^\delta = \int_0^h \frac{1}{\Gamma(n)(\beta_a^\delta/n)^n}\left(\widehat{\beta}^\delta\right)^{n-1}\exp\left(-\frac{\widehat{\beta}^\delta}{\beta_a^\delta/n}\right)d\widehat{\beta}^\delta, \quad (108)$$

and

$$\Pr\left(\widehat{\beta}^\delta > h \mid \beta_r^\delta, n\right) = \int_h^\infty g_{\theta_r}(\widehat{\beta}^\delta \mid n) d\widehat{\beta}^\delta = \int_h^\infty \frac{1}{\Gamma(n)(\beta_r^\delta/n)^n}\left(\widehat{\beta}^\delta\right)^{n-1}\exp\left(-\frac{\widehat{\beta}^\delta}{\beta_r^\delta/n}\right)d\widehat{\beta}^\delta. \quad (109)$$

The possible situations determining the corresponding acceptance schemes are illustrated below.

Situation 1

Consider the situation when the size "n" of a random sample $(X_1 \leq \ldots \leq X_n)$ of product items to test the Weibull MTTF (mean time to failure) "μ" is equal to 20 (see Table 1), but a product lot misclassification probability,

Table 1. The data of lifetime of control relays

Observations (in terms of 10^3 hours)									
X_1	X_2	X_3	X_4	X_5	X_6	X_7	X_8	X_9	X_{10}
4.012	7.867	4.542	5.965	14.48	5.317	4.799	4.228	11.186	9.378
X_{11}	X_{12}	X_{13}	X_{14}	X_{15}	X_{16}	X_{17}	X_{18}	X_{19}	X_{20}
8.636	6.893	8.000	9.198	8.686	8.333	3.688	10.589	9.219	6.544

$$\alpha(h \mid n) = \alpha_p(h \mid n) + \alpha_c(h \mid n), \tag{110}$$

is not prespecified. The problem is to find the separation threshold "*h*" which minimized the product lot misclassification probability (110).

Goodness-of-fit testing. Let $(X_1, \ldots, X_n)$ be a random sample from the two-parameter Weibull distribution (1), and let $\widehat{\beta}$, $\widehat{\delta}$ be the maximum likelihood estimates of β, δ, respectively, computed on the basis of $(X_1, \ldots, X_n)$:

$$\widehat{\delta} = \left[\left(\sum_{i=1}^{n} x_i^{\widehat{\delta}} \ln x_i \right) \left(\sum_{i=1}^{n} x_i^{\widehat{\delta}} \right)^{-1} - \frac{1}{n} \sum_{i=1}^{n} \ln x_i \right]^{-1} = 3.004644, \tag{111}$$

and

$$\widehat{\beta} = \left(\sum_{i=1}^{n} x_i^{\widehat{\delta}} \Big/ n \right)^{1/\widehat{\delta}} = 8.494792, \tag{112}$$

We assess the statistical significance of departures from the Weibull model (1) by performing the Anderson–Darling goodness-of-fit test. There are many goodness-of-fit tests, for example: Kolmogorov-Smirnov and Anderson-Darling tests. The Anderson-Darling test is more sensitive to deviations in the tails of a distribution than the older Kolmogorov-Smirnov test. The Anderson-Darling test statistic value is determined by (e.g. Dodson, 1994):

$$A^2 = -\left[\sum_{i=1}^{n} (2i-1) \left(\ln F_\theta(x_i) + \ln\left(1 - F_\theta(x_{n+1-i})\right) \right) \right] \Big/ n - n, \tag{113}$$

where $F_\theta(x)$ is the cumulative distribution function of X,

$$\theta = (\beta = \widehat{\beta} = 8.494792, \ \delta = \widehat{\delta} = 3.004644), \tag{114}$$

and n is the number of observations. The result from (113) needs to be modified for small sampling values. For the Weibull distribution, the modification of A^2 is

$$A_{\mathrm{mod}}^2 = A^2\left(1+\frac{0.2}{\sqrt{n}}\right). \tag{115}$$

The A_{mod}^2 value must then be compared with critical values, A_{α}^2, which depend on the significance level α and the distribution type. As an example, for the Weibull distribution the determined A_{mod}^2 value has to be less than the following critical values for acceptance of goodness-of-fit (see Table 2):

For this example, α=0.05, $A_{\alpha=0.05}^2 = 0.757$,

$$A^2 = -\left[\sum_{i=1}^{10}(2i-1)\left(\ln F_{\theta}(x_i)+\ln\left(1-F_{\theta}(x_{n+1-i})\right)\right)\right]\Big/20 - 20 = 0.295964, \tag{116}$$

$$A_{\mathrm{mod}}^2 = A^2\left(1+\frac{0.2}{\sqrt{n}}\right) = 0.3092 < A_{\alpha=0.05}^2 = 0.757. \tag{117}$$

Thus, there is not evidence to rule out the Weibull lifetime model (1).

To minimize the probability of product lot misclassification, the separation threshold "*h*" is determined via (108) and (109) as follows:

$$h = \arg\min_h \alpha(h \mid n) = \arg\min_h[\alpha_p(h \mid n)+\alpha_c(h \mid n)]$$

$$= \arg\min_h\left(\Pr\left(\widehat{\beta}^{\delta} \le h \mid \beta_a^{\delta}, n\right)+\Pr\left(\widehat{\beta}^{\delta} > h \mid \beta_r^{\delta}, n\right)\right) = 278.8006, \tag{118}$$

where n=20, δ=3, $\beta_a^{\delta} = 481.692225$, $\beta_r^{\delta} = 175.543814$, and

$$\alpha_p(h \mid n)+\alpha_c(h \mid n) = 0.01524 + 0.010362 = 0.025602. \tag{119}$$

Since

$$\widehat{\beta}^{\delta} = (8.494792)^3 = 612.9969 > h = 278.8006, \tag{120}$$

the product lot is accepted.

Table 2. Critical values for A_{mod}^2

α	0.1	0.05	0.025	0.01
A_{α}^2	0.637	0.757	0.877	1.038

Situation 2

In the case of Situation 1, the costs of product lot misclassification are defined as follows: (1) *zero* for correct classification, (2) c_p=2 (*in terms of money*) when an estimator $\widehat{\beta}^{\delta}$ from R_a is incorrectly classified as R_r, and (3) c_c=3 (*in terms of money*) when an estimator $\widehat{\beta}^{\delta}$ from R_r is incorrectly classified as R_a.

The problem is to find the separation threshold "*h*" which minimizes the *expected cost of product lot misclassification* (ECPLM):

$$\text{ECPLM}(h \mid n) = c_p \alpha_p (h \mid n) + c_c \alpha_c (h \mid n)$$

$$= c_p \Pr\left(\widehat{\beta}^{\delta} \le h \mid \beta_a^{\delta}, n\right) + c_c \Pr\left(\widehat{\beta}^{\delta} > h \mid \beta_r^{\delta}, n\right). \tag{121}$$

To minimize (121), the separation threshold "*h*" is determined as follows:

$$h = \arg\min_h \text{ECPLM}(h \mid n) = \arg\min_h [c_p \alpha_p (h \mid n) + c_c \alpha_c (h \mid n)]$$

$$= \arg\min_h \left(c_p \Pr\left(\widehat{\beta}^{\delta} \le h \mid \beta_a^{\delta}, n\right) + c_c \Pr\left(\widehat{\beta}^{\delta} > h \mid \beta_r^{\delta}, n\right)\right) = 284.4, \tag{122}$$

where n=20, δ=3, $\beta_a^{\delta} = 481.692225$, $\beta_r^{\delta} = 175.543814$, c_p=2, c_c=3, and

$$c_p \alpha_p (h \mid n) + c_c \alpha_c (h \mid n) = 0.036713 + 0.023437 = 0.06015. \tag{123}$$

Since

$$\widehat{\beta}^{\delta} = (8.494792)^3 = 612.9969 > h = 284.4, \tag{124}$$

the product lot is accepted.

Situation 3

Consider Situation 2 when there is the cost c=0.02 (*in terms of money*) of sampling and putting a product item on test, but the size "*n*" of a random sample ($X_1 \le \ldots \le X_n$) of product items to test the Weibull MTTF "μ", the producer's risk, which is denoted by α_p, and the consumer's risk, which is denoted by α_c, are not preassigned.

The problem is to find the optimal separation threshold "*h*" and integer "*n*", which minimize the performance index:

$$cn + c_p \alpha_p (h, n) + c_c \alpha_c (h, n)$$

$$= cn + c_p \Pr\left(\widehat{\beta}^\delta \le h \mid \beta_a^\delta, n\right) + c_c \Pr\left(\widehat{\beta}^\delta > h \mid \beta_r^\delta, n\right). \quad (125)$$

The optimal separation threshold "h" for $\widehat{\beta}^\delta$ and integer "n" are given by

$$(h,n) = \arg\min_n \left[cn + \left(\min_h [c_p \alpha_p (h \mid n) + c_c \alpha_c (h \mid n)] \right) \right]$$

$$= \arg\min_n \left[cn + \min_h \left(c_p \Pr\left(\widehat{\beta}^\delta \le h \mid \beta_a^\delta, n\right) + c_c \Pr\left(\widehat{\beta}^\delta > h \mid \beta_r^\delta, n\right) \right) \right]$$

$$= (286.2665,\ 15), \quad (126)$$

where $\delta = 3$, $\beta_a^\delta = 481.692225$, $\beta_r^\delta = 175.543814$, $c_p = 2$, $c_c = 3$, $c = 0.02$, and

$$cn + c_p \alpha_p (h \mid n) + c_c \alpha_c (h \mid n) = 0.3 + 0.077519 + 0.048021 = 0.42554. \quad (127)$$

Since

$$\widehat{\beta}^\delta = \sum_{i=1}^{n=15} X_i^\delta / n = 624.7586 > h = 286.2665, \quad (128)$$

the product lot is accepted.

Situation 4

Consider the situation when the size "n" of a random sample ($X_1 \le \ldots \le X_n$) of product items to test the Weibull MTTF "μ" is not preassigned, but a product lot misclassification probability is preassigned:

$$\alpha(h,n) = \alpha = 0.1. \quad (129)$$

The problem is to find the optimal separation threshold "h" and the required smallest integer "n", which minimize the performance index:

$$[\alpha(h,n) - \alpha]^2$$

$$= \left[\Pr\left(\widehat{\beta}^\delta \le h \mid \beta_a^\delta, n\right) + \Pr\left(\widehat{\beta}^\delta > h \mid \beta_r^\delta, n\right) - \alpha \right]^2. \quad (130)$$

The optimal separation threshold "h" for $\widehat{\beta}^\delta$ and the required smallest integer "n" are given by

$$(h,n) = \arg\min_n \left(\min_h [\alpha(h \mid n) - \alpha]^2 \right)$$

$$= \arg\min_{n}\left[\min_{h}\left(\Pr\left(\widehat{\beta}^{\delta} \le h \mid \beta_{a}^{\delta}, n\right) + \Pr\left(\widehat{\beta}^{\delta} > h \mid \beta_{r}^{\delta}, n\right) - \alpha\right)^{2}\right]$$

$$= (285.5056,\ 11), \tag{131}$$

where δ=3, $\beta_{a}^{\delta} = 481.692225$, $\beta_{r}^{\delta} = 175.543814$, $\alpha = 0.1$, and

$$\alpha(h,n) = \alpha_{p}(h,n) + \alpha_{c}(h,n) = 0.067952 + 0.032048 = 0.1. \tag{132}$$

Since

$$\widehat{\beta}^{\delta} = \sum_{i=1}^{n=11} X_{i}^{\delta} / n = 645.3054 > h = 285.5056, \tag{133}$$

the product lot is accepted.

Situation 5

Consider the situation when the size "n" of a random sample ($X_1 \le ... \le X_n$) of product items to test the Weibull MTTF "μ" is not preassigned, but the producer's risk, which is denoted by α_p, and the consumer's risk, which is denoted by α_c, are preassigned:

$$\alpha_{p} = 0.04, \quad \alpha_{c} = 0.06. \tag{134}$$

The problem is to find the optimal separation threshold "h" and the required smallest integer "n", which minimize the performance index:

$$(\alpha_{p}(h,n) - \alpha_{p})^{2} + (\alpha_{c}(h,n) - \alpha_{c})^{2}$$

$$= \left(\Pr\left(\widehat{\beta}^{\delta} \le h \mid \beta_{a}^{\delta}, n\right) - \alpha_{p}\right)^{2} + \left(\Pr\left(\widehat{\beta}^{\delta} > h \mid \beta_{r}^{\delta}, n\right) - \alpha_{c}\right)^{2}. \tag{135}$$

The optimal separation threshold "h" for $\widehat{\beta}^{\delta}$ and the required smallest integer "n" are given by

$$(h,n) = \arg\min_{n}\left(\min_{h}[(\alpha_{p}(h \mid n) - \alpha_{p})^{2} + (\alpha_{c}(h \mid n) - \alpha_{c})^{2}]\right)$$

$$= \arg\min_{n}\left(\min_{h}\left[\left(\Pr\left(\widehat{\beta}^{\delta} \le h \mid \beta_{a}^{\delta}, n\right) - \alpha_{p}\right)^{2} + \left(\Pr\left(\widehat{\beta}^{\delta} > h \mid \beta_{r}^{\delta}, n\right) - \alpha_{c}\right)^{2}\right]\right)$$

$$= (263.3863,\ 11), \tag{136}$$

where δ=3, $\beta_a^\delta = 481.692225$, $\beta_r^\delta = 175.543814$, α_p = 0.04, α_c = 0.06, and

$$\alpha_p(h,n) = 0.043232, \tag{137}$$

$$\alpha_c(h,n) = 0.061749, \tag{138}$$

$$\alpha_p(h,n) + \alpha_c(h,n) = 0.10498187. \tag{139}$$

Since

$$\widehat{\beta}^\delta = \sum_{i=1}^{n=11} X_i^\delta / n = 645.3054 > h = 263.3863, \tag{140}$$

the product lot is accepted.

CONCLUSION

The main aim of this chapter is to promote the novel ideas for effective optimization (constrained or unconstrained) of product acceptance processes in terms of misclassification probability under parametric uncertainty of underlying models with real practical applications. It is expected that these ideas will give interesting and novel contributions to statistical theory and its applications at a good mathematical level, where the theoretical results are obtained through the frequentist (non-Bayesian) statistical approach. Frequentist probability interpretations of the methods proposed are clear. Bayesian methods are not considered here. It will be noted, however, that although subjective Bayesian approach has a clear personal probability interpretation, it is not generally clear how this should be applied to non-personal prediction or decisions. Objective Bayesian methods, on the other hand, do not have clear probability interpretations in finite samples.

The methodology described here can be extended in several different directions to handle various problems that arise in practice. It is based on the pivotal quantity averaging (PQA) technique which allows one to eliminate the unknown parameters from the problem and to use available statistical information as completely as possible.

While the details of problems considered in the chapter can change significantly from one industry to the next, the focus is always on making better statistical decisions and predictive inferences.

REFERENCES

Baklizi, A. (2003). Acceptance sampling based on truncated life tests in the Pareto distribution of the second kind. *Advances and Applications in Statistics*, *3*, 33–48.

Balakrishnan, N., Leiva, V., & Lopez, J. (2007). Acceptance sampling plans from truncated life tests based on the generalized Birnbaum-Saunders distribution. *Communications in Statistics. Simulation and Computation*, *36*(3), 643–656. doi:10.1080/03610910701207819

Dodson, B. (1994). *Weibull Analysis*. Milwaukee, WI: ASQ Quality Press.

Epstein, B. (1954). Truncated life tests in the exponential case. *Annals of Mathematical Statistics*, *68*(3), 555–564. doi:10.1214/aoms/1177728723

Fertig, F. W., & Mann, N. R. (1980). Life-test sampling plans for two-parameter Weibull populations. *Technometrics*, *22*(2), 165–177. doi:10.1080/00401706.1980.10486131

Goode, H. P., & Kao, J. H. K. (1961). Sampling plans based on the Weibull distribution. In *Proceedings of Seventh National Symposium on Reliability and Quality Control* (pp. 24–40). Philadelphia: Academic Press.

Gupta, S. S. (1962). Life test sampling plans for normal and lognormal distribution. *Technometrics*, *4*(2), 151–175. doi:10.1080/00401706.1962.10490002

Gupta, S. S., & Groll, P. A. (1961). Gamma distribution in acceptance sampling based on life tests. *Journal of the American Statistical Association*, *56*(296), 942–970. doi:10.1080/01621459.1961.10482137

Kantam, R. R. L., & Rosaiah, K. (1998). Half logistic distribution in acceptance sampling based on life tests. *IAPQR Transactions*, *23*, 117–125.

Kantam, R. R. L., Rosaiah, K., & Srinivasa Rao, G. (2001). Acceptance sampling based on life tests: Log-logistic model. *Journal of Applied Statistics*, *28*(1), 121–128. doi:10.1080/02664760120011644

Nechval, N. A., Berzins, G., Purgailis, M., & Nechval, K. N. (2008). Improved estimation of state of stochastic systems via invariant embedding technique. *WSEAS Transactions on Mathematics*, *7*, 141–159.

Nechval, N. A., Purgailis, M., Berzins, G., Cikste, K., Krasts, J., & Nechval, K. N. (2010). Invariant embedding technique and its applications for improvement or optimization of statistical decisions. In K. Al-Begain, D. Fiems, & W. Knottenbelt (Eds.), *Analytical and Stochastic Modeling Techniques and Applications, (LNCS)* (Vol. 6148, pp. 306–320). Berlin: Springer-Verlag. doi:10.1007/978-3-642-13568-2_22

Nechval, N. A., & Vasermanis, E. K. (2004). *Improved Decisions in Statistics*. Riga: Izglitibas soli.

Rosaiah, K., & Kantam, R. R. L. (2005). Acceptance sampling based on the inverse Rayleigh distribution. *Economic Quality Control*, *20*(2), 277–286. doi:10.1515/EQC.2005.277

Rosaiah, K., Kantam, R. R. L., & Santosh Kumar, Ch. (2006). Reliability test plans for exponentiated log-logistic distribution. *Economic Quality Control*, *21*(2), 165–175. doi:10.1515/EQC.2006.279

Sobel, M., & Tischendrof, J. A. (1959). Acceptance sampling with life test objective. In *Proceedings of Fifth National Symposium on Reliability and Quality Control* (pp. 108–118). Philadelphia: Academic Press.

Squeglia, N. L. (2008). *Zero Acceptance Number Sampling Plans* (5th ed.). Milwaukee, WI: ASQ Quality Press.

Srinivasa, R. G., & Kantam, R. R. L. (2010). Acceptance sampling plans from truncated life tests based on the log-logistic distribution for percentiles. *Economic Quality Control*, *25*, 153–167.

Stephens, K. S. (2001). *The Handbook of Applied Acceptance Sampling Plans, Procedures and Principles*. Milwaukee, WI: ASQ Quality Press.

Tsai, T. R., & Wu, S. J. (2006). Acceptance sampling based on truncated life tests for generalized Rayleigh distribution. *Journal of Applied Statistics*, *33*(6), 595–600. doi:10.1080/02664760600679700

Wu, C. J., & Tsai, T. R. (2005). Acceptance sampling plans for Birnbaum-Saunders distribution under truncated life tests. *International Journal of Reliability Quality and Safety Engineering*, *12*(06), 507–519. doi:10.1142/S0218539305001987

Chapter 2
Combination of Stochastic Methods for Solving ELD Problem of Thermal Power Generation System

Dipankar Santra
RCC Institute of Information Technology, India

Krishna Sarker
Saroj Mohan Institute of Technology, India

Jayanti Sarker
Techno India, India

Anirban Mukherjee
RCC Institute of Information Technology, India

Subrata Mondal
Jadavpur University, India

ABSTRACT

This chapter reports a hybrid optimization technique, a combination of stochastic methods – particle swarm optimization (PSO) and ant colony optimization (ACO), which is applied to find economic dispatch schedule and minimum generation cost for convex and non-convex power generation system simulated in MATLAB. A 40-generator system is considered here with combinations of valve point loading, ramp rate limit, and prohibited operating zone. The output is also noted when transmission loss is taken into consideration. The results are found better than those of many other hybrid methods. Considering the quality of the solution obtained and nature of convergence, PSO-ACO may be accepted as a good alternative for solving ELD problems of varying complexity. Though PSO has been extensively used in ELD problems for its flexibility, robustness, and fast convergence, it often produces suboptimal solution due to its premature convergence. ACO, on the other hand, known for its good global exploration feature, imparts better balance between local and global search when combined with PSO.

DOI: 10.4018/978-1-5225-5045-7.ch002

INTRODUCTION

Economic load dispatch (ELD) has become an indispensable function in the operation and control of modern power generation system. One objective of ELD is to find the optimum generation schedule for the generating units whereby the power demand can be met satisfying the transmission loss at minimum generation cost. Improvements in scheduling the unit outputs can lead to significant cost savings. The fuel cost being the major component of the total cost of power generation in thermal power plants, the fuel cost function is considered as the objective function of the ELD problem. Grossly representing, the fuel cost function of a generator is a simple quadratic function while the constraints are linear in nature. However, in large thermal power plants with higher capacity turbines, the fuel cost function is highly non-smooth and non-convex owing to the effect of valve-point loading (VPL) and non-linear constraints like ramp rate limit (RRL) and prohibited operating zone (POZ). For realistic modeling the fuel cost function is taken as a segmented piecewise nonlinear function; this renders complexity to the load optimization problem since the cost function contains discontinuous values at each boundary forming multiple local optima which prevents the classical optimization methods (Grigsby, 2007) from obtaining the global optima. Also the complication of the problem increases significantly with increase in the number of generating units because of their combinatorial nature. This paper presents a novel solution approach to complex ELD problems of 40-generating units with non smooth fuel cost function i.e. with VPL effect together with singular or multiple constraints (generator constraint, RRL and POZ) and with and without transmission loss.

Over the years different soft computing approaches have been employed to find optimal solution of ELD problem. Neural Network (NN) (Mohammadi & Varahram, 2006) and Abdelaziz et al., 2008), Firefly Algorithm (FA) (Yang et al., 2012), Genetic Algorithm (GA) (Won & Park, 2003; Chiang, 2005), Bacterial Foraging Optimization (BFO) (Hazra & Sinha, 2008), Particle Swarm Optimization (PSO) (Gaing, 2003; Park et al., 2005), Ant Colony Optimization (ACO) (Swarup, 2005; Pothiya et al., 2010), Biogeography-Based Optimization (BBO) (Bhattacharya & Chattopadhyay, 2010), Evolutionary Programming (EP) (Sinha et al., 2003; Jayabarathi et al., 2005), Differential Evolution (DE) (Nomana & Iba. 2008), Gravitational Search Algorithm (GSA) (Duman et al., 2010) and Pattern Search (PS) (Al-Sumait et al., 2007) are some of the well known approaches tried. Predominantly in the 21st century, a new trend has been the development and application of various hybrid soft computing approaches (in ELD problem) producing interesting and improved results. A detailed account of research on various original and hybrid techniques to solve ELD problems of different complexity and for small to large power generation systems has been provided by Santra et al. (2014).

The focus of the present study is hybrid soft computing approach in 40-generator ELD problem. Out of the many studies found in literature, two distinct trends are observed in addressing 40-generator ELD – one is without transmission loss and the other one, with transmission loss. Very few have addressed both which the present study does. Some of the well documented hybrid approaches that do not consider loss are: Chaotic Differential Evolution and Sequential Quadratic Programming (DEC-SQP) developed by Coelho & Mariani (2006) uses DE with chaos sequences as the global optimizer and the SQP to sequentially fine-tune the DE run. Bhattacharya & Chattopadhyay (2010) proposed combination DE and BBO that improves the global search capability of DE by employing BBO effectively to generate promising candidate solutions of ELD. The migration operator of BBO is combined with mutation, crossover and selection operators of DE to enhance the convergence speed and improve quality of solution. Al-sumait et al. (2010) proposed hybrid GA–PS–SQP by combining GA, PS and SQP techniques to find effective solution (of 40-generator ELD problem). It uses GA as the global optimizer and PS and SQP to influence GA output to enhance reliability of the solution. The advantage of the hybridization is that the user is no more required to initiate the algorithmic search with a chosen input; the output is more robust and the error bound of all solutions is quite narrow.

In DE-Harmony Search (DE-HS) hybrid introduced by Dinh et al. (2013), the slow convergence and dimensionality problem of classical HS has been overcome by borrowing the mutation strategy of DE algorithm. The use of differential mutation and crossover enhanced the exploitation and exploration capability in DEHS. Hybridization of Chemical Reaction Optimization (CRO) technique with DE was applied by Roy et al. (2014) in ELD problem with VPL effect. In HCRO-DE, mutation operator of DE is combined with collision and decomposition operation of CRO to achieve better exploration and exploitation of search space than that obtained by DE alone. Subathra et al. (2015) proposed a hybrid of Cross-Entropy (CE) method and SQP technique. While CE guides the exploratory search mechanism, SQP fine-tunes the solution candidates by gradient search mechanism. Effectively CE-SQP hybrid achieves faster convergence, reduced computational complexity and balanced interplay between local and global search. Sayah and Hamouda (2013) applied the hybrid DE-PSO algorithm in large scale power system (with valve-point loading). To augment the global exploration property of DE, PSO is integrated as a supplementary mutation operator thereby also avoiding the premature convergence to local minima. The differential information obtained by DE combined with the experiential information generated by PSO creates a viable solution.

Very few studies and methods are found that deal with 40-generator system with transmission loss. Most of the approaches consider non-smooth cost function but non-linear constraints RRL and POZ are not taken into account. Ciornei & Kyriakides (2012) proposed GA-API, a new hybrid of Real Coded GA (RCGA) by combining it with a special class of Ant Colony Optimization called API to successfully test with convex and non-convex ELD where RRL, POZ and Spinning Reserve constraints were considered along with transmission loss. In this method, the good spreading of GA in the solution space is complemented with good search capability of API around the global solution. Reddy & Vaisakh (2013) presented a hybrid Shuffled Differential Evolution (SDE) algorithm which blended the strong features of Shuffled Frog Leaping Algorithm (SFLA) and DE to solve ELD problem with combination of valve point effect and transmission loss. SDE, characterized by a novel mutation operator, combines the exploration and exploitation mutation capabilities of DE and thus helps overcome the intrinsic limitation of DE and SFLA in solving large scale non-convex ELD.

Hazra et al. (2015) have presented Oppositional Chemical Reaction Optimization (OCRO) algorithm where OBL is integrated with the Chemical Reaction Optimization (CRO) (developed by Lam & Li (2010)) based on dynamics of chemical reaction. OCRO is characterized by the fitness (or closeness to the global optima) of an opposite candidate solution and also an improved diversity among the population. In effect better convergence is achieved as OCRO is applied to ELD involving valve point effect and transmission loss. Barisal & Prusty (2015) introduced Oppositional Invasive Weed Optimization (OIWO) by hybridizing IWO, characterized by the encompassing nature of weeds, and OBL featuring quasi-opposite numbers. IWO is a novel population based stochastic, derivative-free optimization algorithm developed by Mehrabian & Lucas (2006). The algorithm exploits some characteristic features of weeds namely, fast replication and spreading, distribution and self adjustment to the climatic changes. OBL helps improve the convergence rate of IWO thereby producing satisfying results of large scale ELD problem that takes into account valve point loading effect and transmission loss.

From the above citations and relevant literature it is observed that PSO-ACO hybrid soft computing approach has never been tried to optimize large scale load dispatch though mostly PSO and to some extent ACO have been applied individually or in combination with other methods in different ELD problems quite successfully. Invented by Kennedy and Eberhartin in 1995, PSO has been found to yield better and faster solution, satisfying all the constraints for both convex and non-convex ELD. It is found that execution of PSO is very fast towards the convergence point, so number of iteration is less compared to other meta-heuristic techniques. But at later stages when fine tuning operation is required to reach global optima, PSO gives unsatisfactory result due to large accumulation of particles at 'gbest' position. ACO is another probabilistic optimization technique that was introduced by Marco Dorigo in 1992. It is observed that ACO is a powerful yet flexible heuristic algorithm with a good downhill behaviour to better explore around the global solution region than PSO. Again for complex problems, ACO suffers from stagnation and slow convergence. Taking advantage of the mutually complementary contrasting character of PSO (with good local search ability) and ACO (with global search capability) the hybridization of PSO with ACO has been effectively used by the present authors while applying the hybrid method to simulated 40-generator system with and without transmission loss and the different test cases along with comparison of output with other hybrid approaches are reported in this study.

The paper is organized as follows. The 2nd section of the paper on 'ELD Formulation' provides mathematical formulation of standard ELD problem along with its different associated constraints and their brief description. The standard PSO and ACO algorithms followed by the PSO-ACO hybrid approach are described in the subsequent sections on 'PSO', 'ACO' and 'PSO-ACO' respectively. Results of simulation studies are presented and discussed in the last but one section on 'Experiment, Result and Analysis'. Finally the 'Conclusion' in the last section highlights the salient points of this study and indicates future scope of research.

ELD FORMULATION

The ELD problem is generally formulated as a nonlinear optimization problem as shown in (1) where $C_i(P_i)$ is the cost function of the ith generator, and is usually expressed as a quadratic polynomial. The total generation cost F is the arithmetic sum of the individual generator's cost function and the objective of ELD is to minimize the total cost.

$$\text{Minimize } F = \sum_{i=1}^{n} C_i(P_i) = \sum_{i=1}^{n} a_i + b_i P_i + c_i P_i^2 \tag{1}$$

In (1) a_i, b_i and c_i are the cost coefficients of the ith generator; n is the number of committed generators, in our study $n = 40$; P_i is the power output of the ith generator.

ELD problem comes with following constraints out of which power balance with transmission loss and generator capacity constraints are linear in nature while the rest are non-linear.

1. Real Power Balance Constraint

It is well known the total power output ($\sum_{i=1}^{n} Pi$) from all the (n nos) generating units must equal the total load demand (P_D) and the transmission loss (P_L). Therefore it can be expressed as:

$$\sum_{i=1}^{n} P_i = P_D + P_L \tag{2}$$

2. Transmission Loss

This may be expressed in terms of B-coefficients as:

$$P_L = \sum_{i=1}^{n} \sum_{j=1}^{n} P_i B_{ij} P_j + \sum_{i=1}^{n} P_i B_{0i} + B_{00} \tag{3}$$

Here B_{ij}, B_{0i} and B_{00} are the transmission network power loss B-coefficients; B_{ij} is the ij^{th} element of loss coefficient symmetric matrix B, B_{0i} is the ith element of the loss coefficient vector and B_{00} is the loss coefficient constant.

3. Generator Capacity Constraint

The power generated by each generator should be within their respective lower ($P_i^{\min}$) and upper limits ($P_i^{\max}$) of active power generation as expressed in relation (4).

$$P_i^{\min} \leq P_i \leq P_i^{\max} \tag{4}$$

4. Ramp Rate Limit (RRL)

Rigorous change of power output by any generating unit is restricted by the ramp-up limit (UR_i) and ramp-down limit (DR_i) which are the physical properties of a unit. A generating unit, which used to generate P_{i0} can just increase or decrease its generation (P_i) to some extent as expressed in relations (5) and (6).

As generation increases: $P_i - P_{i0} \leq \mathrm{UR}_i$ (5)

As generation decreases: $P_{i0} - P_i \leq \mathrm{DR}_i$ (6)

Combining relations (4), (5) and (6) the power generation constraint can be expressed as:

$$\max\left(P_i^{\min}, P_{i0} - \mathrm{DR}_i\right) \leq P_i \leq \min\left(P_i^{\max}, P_{i0} + \mathrm{UR}_i\right) \quad (7)$$

5. Prohibited Operating Zone (POZ)

A prohibited operating zone of a generator is the span of output power where the procedural operation causes undue disturbance. Normally power adjustment is not allowed in such spans. The allowable operating zones of a generator can be described as follows:

$$P_i^{\min} \leq P_i \leq P_{i,1}^{\text{lower}}; \quad i = 1, 2, \ldots n,$$

$$P_{i,j-1}^{\text{upper}} \leq P_i \leq P_{i,j}^{\text{lower}}; \; j = 2, 3, ..n_i,$$

$$P_{i,n_i}^{\text{upper}} \leq P_i \leq \mathrm{P}_i^{\max} \quad (8)$$

Total number of POZs of the ith unit is n_i. $P_{i,j-1}^{\text{upper}}$ is the upper boundary and $P_{i,j}^{\text{lower}}$ is the lower boundary of the jth POZ of the ith unit.

6. Valve Point Loading

The valve point effect is the ripple effect seen in the otherwise smooth fuel curve when steam admission valves in turbines are released. To incorporate this realistic effect, the objective function F is formulated as the superposition of sinusoidal functions and quadratic functions as shown in (9). A non-smooth curve corresponds to (9) unlike a smooth quadratic polynomial curve representing (1).

$$\text{Minimize } F = \sum_{i=1}^{\mathrm{n}} C_i(P_i)$$

$$= \sum_{i=1}^{n} a_i + b_i P_i + c_i (P_i)^2 + \left| e_i \sin\left(f_i \left(P_i^{\min} - P_i\right)\right)\right| \quad (9)$$

Where e_i and f_i are the valve point coefficient of ith generating unit reflecting valve point effects and $P_i^{\min}$ is the minimum power output of unit i.

PARTICLE SWARM OPTIMIZATION (PSO)

PSO is a population-based heuristic search algorithm that emulates the movement of swarm like a 'flock of birds' in search of food. Birds in a group randomly search for a piece of food in a particular area and in the process some of them come closer to the food source. The other birds follow the bird nearest to the food and thereby in each movement improve their position and form better idea of how far the food is from their respective current position. This natural strategy of the swarm is followed by PSO to solve the non-linear and discontinuous optimization problems.

A "bird" is called a "particle" in PSO and it is analogous to a potential solution in the search space. Each particle has fitness, position and velocity. The fitness of each particle is assessed by the fitness function (i.e. the fuel cost function in case of ELD problem). The particles take flight in the search space with initial random velocity assigned to them and then make parallel searches for optima by updating their velocity and position dynamically. In every iteration the particle swarm optimizer tracks two updated values – one is the '*pbest* or *personal best*' i.e. the best value (fitness) achieved so far by a given particle while the other is the '*gbest* or *global best*' i.e. the best value attained so far by the population. Based on '*pbest*' and '*gbest*', the velocity and position of the particles are updated vide equations (10) and (11) respectively. Each particle progresses in the search space by judging its next position with respect to '*pbest*' and '*gbest*' position.

$$V_i^{t+1} = wV_i^t + c_1\text{rand}_1\left(\right)\left(\text{pbest}_i - X_i^t\right) + c_2\text{rand}_2\left(\right)\left(\text{gbest}_i - X_i^t\right) \tag{10}$$

$$X_i^{t+1} = X_i^{(t)} + V_i^{(t+1)} \tag{11}$$

Here, V_i^t is the velocity of particle i at iteration t and $V_{\text{min}} \leq V_i^t \leq V_{\text{max}}$; w is the inertia weight factor; c_1, c_2 are the acceleration coefficients; $rand_1$, $rand_2$ are random numbers between 0 and 1; X_i^t is the position of particle i at iteration t.

Generally, the inertia weight w that carries the exploration property of PSO is set according to the following equation:

$$w = w_{\text{max}} - \frac{w_{\text{max}} - w_{\text{min}}}{t_{\text{max}}} t \tag{12}$$

Here, w_{max} and w_{min} are the maximum and minimum values of the weight factor. t_{max} is the maximum number of iterations and t is the current number of iteration.

Again the acceleration coefficients c_1 and c_2 are set by following two expressions:

$$c_1 = c_{1\text{max}} - \frac{C_{1\text{max}} - C_{1\text{min}}}{\text{t}_{\text{max}}} t \tag{13}$$

$$c_2 = c_{2max} - \frac{C_{2max} - C_{2min}}{t_{max}} t \quad (14)$$

Here, w_{min}, w_{max} are opening and closing weights; c_{1min}, c_{1max} are initial and final cognitive factors and; c_{2min}, c_{2max} are initial and final social factors. If $c_1 > c_2$, the particle approaches '*pbest*' or the local optima, else it converges to '*gbest*' or the global optima.

ANT COLONY OPTIMIZATION (ACO)

ACO is another swarm-based method which, in search of best solution, follows the strategy of movement of an ant colony towards the source of food through the shortest path. It is found that during movement each ant leaves behind a trail of chemical substance, called pheromone. The other ants are attracted by this chemical essence and breaking their random motion they trace the pheromone trail along the path used by the predecessor ants. Now in search of food, the ant which finds food earlier through the shorter path causes more pheromone deposited in this path compared to what remains deposited in the longer path. This is because, the longer the path the more the pheromone trail gets evaporated, thus reducing its strength. Other ants are most likely to follow the shorter route as the chemical essence is more in this route. In doing so, they will reconstruct the existing pheromone trail by spreading additional amount of pheromone on the way. Conversely, less travelled path will have its pheromone layer evaporated, thus reducing its attraction.

During search the ants use two types of information. The private or local information concerns the ant's own experience of visited nodes and the public or global information concerns other ant's experience accessed through pheromone trail. Though each ant finds a new solution, better solutions are yielded by exchanging information with other ants through the public source (pheromone). Thus, analogous to an ant, the ACO algorithm constructively builds or improves a solution to an optimization problem by moving through nodes (or states) of a neighborhood graph.

The ACO mechanism has three logical parts – 1. initialization, 2. transition rule, and 3. pheromone trail update rule

1. Initialization

In this phase the parameters like the required number of ants, the relative importance of the pheromone trail, visibility, quantity of pheromone deposited on a path, evaporation factor, tuning factor etc. are initialized with suitable values.

2. Transition Rule

The transition probability of an ant k from one state i to next state j is given by:

$$P_{ij}^k(t) = \frac{[\tau_{ij}]^\alpha [\eta_{ij}]^\beta}{\sum_{l \in N_i^k} [\tau_{il}(t)]^\alpha [\eta_{il}]^\beta} \quad \text{if} \quad j \in N_i^k \quad (15)$$

Here, N_i^k is the feasible neighbourhood of ant k at position i, l is the unvisited state or node, η_{ij} is a heuristic function given by the inverse of distance traveled, τ_{ij} is the pheromone trail intensity on edge (i, j), α is the relative importance of the trail $\alpha \geq 0$, β is the relative importance of the visibility $\beta \geq 0$.

3. Pheromone Trail Update Rule

- *Local update rule*

An ant changes the amount of pheromone level τ_{ij} on the visited edge (i, j) as given by:

$$\tau_{ij} = (1 - \varsigma)\, \tau_{ij} + \tau_0 \tag{16}$$

Here ς is a heuristically defined parameter for initial pheromone level.

- *Global update rule*

After all the ants build their individual solutions, global pheromone-update rule (17) is applied only to those paths that are visited by more ants and each ant tends to deposit more pheromone.

$$\tau_{ij} = (1 - \rho)\tau_{ij} + \rho\Delta\tau_{ij} \tag{17}$$

Here, ρ is the pheromone decay factor or evaporation rate $(0 < \rho < 1)$ and $\Delta\tau_{ijk}$ is the amount of pheromone deposited on edge (i,j) by best ant k.

PSO-ACO HYBRID

PSO is good in processing ELD problems with multiple non-linear constraints but it has some limitations. PSO has a tendency to get confined to local optima if, for a number of iterations, the '*gbest*' and '*pbest*' positions remain equal to the particle's position. If the '*gbest*' is not updated with iterations, other particles are drawn towards the '*gbest*' position. Again, If V_{max} is very high particles tends to fly past good local solutions. If V_{max} is very small, the particles may not explore sufficiently beyond good local solutions. Thus based on the velocity the particles are biased towards global or local search.

ACO when applied in ELD suffers from the disadvantage of slow convergence. The pheromone trail of ACO can potentially help diversification and avoid stagnation if ACO can be fed with the '*gbest*' solution found by PSO after a good number of iteration. Taking as input the '*gbest*' particles, the ant's response functions are evaluated, compared and information is exchanged with the best ant to finally arrive at the best solution. By using its memory an ant can retrace the same path backward and update the pheromone trails according to the quality of the solution. Also pheromone trail intensity decreases ('evaporates') over time which is important to avoid a rapid convergence of the algorithm to a sub-optimal region.

Considering the mutually complementary strength and weakness of PSO and ACO, the PSO-ACO hybrid algorithm has been taken up to experiment with multi-constraint ELD problem. The pseudocode and flowchart (Figure 1) of the hybrid PSO-ACO algorithm are given below.

```
Input data for PSO
Initialize particle's position and velocity with random values
```

```
While (terminal condition not satisfied)
          Find local and global fitness values, pbest and gbest
          Update inertia weight
               Update positions and velocities
If gbest solution improves Then
               Save the solution for ACO
End if
End while
Initialize ACO with the gbest found by PSO
While (terminal condition not satisfied)
               Generate ants and make path for each ant
               Evaluate and Compare response function
If response function does not improve Then
               Exchange information with best ant
               Generate path from local position to best ant
End if
End while
Generate optimum result
End
```

Figure 1. Flowchart of hybrid PSO-ACO

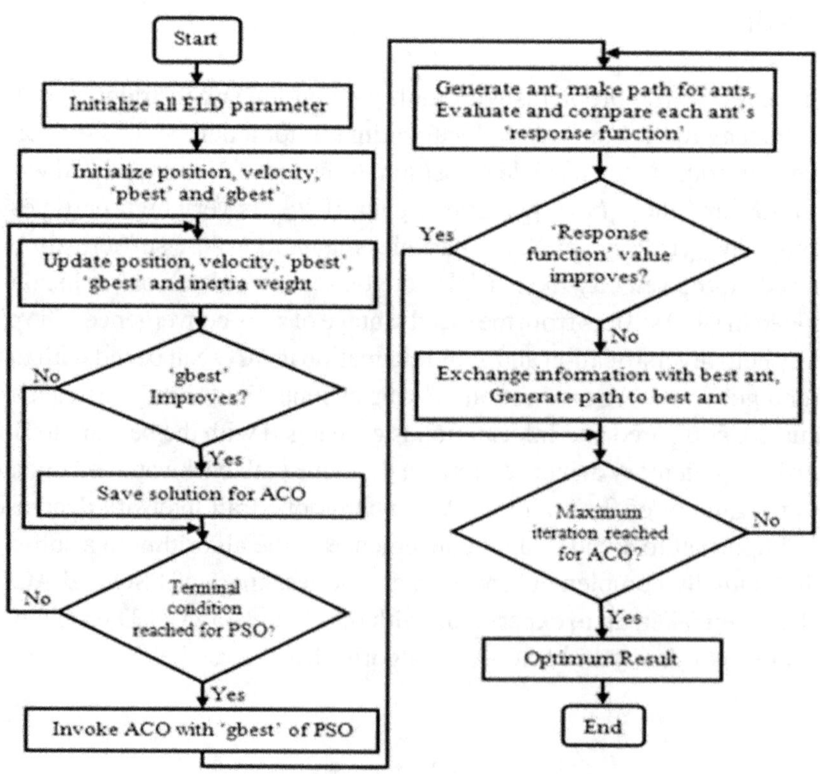

The values of different parameters of PSO-ACO algorithm as used in the MATLAB simulation programs are given below. Through trial and error some of the parameters to which PSO, in particular, is sensitive are fine tuned to obtain better result.

For PSO: Population Size =100, Generations = 500, Inertial constant (w) = 0.3, $iter_{max}$ = 1000, c_1 = 2, c_2 = 2, w_{max} = 0.9, w_{min} = 0.4

For ACO: No. of iterations = 500, No. of ants = 50, Initial pheromone value (τ_0) =0.0088, Pheromone update constant (Q) = 20, Exploration constant (q_0) = 1, $\rho = 0.05$, $\alpha = 1$, $\beta = 5$

EXPERIMENT, RESULT AND ANALYSIS

Proposed PSO-ACO algorithm has been applied to solve simulated 40-generator ELD problems in nine different test cases by varying the constraints and fuel cost function. The program has been written in MATLAB-7.8 and executed on a 64-bit PC with Intel Core i3 2.7 GHz processor, 1 TB HDD, 8 GB RAM and Windows 7 operating system. Following two major cases have been considered:

1. Without Transmission loss: Test Case 1 to 7
2. With Transmission loss: Test Case 8 and 9

50 runs of the PSO-ACO program in MATLAB have been executed for each test case and the best, worst and mean output values out of 50 sets of output are recorded. The success rate (of converging to best output), CPU time and number of iterations (for convergence) are also recorded and all values are listed in Table 4.

- **Test Case 1:** Only linear constraints, without VPL and transmission loss:

The values of basic ELD modeling parameters like cost coefficients a_i, b_i, c_i, and generator capacity constraints (P_{min}, P_{max}) for each generator are given in Table 1. The objective function to be minimized in this case is (1) and the linear constraints considered are (2) and (4). A comparison of output of PSO-ACO and hybrid Interior Point-assisted Differential Evolution (IPM-DE) algorithm tested by Duvvuru & Swarup (2011) is shown in Table 2. The convergence characteristics for this case is depicted in Figure 2.

- **Test Case 2:** With linear constraints plus RRL, without VPL and transmission loss:

In this case cost coefficients (a_i, b_i, c_i), and generator capacity constraints (P_{min}, P_{max}) for each generator are same as that used in Test Case 1. Here RRL is considered as a constraint in addition to the linear constraints (2) and (4). The upper and lower RRLs namely, UR_i and DR_i (refer (5)-(7)) are different for each generator and are listed in Table 1. The objective function to be minimized is again (1) as valve point loading is not considered. The convergence characteristics for this case is depicted in Figure 3.

- **Test Case 3:** With linear constraints plus POZ and RRL, without VPL and transmission loss:

This case considers POZ in addition to the other non-linear constraint RRL. The POZ (refer (8)) falls in the power output zones of 10th to 14th generating units as listed in the Table 1. These prohibited

Table 1. Values of cost coefficients and constraints parameters (Sinha et al., 2003; Wang et al., 2007) used in test cases 1-9

Unit	P_i^{min} (MW)	P_i^{max} (MW)	a_i (\$)	b_i (\$/MW)	c_i (\$/MW²)	e_i	f_i	UR_i (MW/h)	DR_i (MW/h)	P_i^0	Prohibited Zones (MW)
1	36	114	94.705	6.73	0.00690	100	0.084	114	114	100	-
2	36	114	94.705	6.73	0.00690	100	0.084	114	114	100	-
3	60	120	309.54	7.07	0.02028	100	0.084	120	120	90	-
4	80	190	369.03	8.18	0.00942	150	0.063	100	150	150	-
5	47	97	148.89	5.35	0.0114	120	0.077	97	97	80	-
6	68	140	222.33	8.05	0.01142	100	0.084	80	125	120	-
7	110	300	287.71	8.03	0.00357	200	0.042	165	200	280	-
8	135	300	391.98	6.99	0.00492	200	0.042	165	200	200	-
9	135	300	455.76	6.60	0.00573	200	0.042	165	200	230	-
10	130	300	722.82	12.9	0.00605	200	0.042	155	190	240	[130,150] [200,230] [270,299]
11	94	375	635.20	12.9	0.00515	200	0.042	150	185	210	[100,140] [230,280] [300,350]
12	94	375	654.69	12.8	0.00569	200	0.042	150	185	210	[100,140] [230,280] [300,350]
13	125	500	913.40	12.5	0.00421	300	0.035	206	235	230	[150,200] [250,300] [400,450]
14	125	500	1760.4	8.84	0.00752	300	0.035	260	290	355	[200,250] [300,350] [450,490]
15	125	500	1728.3	9.15	0.00708	300	0.035	186	215	350	-
16	125	500	1728.3	9.15	0.00708	300	0.035	186	215	350	-
17	220	500	647.85	7.97	0.00313	300	0.035	240	270	460	-
18	220	500	649.69	7.95	0.00313	300	0.035	240	268	470	-
19	242	550	647.83	7.97	0.00313	300	0.035	290	315	500	-
20	242	550	647.81	7.97	0.00313	300	0.035	290	315	500	-
21	254	550	785.96	6.63	0.00298	300	0.035	335	360	510	-
22	254	550	785.96	6.63	0.00298	300	0.035	335	360	520	-
23	254	550	794.53	6.66	0.00284	300	0.035	335	362	520	-
24	254	550	794.53	6.66	0.00284	300	0.035	350	378	450	-
25	254	550	801.32	7.10	0.00277	300	0.035	350	380	400	-
26	254	550	801.32	7.10	0.00277	300	0.035	350	380	520	-
27	10	150	1055.1	3.33	0.52124	120	0.077	95	145	20	-

continued on following page

Table 1. Continued

Unit	P_i^{min} (MW)	P_i^{max} (MW)	a_i ($)	b_i ($/MW)	c_i ($/MW²)	e_i	f_i	UR_i (MW/h)	DR_i (MW/h)	P_i^0	Prohibited Zones (MW)
28	10	150	1055.1	3.33	0.52124	120	0.077	95	145	20	-
29	10	150	1055.1	3.33	0.52124	120	0.077	98	145	25	-
30	47	97	148.89	5.35	0.01140	120	0.077	97	97	90	-
31	60	190	222.92	6.43	0.00160	150	0.063	90	145	170	-
32	60	190	222.92	6.43	0.00160	150	0.063	90	145	150	-
33	60	190	222.92	6.43	0.00160	150	0.063	90	145	190	-
34	90	200	107.87	8.95	0.0001	200	0.042	105	150	190	-
35	90	200	116.58	8.62	0.0001	200	0.042	105	150	150	-
36	90	200	116.58	8.62	0.0001	200	0.042	105	150	180	-
37	25	110	307.45	5.88	0.0161	80	0.098	110	110	60	-
38	25	110	307.45	5.88	0.0161	80	0.098	110	110	40	-
39	25	110	307.45	5.88	0.0161	80	0.098	110	110	50	-
40	242	550	647.83	7.97	0.00313	300	0.035	290	315	512	-

Figure 2. Convergence characteristics - Test Case 1

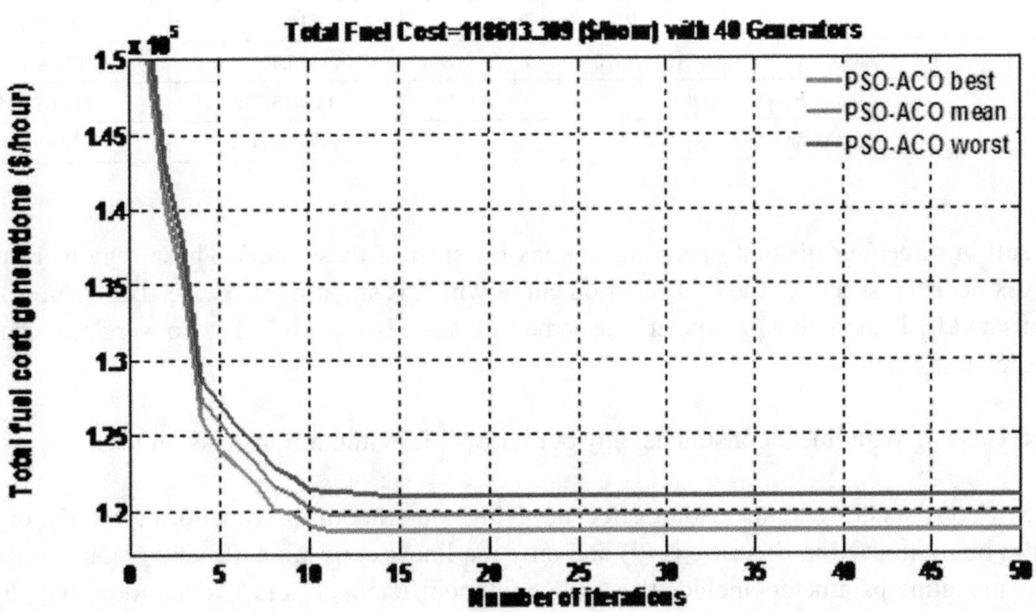

**For a more accurate representation see the electronic version.*

Table 2. Comparison of PSO-ACO with IPM-DE

Unit no.	Generation (MW)		Unit no.	Generation (MW)	
	PSO-ACO	IPM-DE		**PSO-ACO**	IPM-DE
P_1	**114.000**	113.9851	P_{21}	**550.000**	549.9999
P_2	**114.000**	113.9834	P_{22}	**550.000**	549.9970
P_3	**120.000**	119.9999	P_{23}	**550.000**	549.9852
P_4	**190.000**	189.9999	P_{24}	**550.000**	549.9983
P_5	**97.000**	96.9971	P_{25}	**550.000**	549.9983
P_6	**140.000**	139.9256	P_{26}	**550.000**	549.9964
P_7	**300.000**	299.9996	P_{27}	**10.000**	10.0000
P_8	**300.000**	299.9989	P_{28}	**10.000**	10.0007
P_9	**300.000**	299.9621	P_{29}	**10.000**	10.0011
P_{10}	**130.000**	130.0000	P_{30}	**97.000**	96.9939
P_{11}	**94.000**	94.0010	P_{31}	**190.000**	189.9953
P_{12}	**94.000**	94.0000	P_{32}	**190.000**	189.9998
P_{13}	**125.000**	125.1783	P_{33}	**190.000**	189.9987
P_{14}	**272.647**	271.6294	P_{34}	**200.000**	199.9836
P_{15}	**266.176**	266.5286	P_{35}	**200.000**	199.9987
P_{16}	**266.176**	267.3031	P_{36}	**200.000**	199.9927
P_{17}	**500.000**	499.9648	P_{37}	**110.000**	109.8171
P_{18}	**500.000**	499.9788	P_{38}	**110.000**	110.0000
P_{19}	**550.000**	549.9739	P_{39}	**110.000**	109.9957
P_{20}	**550.000**	549.9670	P_{40}	**550.000**	549.8707
Total Best Cost ($/h)				**118613.309**	118 678.15
Mean cost ($/h)				**119799.4421**	118820.97

zones result in three/four disjoint operating regions for each of these units. Those regions result in a non-convex decision space for the cost function curve which is smooth otherwise. The values of other constraints and ELD modeling parameters are same as that of Test Case 2. The convergence characteristics is shown in Figure 4.

- **Test Case 4:** With linear constraints only but with VPL, without transmission loss:

This case considers valve point loading effect implying non-smooth cost function curve. The objective function to be minimized in this case is (9) and only the linear constraints (2) and (4) are considered. The basic modeling parameters include the valve point coefficients e_i and f_i in addition to a_i, b_i, c_i, as listed in Table 1. The values of linear constraints are same as that of Test Case 1. The convergence characteristics is depicted in Figure 5.

Figure 3. Convergence characteristics - Test Case 2

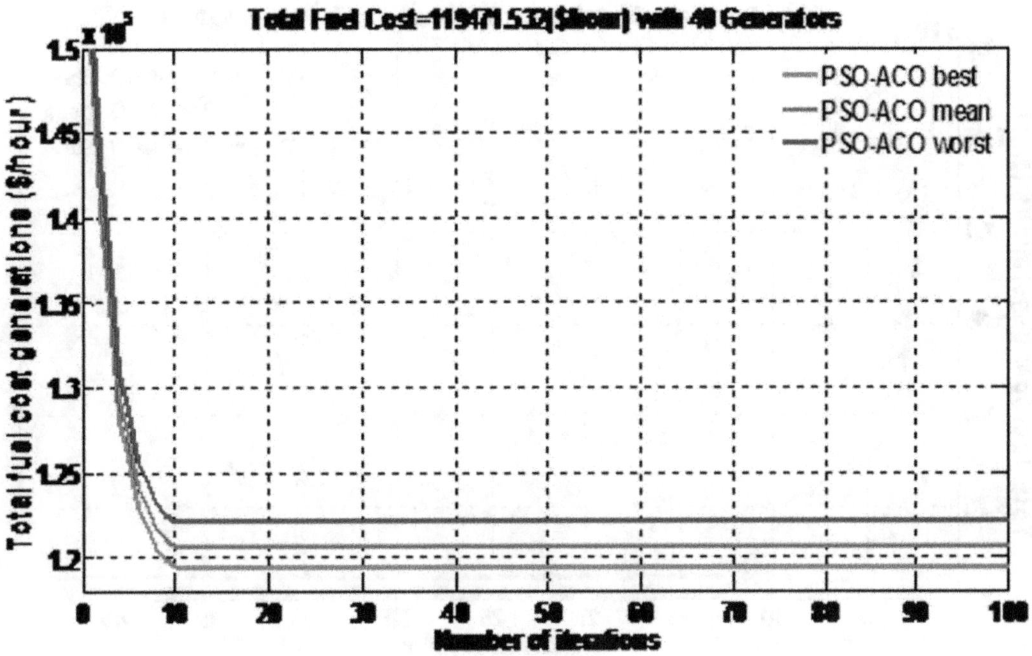

**For a more accurate representation see the electronic version.*

Figure 4. Convergence characteristics - Test Case 3

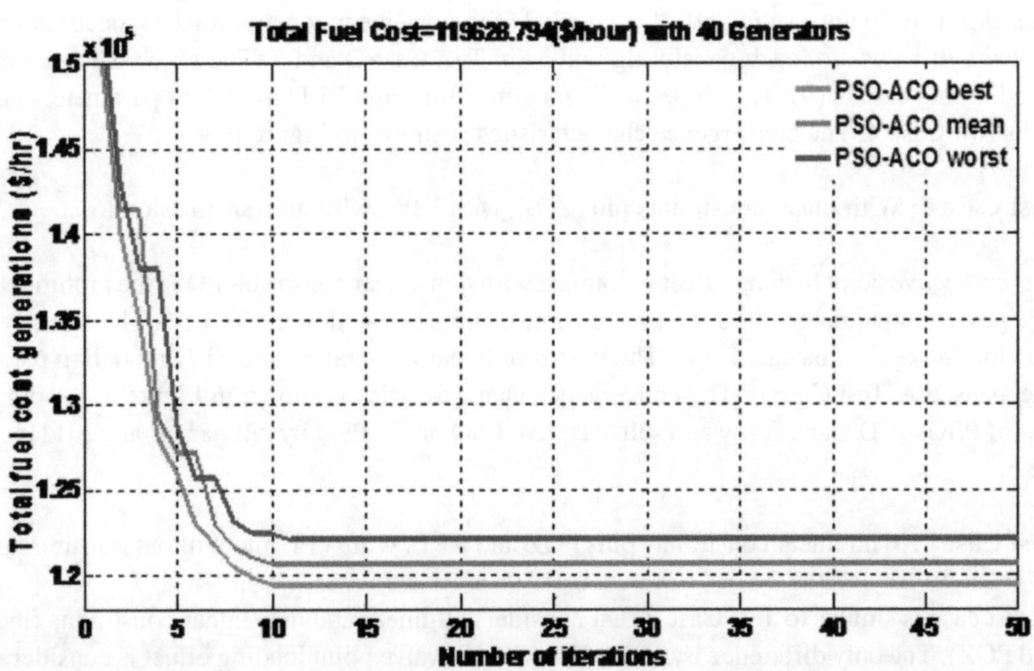

**For a more accurate representation see the electronic version.*

Figure 5. Convergence characteristics - Test Case 4

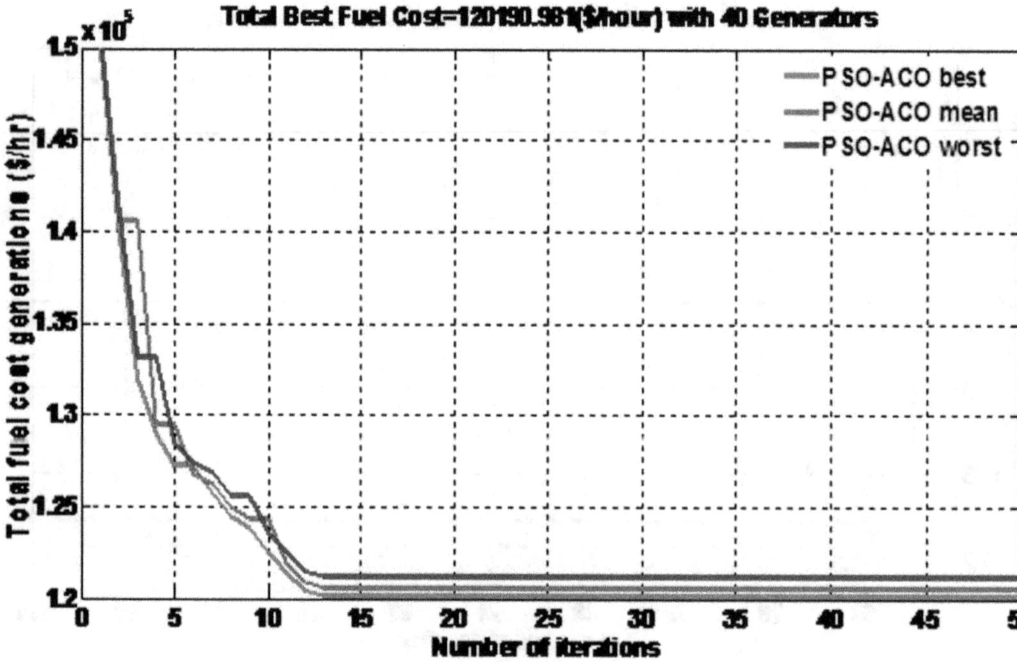

**For a more accurate representation see the electronic version.*

- **Test Case 5:** With linear constraints plus RRL, with VPL, without transmission loss:

In this case valve point loading effect is coupled with non-linear constraint RRL specified with UR_i and DR_i pairs of values for each generating unit as in Test Case 2 and 3. The objective function to be minimized in this case is (9). The values of linear constraints and ELD modeling parameters are same as used in Test Case 4. The convergence characteristics is shown in Figure 6.

- **Test Case 6:** With linear constraints plus POZ, with VPL, without transmission loss:

In this case valve point loading effect is coupled with non-linear constraint POZ; the prohibited zones fall in the power output range of 10th to 14th generating units as listed in the Table 1. The objective function to be minimized in this case is (9). The values of linear constraints and ELD modeling parameters are same as used in Test Case 4. The convergence characteristics is shown in Figure 7. A comparison of output of PSO-ACO and a fuzzy and self-adaptive PSO or NAPSO by Niknam et al. (2011) is shown in Table 3.

- **Test Case 7:** With linear constraints plus POZ and RRL, with VPL and without transmission loss:

This test case is similar to Test Case 3 that considers all linear and non-linear constraints (including RRL and POZ). The only difference is, unlike Test Case 3, valve point loading effect is considered here.

Figure 6. Convergence characteristics -Test Case 5

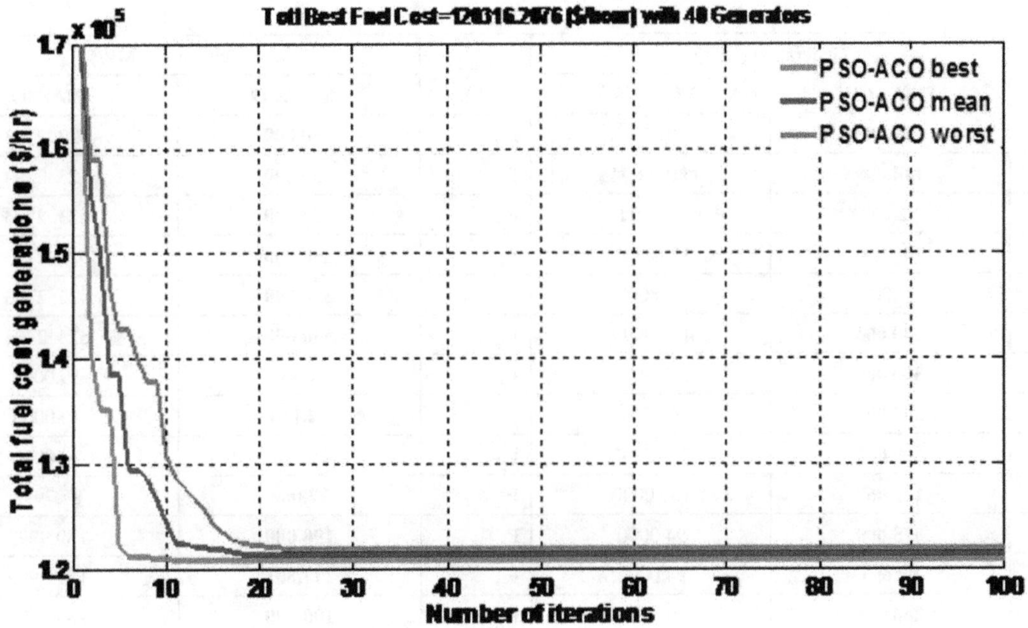

For a more accurate representation see the electronic version.

Figure 7. Convergence characteristics - Test Case 6

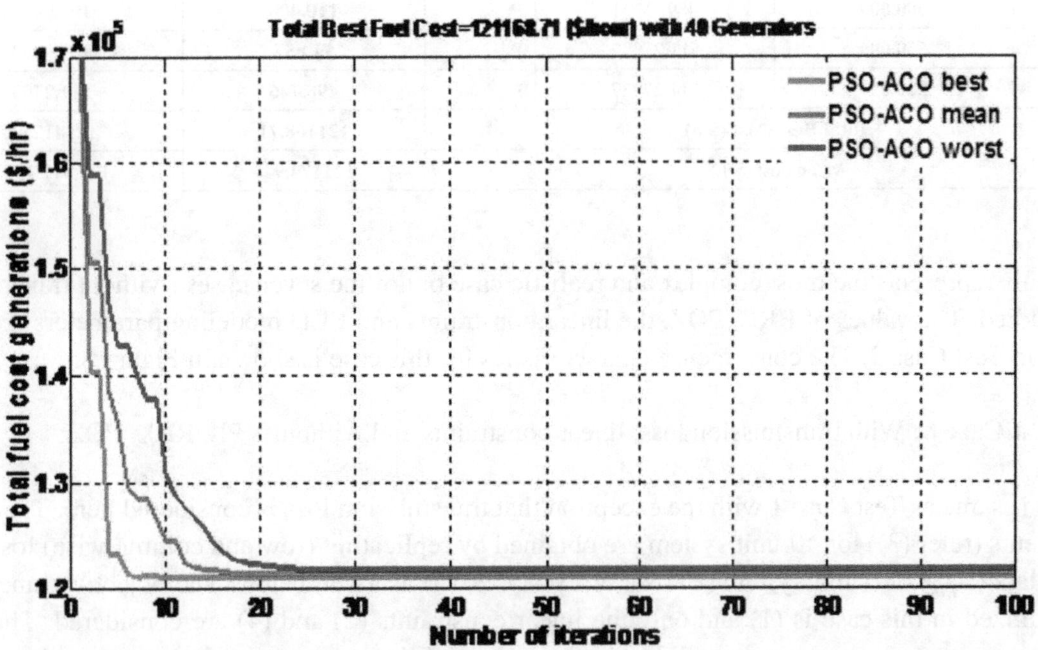

For a more accurate representation see the electronic version.

Table 3. Comparison of PSO-ACO with NAPSO: 40-generator system without VPL for P_D=10500 MW 40-generator system with VPL and POZ for P_D=10500 MW

Unit no.	Generation (MW)		Unit no.	Generation (MW)	
	PSO-ACO	NAPSO		**PSO-ACO**	NAPSO
P_1	**111.883**	110.7997	P_{21}	**550.000**	523. 2793
P_2	**114.000**	110.8004	P_{22}	**550.000**	523. 2793
P_3	**82.269**	97.3971	P_{23}	**550.000**	523. 2793
P_4	**190.000**	179.7331	P_{24}	**550.000**	523. 2793
P_5	**97.000**	87.8009	P_{25}	**550.000**	523. 2793
P_6	**140.000**	139.9999	P_{26}	**550.000**	523. 2793
P_7	**300.000**	259.5996	P_{27}	**21.450**	10.0000
P_8	**300.000**	284.5961	P_{28}	**22.160**	10.0000
P_9	**300.000**	284.5997	P_{29}	**29.481**	10.0000
P_{10}	**129.069**	130.0000	P_{30}	**97.000**	87.7998
P_{11}	**375.000**	94.0000	P_{31}	**190.000**	189.9999
P_{12}	**99.693**	94.0000	P_{32}	**177.801**	189.9999
P_{13}	**240.948**	214.7597	P_{33}	**190.000**	189.9999
P_{14}	**260.551**	394.27937	P_{34}	**200.000**	164.8003
P_{15}	**228.049**	394.27937	P_{35}	**200.000**	194.4019
P_{16}	**255.927**	394. 2794	P_{36}	**200.000**	199.9999
P_{17}	**478.086**	489. 2794	P_{37}	**110.000**	109.9999
P_{18}	**500.000**	489. 2794	P_{38}	**110.000**	109.9999
P_{19}	**550.000**	511.27937	P_{39}	**71.857**	109.9999
P_{20}	**329.031**	511.27937	P_{40}	**498.746**	511.2793
Total Best Cost ($/h)				**121168.71**	121412.62
Mean cost ($/h)				**121174.946**	121,412.7373

Hence this represents the most complex and realistic case out of the seven cases (without transmission loss) studied. The values of RRL, POZ, the linear constraints and ELD modeling parameters are same as used in Test Case 3. The convergence characteristics for this case is shown in Figure 8.

- **Test Case 8:** With transmission loss, linear constraints and without VPL, RRL, POZ:

This is same as Test Case 1 with the exception that transmission loss is considered here. The B-loss coefficients (refer (3)) for 40-unit system are obtained by replicating (row and column wise) loss coefficient data (B_{6x6}) of 6 unit system (Barisal & Prusty, 2015) as given below. The objective function to be minimized in this case is (1) and only the linear constraints (2) and (4) are considered. The basic modeling parameters are taken from Table 1. The values of linear constraints are same as that of Test Case 1.The convergence characteristics for this case is shown in Figure 9.

Figure 8. Convergence characteristics - Test Case 7

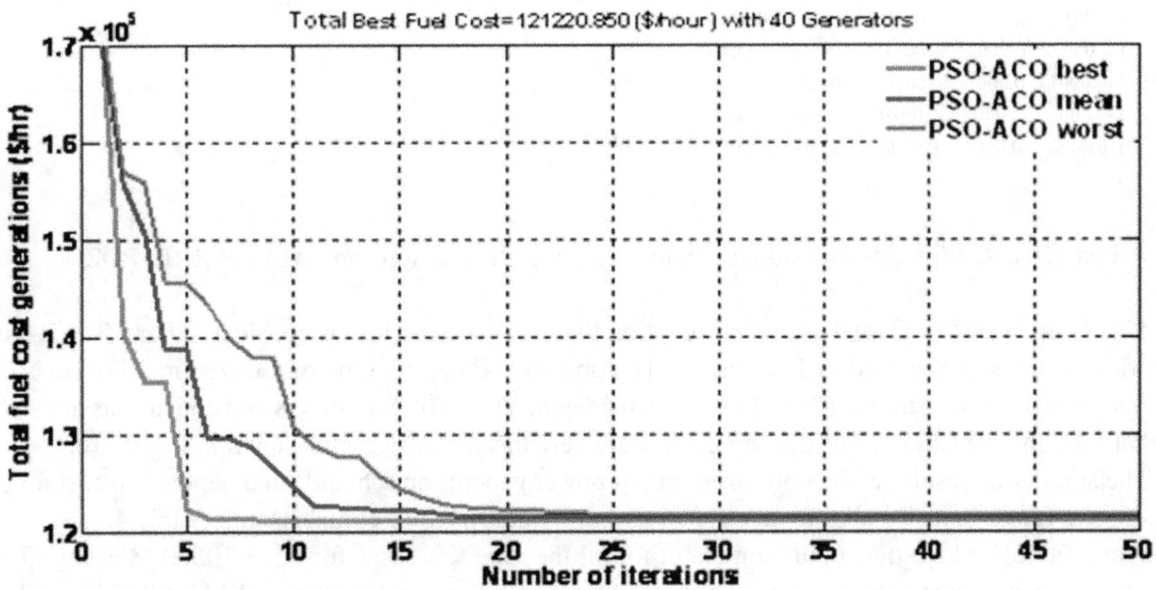

**For a more accurate representation see the electronic version.*

Figure 9. Convergence characteristics - Test Case 8

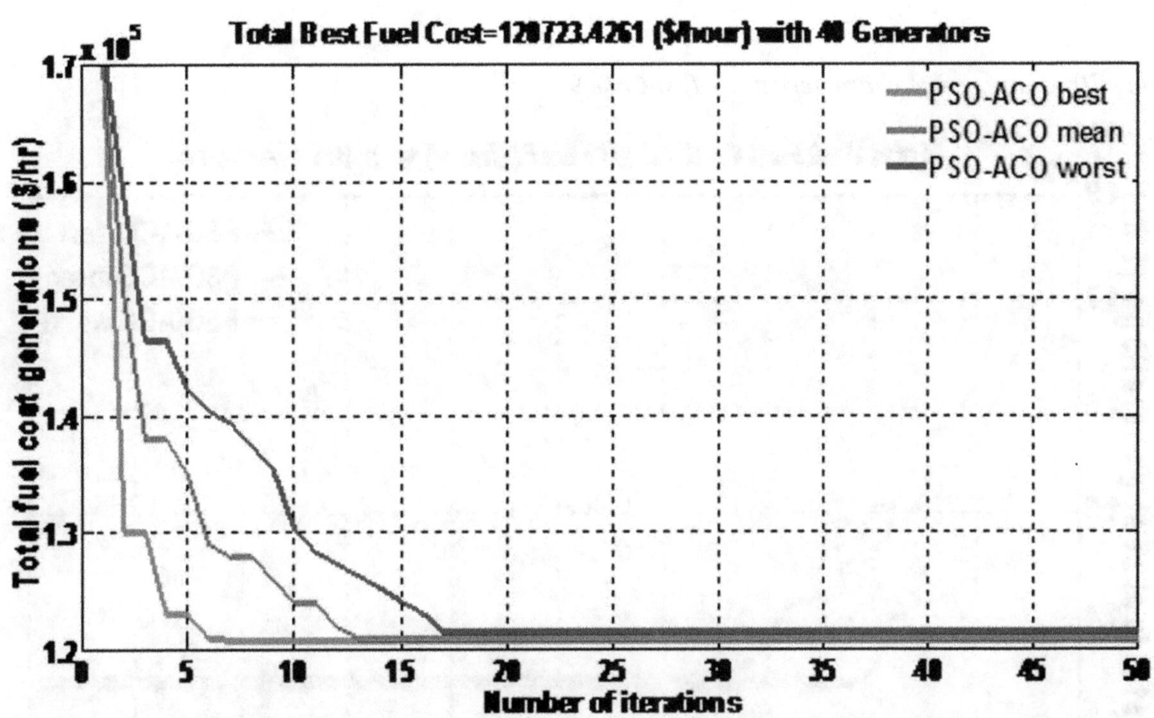

**For a more accurate representation see the electronic version.*

$$B_{6x6} = \begin{bmatrix} 0.0017 & 0.0012 & 0.0007 & -0.0001 & 0.0005 & -0.0002 \\ 0.0012 & 0.0014 & 0.0009 & 0.0001 & -0.0006 & -0.0001 \\ 0.0007 & 0.0009 & 0.0031 & 0.0000 & -0.0010 & -0.0006 \\ -0.0001 & 0.0001 & 0.0000 & 0.0024 & -0.0008 & -0.0006 \\ -0.0005 & -0.0006 & -0.0010 & -0.0006 & 0.0129 & -0.0002 \\ -0.0002 & -0.0001 & -0.0006 & -0.0008 & -0.0002 & 0.0150 \end{bmatrix}$$

- **Test Case 9:** With transmission loss and VPL, linear constraints and without RRL, POZ:

This is same as Test Case 4 with the exception that transmission loss is considered here. The B-loss coefficients are same as used in Test Case 8. The objective function to be minimized in this case is (9) and only the linear constraints (2) and (4) are considered. Values of parameters and constraints are same as that used in Test Case 4. The convergence characteristics for this case is shown in Figure 10.

The simulation result i.e. the optimized output power generation schedule and generation cost for all the 40 generating units for all the nine test cases are summarized for comparison in Table 4.

From Table 4 where the simulation result for all the nine test cases for P_D = 10500 MW for all 40 generating units is summarized, it is seen that the total best cost is minimum (**118613.309** \$/h) for Test Case 1 which is the simplest of the ELD cases considered here, whereas the maximum cost (**133671.6698** \$/h) corresponds to Test Case 9, which is the most realistic case (with loss and VPL). In cases 4, 5, 6, 7 and 9, VPL has been considered and it is observed that there is a distinct positive shift in cost as VPL

Figure 10. Convergence characteristics - Test Case 9

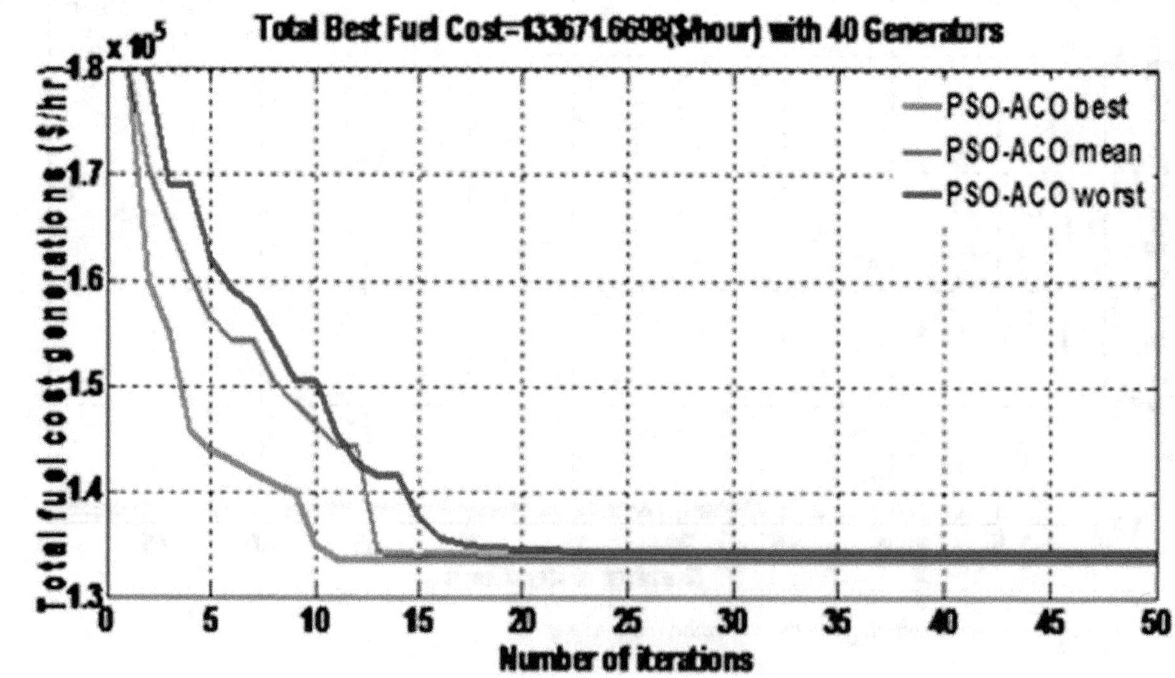

**For a more accurate representation see the electronic version.*

Table 4. Summary of Optimum Power Generation Schedule (MW), Loss (MW) and Total Generated Cost ($/hr) for 40-generator ELD Test Cases 1-9 for P_D = 10500 MW using PSO-ACO hybrid method; POZ region values corresponding to output of $P_{10} - P_{14}$ are in bold

Unit No.	P_i^{min} (MW)	P_i^{max} (MW)	Without Transmission Loss							With Transmission Loss	
			Without VPL, POZ & RRL (Test Case 1)	Without VPL, POZ but with RRL (Test Case 2)	Without VPL but with POZ & RRL (Test Case 3)	Without POZ, RRL but with VPL (Test Case 4)	Without POZ but with RRL & VPL (Test Case 5)	Without RRL but with POZ & VPL (Test Case 6)	With POZ, RRL & VPL (Test Case 7)	Without VPL, POZ & RRL (Test Case 8)	Without POZ, RRL but with VPL (Test Case 9)
P_1	36	114	114.000	111.017	106.158	103.508	113.80560	111.883	104.893	110.98320	110.98320
P_2	36	114	114.000	96.332	114.000	103.253	114.00000	114.000	96.107	114.00000	114.00000
P_3	60	120	120.000	120.000	120.000	105.850	100.34250	82.269	120.000	120.00000	120.00000
P_4	8`0	190	190.000	190.000	176.268	190.000	179.76300	190.000	174.866	190.00000	190.00000
P_5	47	97	97.000	97.000	97.000	97.000	93.01540	97.000	97.000	97.00000	97.00000
P_6	68	140	140.000	140.000	140.000	140.000	139.21740	140.000	75.722	110.44730	110.44730
P_7	110	300	300.000	300.000	300.000	300.000	260.10320	300.000	300.000	300.00000	300.00000
P_8	135	300	300.000	300.000	300.000	300.000	287.01230	300.000	300.000	300.00000	300.00000
P_9	135	300	300.000	300.000	300.000	300.000	276.21730	300.000	284.273	300.00000	300.00000
P_{10}	130	300	130.000	142.509	**162.932**	170.983	130.00000	**129.069**	**258.450**	139.43410	139.43410
P_{11}	94	375	94.000	105.246	**171.327**	165.736	170.10320	**375.000**	**142.608**	98.69900	98.69900
P_{12}	94	375	94.000	151.133	**147.707**	105.218	164.14260	**99.693**	**140.363**	145.37010	145.37010
P_{13}	125	500	125.000	140.737	**128.870**	163.618	221.21500	**240.948**	**126.275**	155.33840	155.33840
P_{14}	125	500	272.647	165.603	**258.143**	300.065	306.18280	**260.551**	**500.000**	230.75650	230.75650
P_{15}	125	500	266.176	286.276	270.051	218.845	394.21530	228.049	133.960	352.23800	352.23800
P_{16}	125	500	266.176	349.397	229.006	276.409	390.29000	255.927	335.589	276.16410	276.16410
P_{17}	220	500	500.000	500.000	500.000	500.000	479.42640	478.086	462.716	467.15480	467.15480
P_{18}	220	500	500.000	500.000	381.075	500.000	489.23100	500.000	500.000	500.00000	500.00000
P_{19}	242	550	550.000	550.000	550.000	550.000	509.17400	550.000	550.000	550.00000	550.00000
P_{20}	242	550	550.000	550.000	550.000	533.591	512.02310	329.031	501.465	550.00000	550.00000
P_{21}	254	550	550.000	550.000	540.484	550.000	520.32650	550.000	550.000	550.00000	550.00000
P_{22}	254	550	550.000	550.000	550.000	550.000	522.32750	550.000	550.000	550.00000	550.00000
P_{23}	254	550	550.000	550.000	550.000	550.000	526.34200	550.000	550.000	550.00000	550.00000
P_{24}	254	550	550.000	491.249	550.000	413.535	519.21000	550.000	550.000	550.00000	550.00000
P_{25}	254	550	550.000	550.000	550.000	550.000	523.19300	550.000	550.000	550.00000	550.00000
P_{26}	254	550	550.000	550.000	550.000	550.000	519.46300	550.000	519.430	550.00000	550.00000
P_{27}	10	150	10.000	28.485	12.848	10.972	14.22740	21.450	26.766	16.05540	16.05540
P_{28}	10	150	10.000	18.837	26.051	29.107	12.93200	22.160	25.401	14.12960	14.12960
P_{29}	10	150	10.000	10.380	30.418	38.119	11.95240	29.481	27.095	13.76300	13.76300
P_{30}	47	97	97.000	97.000	97.000	89.426	90.00000	97.000	96.008	87.04150	87.04150
P_{31}	60	190	190.000	190.000	190.000	190.000	186.13200	190.000	190.000	190.00000	190.00000
P_{32}	60	190	190.000	190.000	190.000	190.000	187.98500	177.801	190.000	190.00000	190.00000
P_{33}	60	190	190.000	190.000	190.000	184.766	190.00000	190.000	190.000	190.00000	190.00000
P_{34}	90	200	200.000	200.000	200.000	200.000	173.54370	200.000	200.000	200.00000	200.00000
P_{35}	90	200	200.000	200.000	200.000	200.000	169.10340	200.000	200.000	200.00000	200.00000

continued on following page

Table 4. Continued

Unit No.	P_i^{min} (MW)	P_i^{max} (MW)	Without Transmission Loss							With Transmission Loss	
			Without VPL, POZ & RRL (Test Case 1)	Without VPL, POZ but with RRL (Test Case 2)	Without VPL but with POZ & RRL (Test Case 3)	Without POZ, RRL but with VPL (Test Case 4)	Without POZ but with RRL & VPL (Test Case 5)	Without RRL but with POZ & VPL (Test Case 6)	With POZ, RRL & VPL (Test Case 7)	Without VPL, POZ & RRL (Test Case 8)	Without POZ, RRL but with VPL (Test Case 9)
P_{36}	90	200	200.000	200.000	200.000	200.000	162.43200	200.000	147.469	200.00000	200.00000
P_{37}	25	110	110.000	110.000	100.664	110.000	110.00000	110.000	90.050	110.00000	110.00000
P_{38}	25	110	110.000	110.000	110.000	110.000	110.00000	110.000	110.000	110.00000	110.00000
P_{39}	25	110	110.000	110.000	110.000	110.000	110.00000	71.857	82.330	110.00000	110.00000
P_{40}	242	550	550.000	508.801	550.000	550.000	511.35000	498.746	451.163	550.00000	550.00000
Total Power			10500.000	10500.000	10500.000	10500.000	10500.000	10500.000	10500.000	10589.08360	11444.07530
Loss (MW)			0.000	0.000	0.000	0.000	0.00	0.000	0.000	89.08360	944.0733
Total Best cost ($/h)			118613.309	119471.532	119628.794	120190.981	120316.2076	121168.712	121220.850	120723.4261	133671.6698
Worst cost ($/h)			120985.5752	121860.9626	122712.268	121272.6998	121411.879	121215.285	121311.456	120853.5347	134028.5477
Mean cost ($/h)			119799.4421	120666.2473	121231.788	120538.458	121321.547	121175.574	121238.436	120726.5327	133852.5215
Success rate (%)			92	94	96	94	94	93	93	96	94
CPU Time (s)			9.0153	9.0235	8.342	8.6748	9.124	8.47637	9.3254	9.978	10.023
Iteration			9	7	7	11	9	7	7	7	11
Trial			50	50	50	50	50	50	50	50	50

effect i.e. non-smooth cost function is introduced in the ELD problem, be it with transmission loss or without loss. Again the generation cost increases as non-linear constraints RRL or POZ are added; this is evident if we compare best costs of Test Case 2 and 3 with that of Test Case 1 and also best costs of Test Case 5, 6 and 7 with that of Test Case 4. Again introducing POZ (in Test Case 6) entails more cost than the cost while considering RRL only (in Test Case 5). As far as loss is concerned, Test Case 9 with VPL effect incurs more loss (**944.0753 MW**) than Test Case 8 without VPL (**89.08360 MW**). Variation of total generation cost ($/hr) with generator power output (MW) pertaining to four test cases with VPL but without loss (Test Case 4, 5, 6 and 7) is depicted in Figure 11. From the convergence characteristics of PSO-ACO shown above (in Figure 2 to10), it is found that the convergence time and the number of iterations to reach the optimum (best cost) value are maximum for Test Case 9 while the minimum number of iteration and the minimum convergence time is consumed in Test Case 3. Test Case 9 being the most realistic case (with loss and VPL) and Test Case 7 having the most complex character (with VPL, RRL and POZ), PSO-ACO handles them well as the convergence is achieved in relatively lesser time.

In the 'without loss' category the total generation cost outputted by the PSO-ACO hybrid in **Test Case 4** corresponding to P_D = 10500 MW have been compared (in Table 5) with those obtained under same test condition for four other hybrid methods discussed in the Introduction. The generation cost in $/hr as obtained for the hybrids HQPSO (Chakraborty et al., 2011), CE-SQP (Subathra et al., 2015), HCRO-DE (Roy et al., 2014), DE-HS (Dinh et al. 2013), DE-PSO (Sayah & Hamouda, 2013), GA-PS-SQP (Alsumait et al., 2010), DE-BBO (Bhattacharya & Chattopadhyay, 2010), DEC-SQP (Coelho & Mariani, 2006) are**121318.60, 121412.88, 121412.55, 121412.5357, 121412.56, 121458.14, 121420.8948** and **121741.9793** respectively whereas the (best) generation cost for PSO-ACO is **120190.981** $/hr. Also

Figure 11. Comparative study for 40 generator power output vs total generation cost for different demands - (a) Without loss, POZ, RRL but with VPL (Case 4), (b) Without loss, POZ but with RRL & VPL (Case 5), (c) Without loss, RRL but with POZ & VPL (Case 6), (d) Without loss but with POZ, RRL & VPL (Case 7).

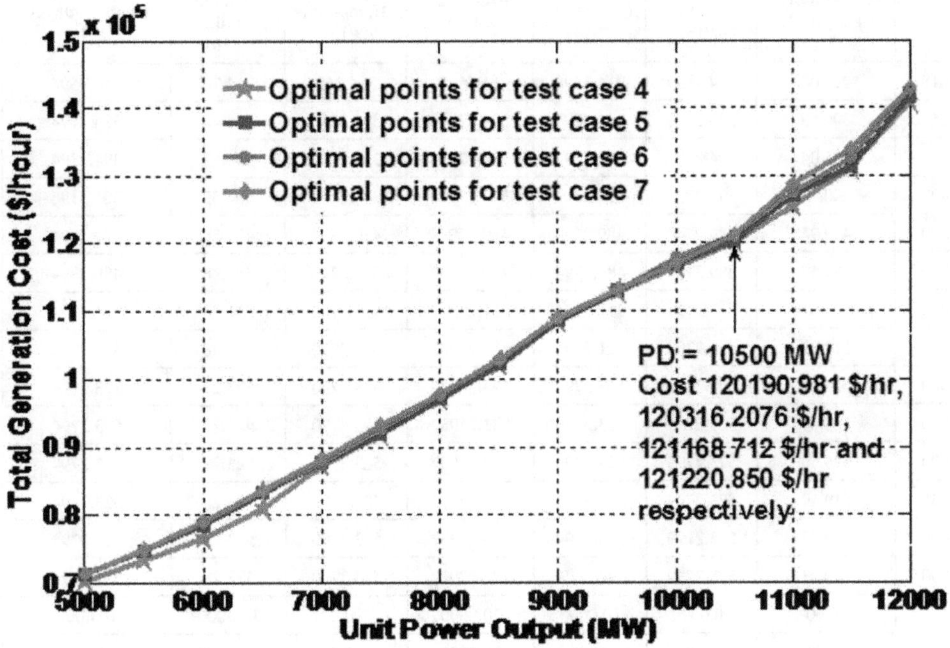

Table 5. Comparison of output of PSO-ACO with different hybrid methods with VPL only and without transmission loss for 40-unit system, P_D=10,500 MW.

Unit No.	PSO-ACO	Other Hybrid Methods							
		HQPSO (Chakraborty et al., 2011)	CE-SQP (Subathra et al., 2015)	HCRO-DE (Roy et al., 2014)	DEHS (Dinh et al., 2013)	DEPSO (Sayah & Hamouda, 2013)	GA–PS–SQP (Alsumait et al., 2010)	DE/BBO (Bhattacharya & Chattopadhyay, 2010)	DEC-SQP (Coelho & Mariani, 2006)
P_1	103.508	114.00	110.7998	110.8015	110.79983	110.8020	110.97	110.7998	111.7576
P_2	103.253	114.00	110.8002	110.7998	110.79983	110.8020	111.02	110.7998	111.5584
P_3	105.850	104.88	97.3999	97.3999	97.39991	97.400	120.00	97.3998	97.3999
P_4	190.000	179.80	179.7331	179.7331	179.73310	179.7330	179.73	179.7331	179.7331
P_5	97.000	93.04	87.7999	87.7999	87.79990	87.80	88.27	87.9576	91.6560
P_6	140.000	139.84	140.0000	140.0000	140.0000	140.0	140.00	140.0000	140.0000
P_7	300.000	260.06	259.5996	259.5997	259.59965	259.6	259.60	259.5997	300.0000
P_8	300.000	287.04	284.5998	284.5997	284.59965	284.6	284.60	284.5997	300.0000
P_9	300.000	276.10	284.5998	284.5997	284.59965	284.6	284.60	284.5997	284.5997
P_{10}	170.983	130.00	130.0000	130.0000	130.0000	130.0	130.00	130.0000	130.0000
P_{11}	165.736	170.42	94.0000	94.0000	94.0000	94.0	168.80	168.7998	168.7998
P_{12}	105.218	164.77	94.0000	94.0000	94.0000	94.0	168.80	94.0000	94.0000

continued on following page

Table 5. Continued

Unit No.	PSO-ACO	Other Hybrid Methods							
		HQPSO (Chakraborty et al., 2011)	CE-SQP (Subathra et al., 2015)	HCRO-DE (Roy et al., 2014)	DEHS (Dinh et al., 2013)	DEPSO (Sayah & Hamouda, 2013)	GA–PS–SQP (Alsumait et al., 2010)	DE/BBO (Bhattacharya & Chattopadhyay, 2010)	DEC-SQP (Coelho & Mariani, 2006)
P_{13}	163.618	223.00	214.7599	214.7598	214.75979	214.760	214.76	214.7598	214.7598
P_{14}	300.065	306.58	394.2798	394.2794	394.27938	394.2790	394.28	394.2794	394.2794
P_{15}	218.845	395.04	394.2799	394.2794	394.27938	394.2790	304.52	394.2794	304.5196
P_{16}	276.409	390.29	394.2799	394.2794	394.27938	394.2790	304.52	304.5196	304.5196
P_{17}	500.000	479.94	489.2749	489.2794	489.27937	489.2790	489.28	489.2794	489.2794
P_{18}	500.000	489.67	489.2799	489.2794	489.27937	489.2790	489.28	489.2794	489.2794
P_{19}	550.000	509.01	511.2799	511.2794	511.27937	511.2790	511.28	511.2794	511.2794
P_{20}	533.591	512.07	511.2798	511.2794	511.27937	511.2790	511.28	511.2794	511.2794
P_{21}	550.000	521.22	523.2793	523.2794	523.27937	523.2790	523.28	523.2794	523.2794
P_{22}	550.000	523.60	523.2799	523.2794	523.27937	523.2790	523.28	523.2794	523.2853
P_{23}	550.000	526.50	523.2799	523.2794	523.27937	523.2790	523.28	523.2794	523.2847
P_{24}	413.535	519.19	523.2799	523.2794	523.27937	523.2790	523.28	523.2794	523.2794
P_{25}	550.000	523.27	523.2799	523.2794	523.27937	523.2790	523.28	523.2794	523.2794
P_{26}	550.000	519.23	523.2799	523.279	523.27937	523.2790	523.28	523.2794	523.2794
P_{27}	10.972	10.00	10.0000	10.0000	10.0000	10.0	10.0000	10.0000	10.0000
P_{28}	29.107	10.00	10.0000	10.0000	10.0000	10.0	10.0000	10.0000	10.0000
P_{29}	38.119	10.00	10.0000	10.0000	10.0000	10.0	10.0000	10.0000	10.0000
P_{30}	89.426	89.01	87.8009	87.7999	87.7999	87.80	88.66	97.000	90.3329
P_{31}	190.000	186.03	189.9999	190.0000	190.0000	190.0	190.000	190.000	190.000
P_{32}	190.000	188.50	189.9999	190.0000	190.0000	190.0	190.000	190.000	190.000
P_{33}	184.766	190.00	190.0000	190.0000	190.0000	190.0	190.000	190.000	190.000
P_{34}	200.000	174.80	164.7999	164.7998	164.7998	164.80	164.80	164.7998	200.000
P_{35}	200.000	169.07	194.4138	194.3956	194.3977	194.395	200.000	200.000	200.000
P_{36}	200.000	163.56	200.000	200.000	200.0000	200.0	200.000	200.000	200.000
P_{37}	110.000	100.88	109.9999	110.0000	110.0000	110.0	110.000	110.000	110.000
P_{38}	110.000	110.00	109.9999	110.0000	110.0000	110.0	110.000	110.000	110.000
P_{39}	110.000	110.00	109.9999	110.0000	110.0000	110.0	110.000	110.000	110.000
P_{40}	550.000	515.59	511.2799	511.2794	511.2793	511.279	511.28	511.2794	511.2794
Total Best Cost ($/h)	**120190.981**	**121318.60**	**121412.88**	**121412.55**	**121412.5357**	**121412.56**	**121458.14**	**121420.8948**	**121741.9793**
Mean Cost ($/h)	**120538.458**	**121427.47**	**121423.65**	**121,413.11**	**121414.9488**	**121419.31**	**122039**	**121420.8952**	**122295.1278**

looking at the generation schedule, it is evident from Table 5 that PSO-ACO yields the most optimum and economic solution as far as 40-generator convex ELD problem without transmission loss is concerned. Again PSO-ACO could be compared with IPM-DE for **Test Case 1** and with NAPSO for **Test Case 6** in Table 2 and Table 3 respectively; for both the cases PSO-ACO appears to be more economic. For Test Case 2, 3, 5 and 7 performance of PSO-ACO could not be compared as no other hybrid method's performance have been reported for similar cases with enhanced complexity. Test Case 3 and 7 are particularly interesting as POZ and RRL are both considered here with or without VPL effect.

Now, in the 'with loss' category the total generation cost and transmission loss outputted by the PSO-ACO hybrid in **Test Case 9** corresponding to P_D = 10500 MW have been compared (in Table 6) with those obtained under same test condition for different other hybrid methods discussed in the Introduction. The generation cost in $/hr as obtained for hybrids OIWO (Barisal & Prusty, 2015), OCRO (Hazra et al., 2015), SDE (Reddy & Vaisakh, 2013), GA-API (Ciornei & Kyriakides, 2012) are **136452.68, 136563.48, 138157.46** and **139864.96** respectively whereas the generation cost for PSO-ACO **is 133671.6698** $/hr which implies PSO-ACO is most economic. As far as transmission loss is concerned, PSO-ACO is far more effective than the other methods.

CONCLUSION

In this study, the swarm intelligence-based hybrid search algorithm PSO-ACO is applied for the first time in finding optimal solution of 40-generator ELD problem with and without transmission loss. The method has been tested with wide variation of constraints and cost curve characteristics and the test results strongly attests the potential of PSO-ACO hybrid as an effective multi-constraint optimization method to handle the possible variations of large scale ELD problem. The premature convergence tendency of PSO has been successfully overcome with the influence of ACO whereby the solutions tend to stretch beyond local optima particularly in cases where VPL effect has been considered. However, the convergence characteristics indicate that there is still scope for improvement and rigorous trials are needed to fine tune the parameters. In future PSO-ACO hybrid can be tried for ELD solution of 3, 13

Table 6. Comparison of Loss (MW) and Total Generation Cost ($/hr) for 40-generator ELD with Transmission Loss; Test Case 9 with VPL but without POZ and RRL and with P_D = 10500 MW

Unit (MW)	PSO-ACO	OIWO (Barisal & Prusty, 2015)	OCRO (Hazra et al., 2015)	SDE (Reddy & Vaisakh, 2013)	GAAPI (Ciornei & Kyriakides, 2012)
Total power (MW)	**11444.07530**	11457.2965	11468.9607	11474.43	11545.06
P_{loss} (MW)	**944.0753**	957.2965	968.9607	974.43	1045.06
Total Cost ($/h)	**133671.6698**	136452.68	136563.48	138157.46	139864.96

and 15 generator systems or even more complex power system optimization problems like Dynamic Economic Dispatch (DED) problem and Combined Economic and Emission Dispatch (CEED) problem. Also it remains to be seen how this method handles other complex cases of 40-generator ELD when transmission loss is considered.

REFERENCES

Abdelaziz, A. Y., Mekhamer, S. F., Badr, M. A. L., & Kamh, M. Z. (2008). Economic dispatch using an enhanced Hopfield neural network. *Electric Power Components and Systems*, *36*(7), 719–732. doi:10.1080/15325000701881969

Al-Sumait, J.S., AL-Othman, A.K., & Sykulski, J.K. (2007). Application of pattern search method to power system valve-point economic load dispatch. *International Journal of Electrical Power & Energy Systems, 29*(10), 720–730.

Alsumait, J. S., Sykulski, J. K., & Al-Othman, A. K. (2010). A hybrid GA–PS–SQP method to solve power system valve-point economic dispatch problems. *Applied Energy*, *87*(5), 1773–1781. doi:10.1016/j.apenergy.2009.10.007

Barisal, A. K., & Prusty, R. C. (2015). Large scale economic dispatch of power systems using oppositional invasive weed optimization. *Applied Soft Computing*, *29*, 122–137. doi:10.1016/j.asoc.2014.12.014

Bhattacharya, A., & Chattopadhyay, P. K. (2010). Hybrid differential evolution with biogeography-based optimization for solution of economic load dispatch. *IEEE Transactions on Power Systems*, *25*(4), 1955–1964. doi:10.1109/TPWRS.2010.2043270

Bhattacharya, A., & Chattopadhyay, P. (2010). Biogeography-based optimization for different economic load dispatch problems. *IEEE Transactions on Power Systems*, *25*(2), 1064–1077. doi:10.1109/TPWRS.2009.2034525

Chakraborty, S., Senjyu, T., Yona, A., Saber, A. Y., & Funabashi, T. (2011). Solving economic load dispatch problem with valve-point effects using a hybrid quantum mechanics inspired particle swarm optimization. *IET Generation, Transmission & Distribution*, *5*(10), 1042–1052. doi:10.1049/iet-gtd.2011.0038

Chiang, C. L. (2005). Improved genetic algorithm for power economic dispatch of units with valve-point effects and multiple fuels. *IEEE Transactions on Power Systems*, *20*(4), 1690–1699. doi:10.1109/TPWRS.2005.857924

Ciornei, I., & Kyriakides, E. (2012). A GA-API solution for the economic dispatch of generation in power system operation. *IEEE Transactions on Power Systems*, *27*(1), 233–242. doi:10.1109/TPWRS.2011.2168833

Coelho, L. S., & Mariani, V. C. (2006). Combining of chaotic differential evolution and quadratic programming for economic dispatch optimization with valve-point effect. *IEEE Transactions on Power Systems*, *21*(2), 989–996. doi:10.1109/TPWRS.2006.873410

Dinh, L. L., Ngoc, D. V., Nguyen, T. H., & Anh, D. L. (2013).A hybrid differential evolution and harmony search for nonconvex economic dispatch problems. *Proceedings of the Power Engineering and Optimization Conference (PEOCO)*, 238-243.

Duman, S., Güvenç, U., & Yörükeren, N. (2010). Gravitational search algorithm for economic dispatch with valve-point effects. *International Review of Electrical Engineering*, *5*(6), 2890–2895.

Duvvuru, N., & Swarup, K. S. (2011). A hybrid interior point assisted differential evolution algorithm for economic dispatch. *IEEE Transactions on Power Systems*, *26*(2), 541–549. doi:10.1109/TPWRS.2010.2053224

Gaing, Z. L. (2003). Particle swarm optimization to solving the economic dispatch considering the generator constraints. *IEEE Transactions on Power Systems*, *18*(3), 1187–1195. doi:10.1109/TPWRS.2003.814889

Grigsby, L. L. (2007). *Power System Stability and Control* (2nd ed.). CRC Press. doi:10.1201/9781420009248

Hazra, J., & Sinha, A. K. (2008). Environmental constrained economic dispatch using bacteria foraging optimization. *Proceedings of Power System Technology and IEEE Power India Conference, Joint International Conference*, 1-6. 10.1109/ICPST.2008.4745330

Hazra, S., Roy, P. K., & Sinha, A. (2015). An efficient evolutionary algorithm applied to economic load dispatch problem. *Proceedings of the CCC and IT*, 1-6.

Jayabarathi, T., Jayaprakash, K., Jeyakumar, D. N., & Raghunathan, T. (2005). Evolutionary programming techniques for different kinds of economic dispatch problems. *Electric Power Systems Research*, *73*, 2169–2176.

Lam, A. Y. S., & Li, V. O. K. (2010). Chemical-reaction-inspired metaheuristic for optimization. *IEEE Transactions on Evolutionary Computation*, *14*(3), 381–399. doi:10.1109/TEVC.2009.2033580

Mehrabian, A. R., & Lucas, C. (2006). A novel numerical optimization algorithm inspired from weed colonization. *Ecological Informatics*, *1*(4), 355–366. doi:10.1016/j.ecoinf.2006.07.003

Mohammadi, A., & Varahram, M. H. (2006). Using neural network for solving of on-line economic dispatch problem. *Proceedings of International Conference on Intelligent Agents for Modelling, Control and Automation*, 81 – 87. 10.1109/CIMCA.2006.228

Niknam, T., Mojarrad, H. D., & Meymand, H. Z. (2011). Non-smooth economic dispatch computation by fuzzy and self-adaptive particle swarm optimization. *Applied Soft Computing*, *11*(2), 2805–2817. doi:10.1016/j.asoc.2010.11.010

Nomana, N., & Iba, H. (2008). Differential evolution for economic load dispatch problems. *Electric Power Systems Research*, *78*(8), 1322–1331. doi:10.1016/j.epsr.2007.11.007

Park, J. B., Lee, K. S., Shin, J. R., & Lee, K. Y. (2005). A particle swarm optimization for economic dispatch with nonsmooth cost functions. *IEEE Transactions on Power Systems*, *20*(1), 34–42. doi:10.1109/TPWRS.2004.831275

Sinha, N., Chakrabarti, R., & Chattopadhyay, P. K. (2003). Evolutionary programming techniques for economic load dispatch. *IEEE Transactions on Evolutionary Computation*, *7*(1), 83–94. doi:10.1109/TEVC.2002.806788

Pothiya, S., Ngamroo, I., & Kongprawechnon, W. (2010). Ant colony optimization for economic dispatch problem with non-smooth cost functions. *International Journal of Electrical Power & Energy Systems*, *32*(5), 478–487. doi:10.1016/j.ijepes.2009.09.016

Reddy, A. S., & Vaisakh, K. (2013). Shuffled differential evolution for large scale economic dispatch. *Electric Power Systems Research*, *96*, 237–245. doi:10.1016/j.epsr.2012.11.010

Roy, P. K., Bhui, S., & Paul, C. (2014). Solution of economic load dispatch using hybrid chemical reaction optimization approach. *Applied Soft Computing*, *24*, 109–125. doi:10.1016/j.asoc.2014.07.013

Santra, D., Mondal, A., & Mukherjee, A. (2014). Study of economic load dispatch by various hybrid optimization techniques. In Hybrid Soft Computing: Research and Applications (pp. 37-74). Springer India.

Sayah, S., & Hamouda, A. (2013). A hybrid differential evolution algorithm based on particle swarm optimization for nonconvex economic dispatch problems. *Applied Soft Computing*, *13*(4), 1608–1619.

Subathra, M. S. P., Selvan, S. E., Victoire, T. A. A., Christinal, A. H., & Amato, U. (2015). A hybrid with cross- entropy method and sequential quadratic programming to solve economic load dispatch problem. *IEEE Systems Journal*, *9*(3), 1031–1044. doi:10.1109/JSYST.2013.2297471

Swarup, K. S. (2005). Ant colony optimization for economic generator scheduling and load dispatch. *Proceedings of the 6th WSEAS International Conference on Evolutionary Computing*, 167-175.

Wang, S. K., Chiou, J. P., & Liu, C. W. (2007). Non-smooth/non-convex economic dispatch by a novel hybrid differential evolution algorithm. *IET Generation, Transmission & Distribution*, *1*(5), 793–803. doi:10.1049/iet-gtd:20070183

Won, J. R., & Park, Y. M. (2003). Economic dispatch solutions with piecewise quadratic cost functions using improved genetic algorithm. *International Journal of Electrical Power & Energy Systems*, *25*(5), 355–361. doi:10.1016/S0142-0615(02)00098-4

Yang, X. S., Hosseini, S. S. S., & Gandomi, A. H. (2012). Firefly algorithm for solving non-convex economic dispatch problems with valve loading effect. *Applied Soft Computing*, *12*(3), 1180–1186. doi:10.1016/j.asoc.2011.09.017

Chapter 3
Factors Influencing Patients' Satisfaction

Soraia Oueida
American University of Middle East, Kuwait

Seifedine Kadry
Beirut Arab University, Lebanon

Sorin Ionescu
Polytechnic University of Bucharest, Romania

ABSTRACT

In complex systems, such as healthcare, patient satisfaction is identified as the patient experience of care and has been referred to as the "indispensable outcome." The main goals of ED practitioners are the patient satisfaction along with maintaining optimal outcomes. Patient satisfaction has become a very important outcome measure when assessing healthcare systems performance. Nevertheless, it is a complex confusing concept. Some providers suggest several activities in order to enhance the satisfaction without being sure if these actions really improve satisfaction or no. Also, patient satisfaction enhancement activities should not conflict with cost efficacy. Several factors fall under patient satisfaction. Interviews with physicians prove that patients have specific expectations during a clinical encounter; being aware of these expectations, physicians can fulfill patient satisfaction. The goal of this chapter is to determine the contribution and effect of these factors in influencing patient satisfaction.

INTRODUCTION

Patient satisfaction started to be more highlighted and discussed in 1980's. Between 1980 and 1996, the researches on this topic were highly increased. The approach of considering patient satisfaction as an important measure plays an effective role in making the healthcare providers accountable to their patients and thus is an important healthcare service provided by the hospital to its patients. Patient satisfaction can be referred to as a general point of view towards an experience of healthcare, may comprises emotional facet and relates to some desired expectations or social networks (Keegan et al.,

DOI: 10.4018/978-1-5225-5045-7.ch003

1995). It is described by some authors as emergent and fluid (The development of the Royal College of Surgeons of England), or a passive form of establishing consumer's views (Mclvor and Carr-Hill, 1989). A good system must address patient satisfaction as the point of care for each patient's medical care and service required before exiting the ED. Patient needs are related to the healthcare resources required to accomplish a service. To ensure patient satisfaction, each patient must be involved in the evaluation of his/her own experience during his/her visit to this facility. This will help in ensuring continuous quality and improvement. Nowadays, patient satisfaction has become a high index representing the status of a hospital/facility and a desirable outcome attracting decision makers.

The patient's experience is classified under the satisfaction or dissatisfaction to a certain medical service. It is the patient's judgment on the quality of care provided during a visit. Therefore, patient satisfaction should be considered an important factor in assessing the quality of care in hospitals (Torcson, 2005). Other authors acknowledged in their studies that patient's reports of their health experience and their satisfaction with the quality of service are very important health measures (Bolus and Pitts, 1999). It is obvious that focusing on the quality of care provided in health services is a priority for any healthcare system. However, the high rising costs of treatments, equipment, etc.., the shortage in medical resources and the variations in clinical practice have affected the notion of quality. Thus, a quality healthcare system is the one which guarantees a continuous quality of care despite all the shortages and limitations faced. The assessment of this quality is usually based on technical perspectives and the process in which the care is delivered. This assessment becomes more authentic of it is centered on patient's views, experiences and perceptions. Thus, the system should assess these perceptions and monitor its quality based on these feedbacks in order to keep operating properly. Another goal of patient satisfaction research is to evaluate patient's views and associate available values with different outcomes.

The human connection between patients and physicians has a huge impact. The annals of medicine throughout the years stated that some excellent care delivery failed to have a good patient satisfaction because of the unapproachable physician and some other poor care delivery outcome an excellent patient satisfaction because of the good relationship with the physician. Therefore, it can be said that the quality of medical care is highly influenced by the relationship between patients and physicians.

In the 80's businesses approached customer satisfaction surveys in order to monitor services they provide. Nowadays, healthcare organizations started to adopt this concept where survey results are analyzed in order to evaluate the quality of the medical service provided. The sections below present the survey conducted in our studied ED in order to assess the main factors affecting patient satisfaction.

Quintana et al., 2006, discussed patient satisfaction and its relation to some areas in the hospital. They found that the lower the length of stay was, the higher the satisfaction rate was related to specific domains. Actually, the quality of care delivered to patients affects the overall satisfaction and indirectly the length of stay in the hospital. Majority of hospitals over the last years tend to reduce the length of stay in order to increase the rate of patient satisfaction (Borghans et al., 2012).

METHODS AND MATERIALS

This study was conducted in March 2017, in a hospital located in Lebanon. The hospital has two emergency rooms in its emergency department. The two emergency rooms (ERs) were considered for the survey: the public emergency room (ER A) and the private emergency room (ER B). The patients were selected randomly and a pre-structured survey questionnaire was provided to the patients. Data was collected based on the patient's experience in the ED and patient satisfaction with regard to services and management at the ED. For the purpose of obtaining accurate results 200 patients were given the questionnaires: 100 patients from each ER. Descriptive statistical analysis will be carried out on the data collected in order to evaluate the patient satisfaction and will be presented in below.

Selection of Participants

The participants consisted of consecutive patients treated in the ED during March 2017 and April 2017 through different period of times (day, evening, weekdays, weekends). Some patients are excluded from the survey; such as severe cases, children, patients not willing to participate and those unable to communicate because of language. The survey was designed in a way to be easy for patients to undertake it. Questions are clear and straight forward in order to ensure accuracy. This will ensure that results are trustworthy and patients are not obliged to skip a question or answer anything just to finalize the survey. The design of the survey and choice of questions was based on the several visits to the hospital and the overview on the problems usually faced by patients during their clinical visit.

Data Collection and Processing

Data collection was performed before patient's discharge from ED. The questions can be divided into 3 phases: the first phase of questions were related to his impression and rating for the service and medical care after completion of his demand, the second phase was related to his opinion on how to improve the operation of the emergency department and the factors affecting his experience during this visit. Finally, a retrospective chart review is approached. Immediately after completion of medical care and before transferring the patient to another unit or discharge, the level of satisfaction with the ED visit was determined including diagnosis, procedures followed in the ED, level of explanation of patient's case, physicians and other resources' skills and level of care, the waiting time spent before receiving care, the total waiting time spent in the ED, the time spent with the physician, any suspicious of unnecessary diagnosis and whether patient will follow the treatment/recommendations. The number of patients enrolled with this survey was 100. The last phase included reviewing the medical records after the patient left the ED in order to determine what exactly was performed with this patient. Other information collected was day of the week, shift time, the time spent from triage to examination, the final patient discharge and the category of diagnostic.

Data Analysis

The survey was prepared using the website www.surveymonkey.com, which is an academic and student surveys database for projects of all types. The main concern and outcome measure is the level of satisfaction with the ED visit. The survey was prepared using the website then printed and provided to patients at their arrival to the ED. Another way to collect data from the survey responses was to place the link *(*https://www.surveymonkey.com/r/XGPMF5Y*)* on the hospital's website where patients are requested to access and place their answers as well as sharing the link on social media to grab more general responses. The major outcome measure from this survey is to determine the level of satisfaction with the ED. We focused in our analysis on the factors that mostly affect patient's satisfaction (such as long waiting times, unclear assessment of medical case, ED layout, resources shortage, etc.) in order to figure out ways of improvement based on patients' opinion and experience during their visits. Data responses were collected from the survey website and statistical analysis were performed and presented below.

Results

From all the arriving patients to the ED, only 100 respondents took the survey and results of each question are presented in Figures 1-9 (except question 9 where answers are based on personal view and need text analysis).

Figure 1. Question 1 Chart

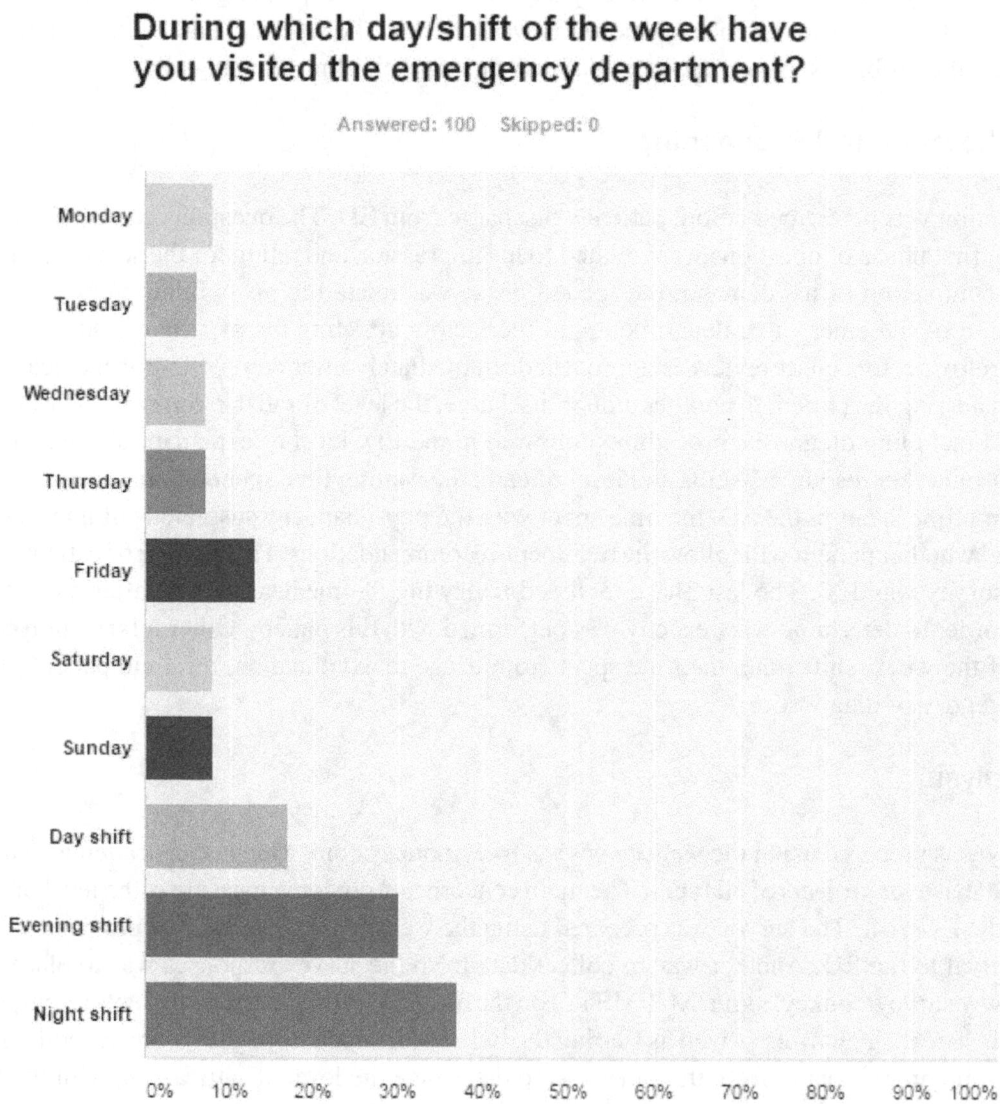

Figure 2. Question 2 Chart

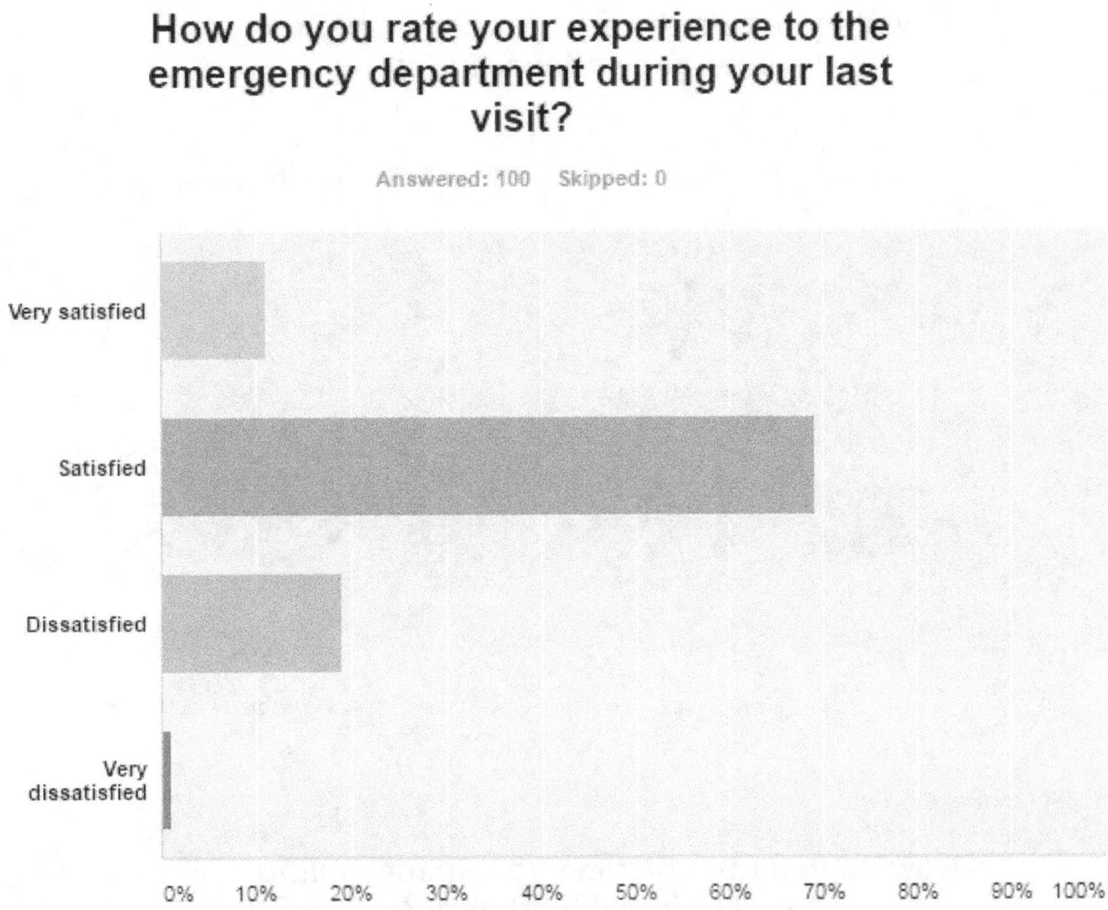

Outputs Statistics

Output Statistics are collected for each question and presented in Tables 1-9.

Discussion

As discussed before the study was performed for one month and data collection showed a sample of 100 patients that took this survey during this time period. From the records collected, the busiest day of the week was Friday, night shifts and this is where improvements should be most likely applied. It is clear that the main issue patients are facing during their visit to the ED is the waiting time (total of 61% of the patients); either the waiting time taken to be initially served (29%), the time a doctor takes to assess patient's condition (20%) or the total time spent in the ED (12%). Another factor that should be taken into consideration while assessing the patient's satisfaction is the level of care the patient receives

Figure 3. Question 3 Chart

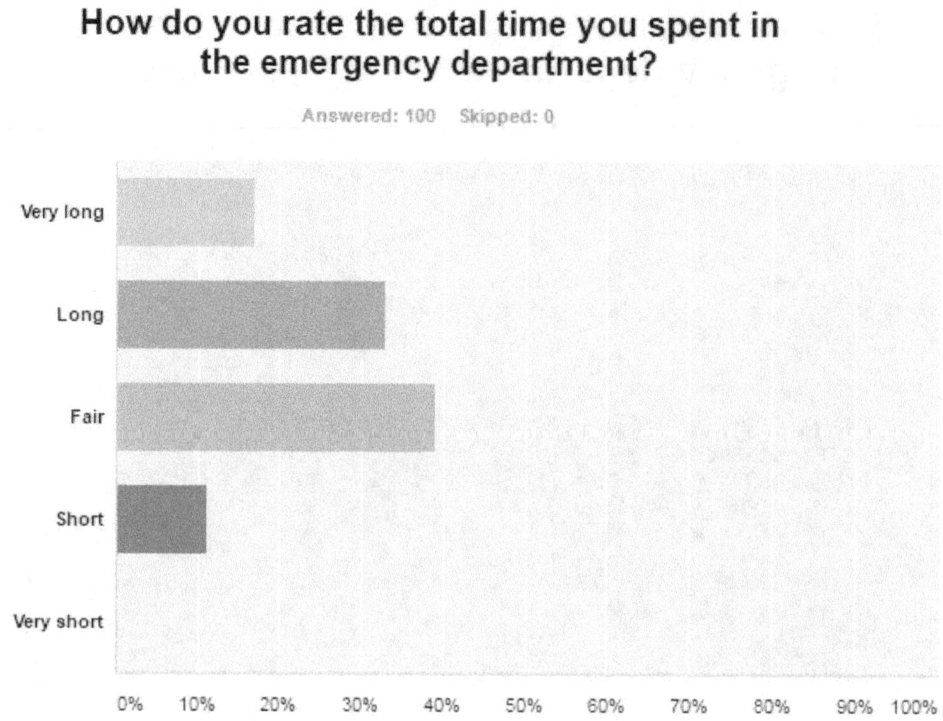

Figure 4. Question 4 Chart

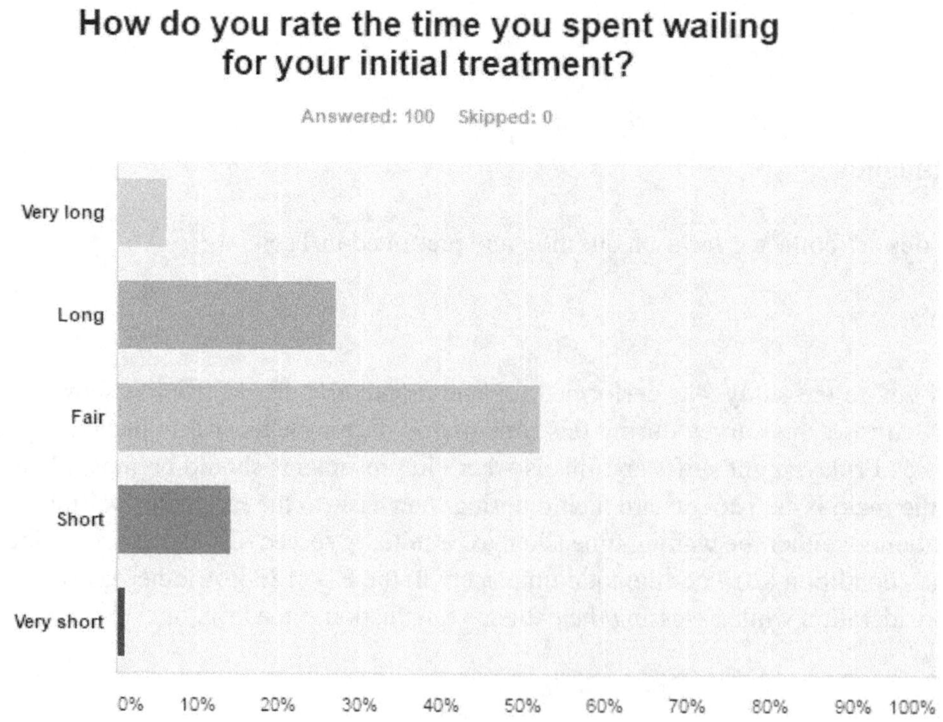

Figure 5. Question 5 Chart

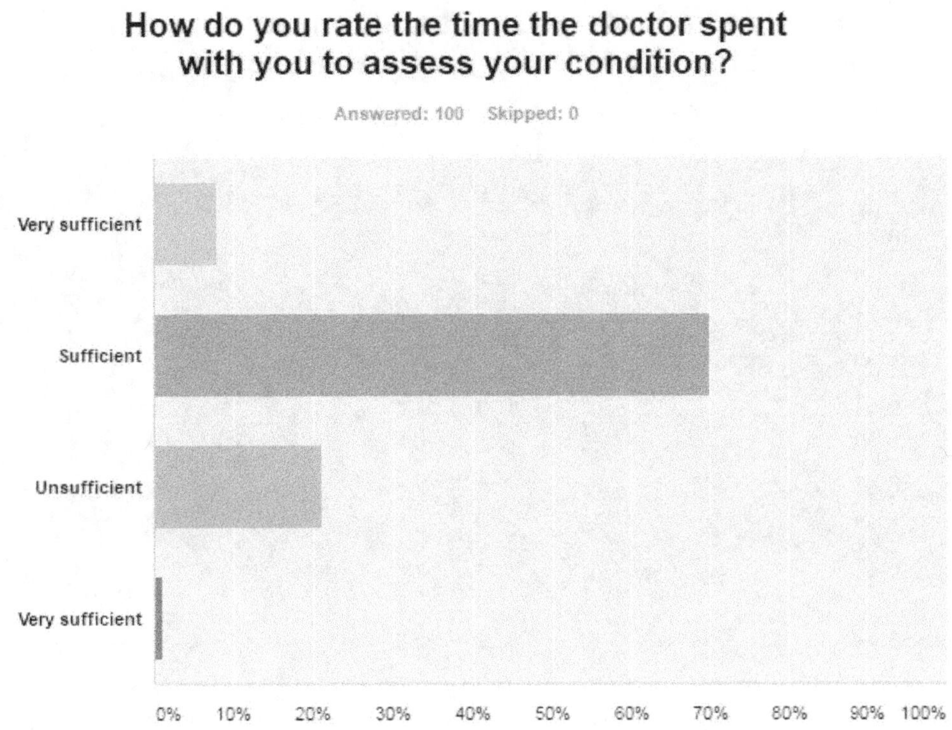

Figure 6. Question 6 Chart

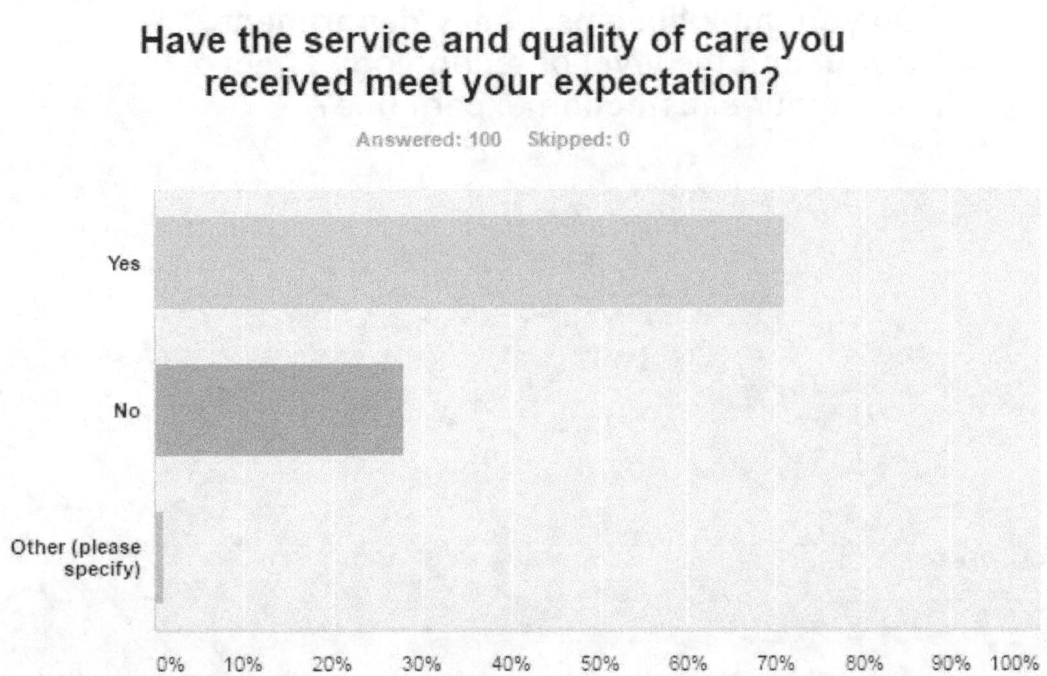

Figure 7. Question 7 Chart

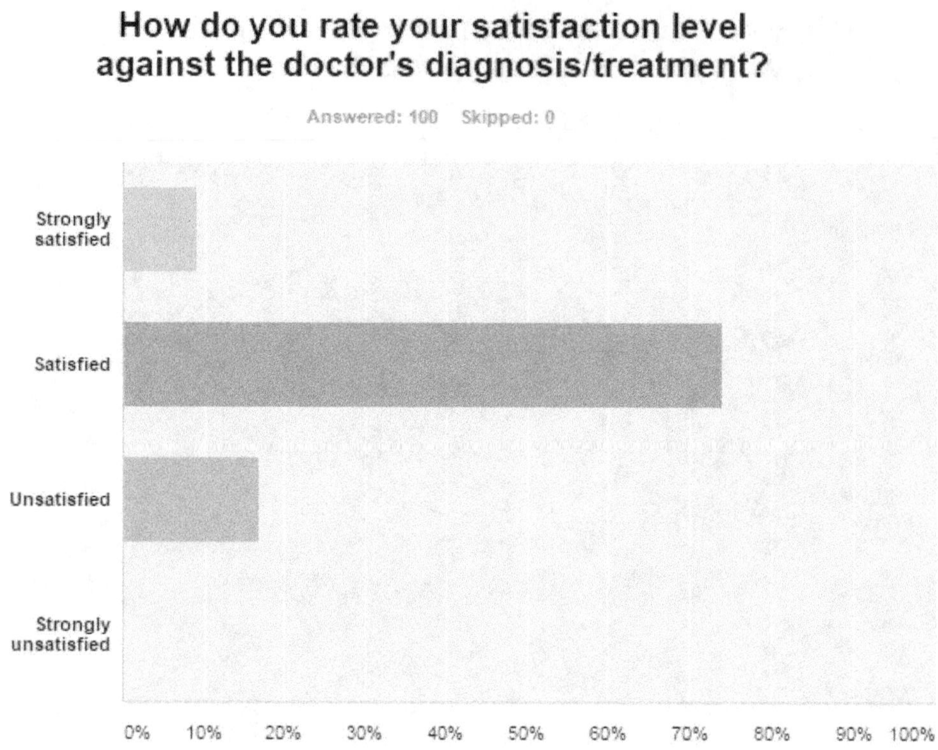

Figure 8. Question 8 Chart

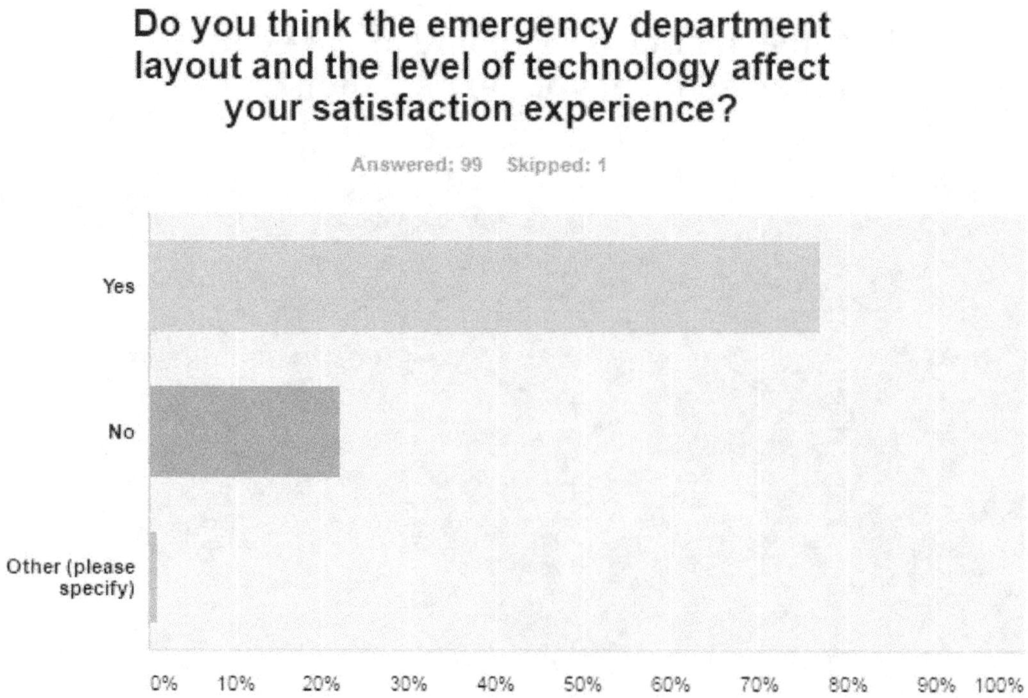

Figure 9. Question 10 Chart

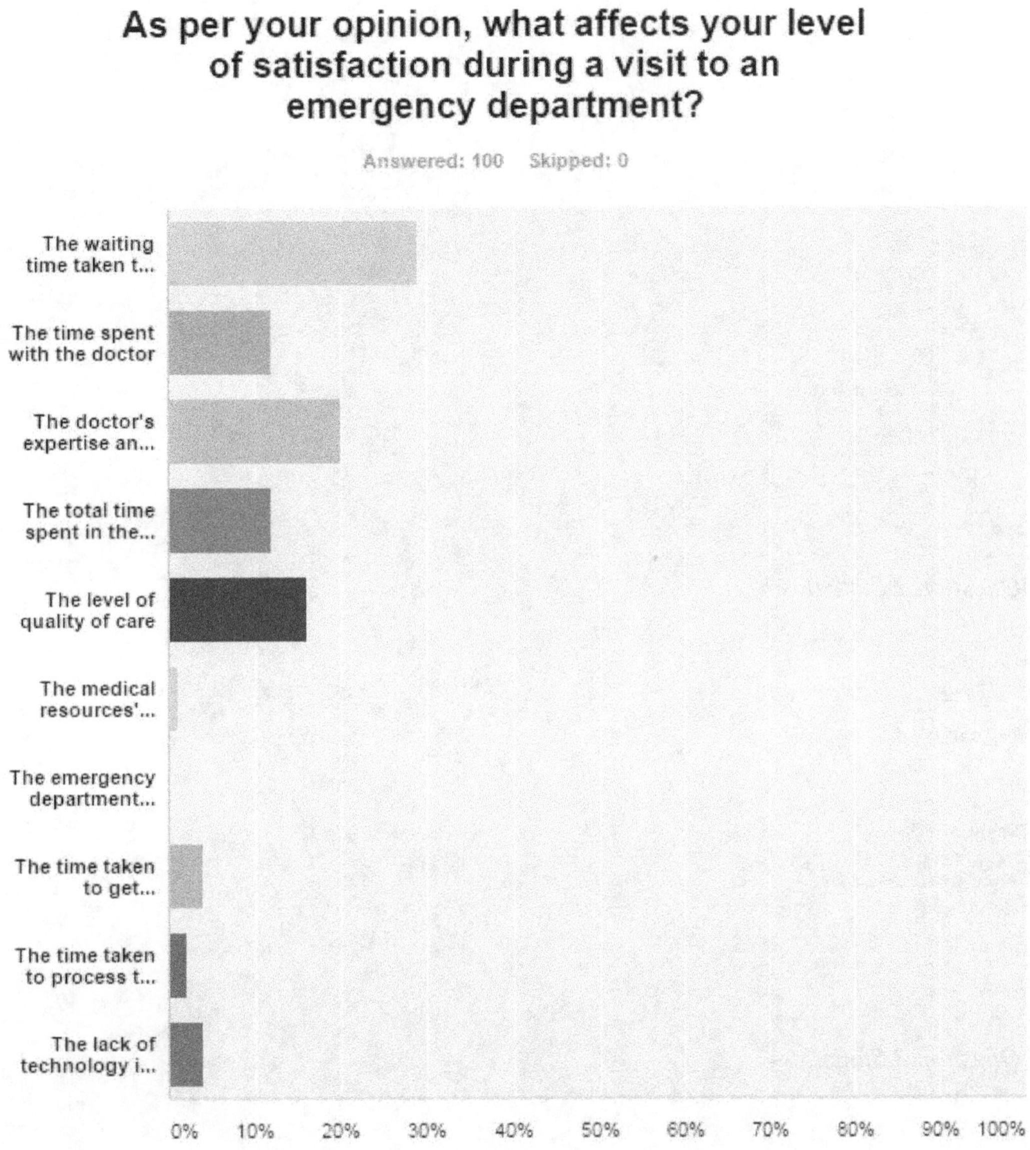

(16%). This study shows also that some patients believe that technology may affect the services in the ED and adding more expertise medical staff will definitely make a difference (answers from question 9).

Limitations

The limitations faced during this data collection were related to the excluded patients: high acuity cases, very old patients or babies and patients not willing to participate. Therefore, the outputs of this survey data collection are invalid for these groups of patients which were not sampled in this work from the

Table 1. Question 1 Statistics

Answer Choices	Responses	
Monday	8.00%	8
Tuesday	6.00%	6
Wednesday	7.00%	7
Thursday	7.00%	7
Friday	13.00%	13
Saturday	8.00%	8
Sunday	8.00%	8
Day shift	17.00%	17
Evening shift	30.00%	30
Night shift	37.00%	37
Total Respondents: 100		

Table 2. Question 2 Statistics

Answer Choices	Responses	
Very satisfied	11.00%	11
Satisfied	69.00%	69
Dissatisfied	19.00%	19
Very dissatisfied	1.00%	1
Total		100

Table 3. Question 3 Statistics

Answer Choices	Responses	
Very long	17.00%	17
Long	33.00%	33
Fair	39.00%	39
Short	11.00%	11
Very short	0.00%	0
Total		100

Table 4. Question 4 Statistics

Answer Choices	Responses	
Very long	6.00%	6
Long	27.00%	27
Fair	52.00%	52
Short	14.00%	14
Very short	1.00%	1
Total		100

Table 5. Question 5 Statistics

Answer Choices	Responses	
Very sufficient	8.00%	8
Sufficient	70.00%	70
Unsufficient	21.00%	21
Very sufficient	1.00%	1
Total		100

Table 6. Question 6 Statistics

Answer Choices		Responses	
Yes		71.00%	71
No		28.00%	28
Other (please specify)	Responses	1.00%	1
Total			100

Table 7. Question 7 Statistics

Answer Choices	Responses	
Strongly satisfied	9.00%	9
Satisfied	74.00%	74
Unsatisfied	17.00%	17
Strongly unsatisfied	0.00%	0
Total		100

Table 8. Question 8 Statistics

Answer Choices		Responses	
Yes		76.77%	76
No		22.22%	22
Other (please specify)	Responses	1.01%	1
Total			99

Table 9. Question 10 Statistics

Answer Choices	Responses	
The waiting time taken to be initially served	29.00%	29
The time spent with the doctor	12.00%	12
The doctor's expertise and time taken to assess your condition	20.00%	20
The total time spent in the emergency department	12.00%	12
The level of quality of care	16.00%	16
The medical resources' attitude	1.00%	1
The emergency department layout	0.00%	0
The time taken to get insurance approval	4.00%	4
The time taken to process the payment before leaving the emergency department	2.00%	2
The lack of technology in the emergency department	4.00%	4
Total		100

beginning. The major reason to exclude these groups from the sampling is because they would generally report lower scores on satisfaction. Measuring patient satisfaction is a very complex endeavor and there is no validated 100% instrument to do so. Thus, our measure of overall satisfaction may not fully capture the real situation. Nevertheless, our surveyed outcome measure resulted in responses and outputs statistics similar to previous studies.

REFERENCES

Bolus, R., & Pitts, J. (1999, April). Patient satisfaction: The indispensable outcome. *Managed Care (Langhorne, Pa.), 8*(4), 24–28. PMID:10387373

Borghans, I., Kleefstra, S. M., Kool, R. B., & Westert, G. P. (2012). Is the length of stay in hospital correlated with patient satisfaction? *International Journal for Quality in Health Care, 24*(5), 443–451. doi:10.1093/intqhc/mzs037 PMID:22789666

Boulding, W., Glickman, S. W., Manary, M. P., Schulman, K. A., & Staelin, R. (2011). Relationship between patient satisfaction with inpatient care and hospital readmission within 30 days. *The American Journal of Managed Care, 17*(1), 41–48. PMID:21348567

Carr-Hill, R., & Mclvor, S. (1989). *The NHS and its Customers*. Centre for Health Economics: University of York.

Cleary, P. D., & McNeil, B. J. (1998). *Patient satisfaction as an indicator of quality care inquiry*. Academic Press.

Hojat, M., Louis, D. Z., Maxwell, K., Markham, F. W., Wender, R. C., & Gonnella, J. S. (2011, June). A Brief Instrument to Measure Patients' Overall Satisfaction With Primary Care Physicians. *Family Medicine, 43*(6), 412–417. PMID:21656396

Keegan, D. P., Eiler, R. G., & Jones, C. R. (1995). Are your performance measures obsolete? *Management Accounting, 70*(12), 45–50.

Llewellyn-Thomas, H. A., Arshinoff, R., Bell, M., Williams, J. I., & Naylor, C. D. (1998). In the queue for total joint replacement: Patients' perspectives on waiting times. *Journal of Evaluation in Clinical Practice, 4*(1), 63–74. doi:10.1046/j.1365-2753.1998.t01-1-00006.x PMID:9524913

Quintana, J. M., Gonzalez, N., Bilbao, A., Aizpuru, F., Escobar, A., & Esteban, C. (2006). Predictors of patient satisfaction with hospital care. *BMC Health Services Research, 6*(102). PMID:16914046

Shesser, R., Kirsch, T., Smith, J., & Kirsch, R. (1991). An analysis of emergency department use by patients with minor illness. *Annals of Emergency Medicine, 20*(7), 743-747.

Toma, G., Triner, W., & McNutt, L.-A. (2009). Patient satisfaction as a function of emergency department previsit expectations. *Annals of Emergency Medicine, 54*(3), 360–367. doi:10.1016/j.annemergmed.2009.01.024 PMID:19282060

Torcson, P. J. (2005). Patient Satisfaction: The Hospitalist's Role. *The Hospitalist*, 27–30.

Chapter 4
Stability of Linear Systems With Multitask Right-Hand Member

G. V. Alferov
Saint Petersburg State University, Russia

G. G. Ivanov
Saint Petersburg State University, Russia

P. A. Efimova
Saint Petersburg State University, Russia

A. S. Sharlay
Saint Petersburg State University, Russia

ABSTRACT

To study the dynamics of mechanical systems and to define the construction parameters and control laws, it is necessary to have computational models accurately describing properties of real mechanisms. From a mathematical point of view, the computational models of mechanical systems are actually the systems of differential equations. These models can contain equations that also describe non-mechanical phenomena. In this chapter, the problems of stability and asymptotic stability conditions for the motion of mechanical systems with holonomic and non-holonomic constraints are under consideration. Stability analysis for the systems of differential equations is given in term of the second Lyapunov's method. With the use of the set-theoretic approach, the necessary and sufficient conditions for stability and asymptotic stability of zero solution of the considered system are formulated. The proposed approaches can be used to study the stability of the motion for robot manipulators, transport, space, and socio-economic systems.

INTRODUCTION

Nowadays mechatronics is a field of knowledge that has a large potential. However, the unpredictability and rapid dynamics of the external environment give rise to a number of problems that are associated with the incompleteness and inconsistency of data on external world conditions. Therefore, one of the most important problems of robotic systems is a mathematical definition of the acting forces and mo-

DOI: 10.4018/978-1-5225-5045-7.ch004

ments in the form of motion equations. These equations are necessary to model the motion with the help of a computer and to choose the laws of motion control. In addition, they are used to assess the quality of the kinematic structure and the construction of mechanical systems. To simplify the description of the mechanical system, the components are replaced by functional converters reducing various kinds of imperfections and so on. Hence, there is an under definition in the description of mechanical systems that contain elements with incomplete information. Thus, in the study of stability of automatic control systems not only one system of differential equations is considered, but a set of systems is under consideration. These systems contain functional parameters that can be arbitrarily varied within the specified limits. In this case differential inclusions are used.

A number of interesting cases, such as the problem of absolute stability of controlled systems or the study of linear nonstationary models

$$\dot{x} = A(t)x,$$

where matrix $A(t)$ element by element satisfies inequality

$$\alpha \le A(t) \le \beta,$$

can be described by linear differential inclusions. This explains the interest in the problem of stability of linear parametrically perturbed systems. Section 1 is devoted to this problem, §§1–3.

In §1 we consider the problem of zero solution stability for a selective linear differential inclusion

$$\dot{x} \in R(x) = conv\left(\bigcup_{k=1}^{N} A_k x\right),$$

where A_k are Hurwitz matrices of dimension $n \times n$.

A.F. Filippov (1985) showed that if solution $x \equiv 0$ of this differential inclusion is asymptotically stable, then it is exponentially stable. A.M. Meilahs showed that asymptotic stability of the considered inclusion implies existence of a function $v(x)$ which is a Lyapunov function. He also determined an analytical form of this function. Based on the results of A.M. Meilakhs (1975), it was proved by A.P. Molchanov and E.S. Pyatnitskii (1986) that for asymptotic stability of equilibrium position $x = 0$ of differential inclusion it is necessary and sufficient a natural number *m*, a number $\gamma > 0$ and symmetric matrices $L_1, \ldots, L_m$ to exist, such that function

$$v(x) = \max_{i=1}^{m} x^T L_i x$$

satisfies the following relation for all *x*

$$\max_{y \in R(x)} \frac{\partial v(x)}{\partial y} \leq -\gamma \| x \|^2,$$

where $\frac{\partial v(x)}{\partial y}$ denotes the derivative of the function $v(x)$ in the direction y.

These results are used in the search of stability for the selector-linear differential inclusion in §1.

In §§2–3 some modifications of the problem of differential inclusions stability are considered (Alferov, Ivanov, Efimova, & Sharlay, 2017; Ivanov, Alferov, Sharlay & Efimova, 2017; Ivanov & Sharlay, 2015; Pichugin, Alferov & Malafeyev, 2016). Namely, a linear homogeneous switched system is considered. Using the direct Lyapunov-function method, the necessary and sufficient conditions are found for the considered system to be stable irrespective of the special realization of the permissible switching modes. Similar problems for second-order systems were considered before, for example, in the papers of A.F. Filippov and other authors (Kulakov, Sokolov, Shalyto & Alferov, 2016; Malafeyev, Alferov & Andreyeva, 2015; Alferov, Malafeyev & Maltseva, 2015; Kulakov, Alferov & Efimova, 2015; Neverova, Malafeyev, Alferov & Smirnova, 2015; Kulakov, Alferov, Efimova, Chernakova & Shymanchuk, 2015; Alferov & Malafeyev, 1996; Alferov & Gorizontov, 1987).

1. STABILITY OF SELECTOR-LINEAR DIFFERENTIAL INCLUSIONS

1.1. Preliminary Remarks

Searching stability of automatic control systems with incomplete information elements implies considering not the only system but a set of differential equations systems containing functional parameters that can arbitrarily vary within given limits. A similar situation arises in the well-known problem of absolute stability of controlled systems while studying linear nonstationary models

$$\dot{x} = A(t)x,$$

where matrix $A(t)$ elementwise satisfies inequality

$$\alpha \leq A(t) \leq \beta,$$

and so on.

To describe systems containing elements with incomplete information the following differential inclusions are used

$$\dot{x} \in R(x), \tag{1.1}$$

where the n-dimensional vector x characterizes the deviation of the system from the mode prescribed by control objective, and $R(x)$ denotes a set of permissible velocities. For the problems given above the set $R(x)$ for each x is a polyhedron of the form

$$R(x) = conv\left(\bigcup_{k=1}^{N} A_k x\right), \tag{1.2}$$

The differential inclusions (1.1), where $R(x)$ has the form (1.2), is named selector-linear, since the multivalued mapping $R(x)$ in (1.2) is the union of linear single-valued mappings (selectors).

Solution of the inclusion (1.1) is generally taken as an absolutely continuous function $x(t)$ satisfying relation

$$\dot{x} \in R(x(t))$$

almost everywhere on the considered interval.

The problem consists in analyzing asymptotic stability of zero solution $x(t) \equiv 0$ of inclusion (1.1).

Solution $x \equiv 0$ of inclusion (1.1) is stable if for each $\varepsilon > 0$ there exists $\delta > 0$, such that for any x_0 for which $\| x_0 \| < \delta$ each solution $\tilde{x}(t)$ with initial condition $\tilde{x}(t_0) = x_0$ exists for $t \geq t_0$ and satisfies inequality

$$\| \tilde{x}(t) \| < \varepsilon.$$

Definition of asymptotic stability, in addition to stability, also requires that $\tilde{x} \to 0$ as $t \to \infty$.

Further assume that all matrices A_k in the expression (1.2) are Hurwitz, since this condition is obviously necessary for asymptotic stability of zero solution (1.1).

1.2. Auxiliary Statements

Before analyzing stability of zero solution of the autonomous selector-linear differential inclusion (1.1) an auxiliary assertion must be proved. In the formulation of this assertion $Z(x)$ denotes a matrix of dimension $N \times n$ in which the k-th row is defined by equality

$$z_k(x) = (A_k x)^T, \qquad k = 1, 2, \ldots, N.$$

Theorem 1.1: If for any $x \in S = \{x : \ \| x \| = 1\}$ there exists an n-dimensional vector $y(x) \in S$, such that

$$Z(x)y(x) < 0,\ x \cdot y(x) > 0,$$

then there exist $\delta > 0$ and a positively definite homogeneous function $v(x)$, such that

$$\frac{d}{dt}v(x) \le -\delta \parallel x \parallel^2,$$

where derivative is taken by any of systems

$$\dot{x} = A_k x, \qquad k = 1, 2, \ldots, N.$$

Proof

Let us take an arbitrary point $x_0 \in S$ and construct a set

$$M(x_0) = \{y : \ Z(x_0)y < 0,\ x_0 \cdot y > 0, y \in S\}.$$

In the set $M(x_0)$ some point y_0 is taken. Since $M(x_0)$ is obviously open, then point y_0 belongs to this set together with some neighborhood of itself.

Using the fact that each point $x \in S$ and that $y \in S$ can be represented as

$$x = x_0 + \Delta x, \qquad y = y_0 + \Delta y,$$

the following equality system is obtained

$$Z(x)y = Z(x_0)y_0 + Z(\Delta x)y_0 + Z(x)\Delta y, x \cdot y = x_0 \cdot y_0 + \Delta x \cdot y_0 + x \cdot \Delta y. \tag{1.3}$$

By construction

$$Z(x_0)y_0 < 0,\ x_0 \cdot y_0 > 0.$$

Then considering (1.3) and the fact that y_0 is an interior point of set $M(x_0)$ there follows existence of constants $\gamma > 0$ and $\theta > 0$, such that for any $x \in \Gamma(x_0)$,

$$\Gamma(x_0) = \{x : \ \parallel x - x_0 \parallel < \gamma, x \in S\},$$

and for any $y \in \Theta(x_0)$,

$$\Theta(x_0) = \{y : \| y - x_0 \| < \theta, y \in S\}$$

$$Z(x)y < 0, \;\; x \cdot y > 0.$$

This means that the set $\Theta(y_0)$ belongs to each of the sets $M(x)$ for $x \in \Gamma(x_0)$, i.e.

$$\Theta(y_0) \subset \bigcap_{x \in \Gamma(x_0)} M(x)$$

Repeating similar arguments, each point $x \in S$ can be associated with some neighborhood $\Gamma(x) \subset S$. It is clear that the collection of neighborhoods $\Gamma(x)$, $x \in S$, forms an open covering of set *S*. Then by virtue of compactness of the sphere, it is possible to extract from the set *S* a finite subcovering

$$\Gamma(x_1), \ldots, \Gamma(x_m).$$

Without loss of generality, the considered subcovering is supposed to be minimal, i.e. no any of the sets $\Gamma(x_j)$, $j = 1, 2, \ldots, m$ can be taken out of this subcovering.

It is clear that sets $\Gamma(x_j)$ can intersect pairwise. Therefore, some open subset $\Gamma'(x_j)$ is taken from each $\Gamma(x_j)$ in such a way that for $j_1 \neq j_2$

$$\Gamma'(x_{j_1}) \cap \Gamma'(x_{j_2}) = \varnothing$$

and that closed sets $\overline{\Gamma'}(x_j)$, $j = 1, 2, m$, form a covering of set S.

Note that the given above procedure for constructing the set $\Gamma(x)$ associates each set $\Gamma(x_j)$ with some set $\Theta(y_j)$, $j = 1, 2, \ldots, m$.

Let an arbitrary continuous vector-function

$$p_j(x) : \Gamma'(x_j) \to \Theta(y_j), \qquad j = 1, 2, \ldots, m.$$

be defined on $\Gamma'(x_j)$.

It is clear that neighborhoods $\Gamma(x_j)$ and $\Theta(y_j)$ can always be taken in such a way that there exists a set of positive numbers $\delta_1, \ldots, \delta_m$, such that for any $x \in \Gamma(x_j)$

$$Z(x)p_j(x) \leq -\delta_j \leq -\delta, \qquad j = 1, 2, \ldots, m,$$

where

$$\delta = \min_{j\in\{1,\ldots,m\}} \delta_j.$$

Functions p_j are taken on the boundaries of sets $\Gamma'(x_j)$ by continuity.

Thus, on the sphere S vector-function $p(x)$ is constructed. If $x \ni S$, then it can be assumed that

$$p(x) = \| x \| \, p\left(\frac{x}{\| x \|}\right).$$

Since the choice of the functions p_j is arbitrary, then these functions can be demanded to satisfy the conditions

$$\omega_j \wedge d\omega_j \equiv 0, \qquad x \in \Gamma'(x_j), \quad j = 1,\ldots,m,$$

where $\omega_j = p_j dx$ is a differential 1-form, and $d\omega_j$ is its external derivative.

According to Frobenius's theorem, the vector-function $p(x)$ constructed above defines a completely integrable field of hyperplanes in the space above. Hence, there exists the function $v(x)$ which the vector-function $p(x)$ is a gradient for.

Condition $x \cdot p(x) < 0$ for $x \neq 0$ implies that the homogeneous function $v(x)$ is positively definite, and condition $Z(x)p(x) \leq -\delta$ for $x \in S$ implies that relation

$$\frac{d}{dt} v(x) \leq -\delta \| x \|^2,$$

is realized due to the choice of the function $p(x)$ and linearity of matrix $Z(x)$ coefficients with respect to *x*. Here the derivative is taken by any of systems

$$\dot{x} = A_k x, \qquad k = 1,\ldots,N.$$

1.3. Asymptotic Stability of the Trivial Solution of Automorphic Selector-Linear Differential Inclusion

Now let the problem of asymptotic stability for the trivial solution $x \equiv 0$ of autonomous selector-linear differential inclusion (1.1) be considered.

Theorem 1.2: For zero solution $x \equiv 0$ of selector-linear differential inclusion (1.1) be asymptotically stable, it is necessary and sufficient an n-dimensional vector $y\left(x\right) \in S$ to exist for any $x \in S$ and

$$Z\left(x\right)y\left(x\right) < 0,\ x \cdot y(x) > 0. \tag{1.4}$$

Proof

Necessity. Let solution $x \equiv 0$ of selector-linear differential inclusion of the form (1.1) be asymptotically stable. Then it follows from the works of A.M. Meilakhs, A.P. Molchanov and E.S. Pyatnitsky that there exist a positive integer m, positive δ and symmetric $n \times n$ -matrices L_i, $i = 1, 2,\ldots, m$, such that function

$$v\left(x\right) = \max_{i \in \{1,2,\ldots,m\}} x^T L_i x$$

is a positively definite quadratic form and due to (1.1) has a derivative $w\left(x\right)$ satisfying the estimate

$$w\left(x\right) \le -\delta \parallel x \parallel^2 . \tag{1.5}$$

It is obvious that if the estimate (1.5) is valid for derivative of the function $v\left(x\right)$ taken by the differential inclusion (1.1), then it is also realized for derivative of the function $v\left(x\right)$ taken by any of the systems

$$\dot{x} = A_k x, \qquad k = 1, 2, \ldots, N.$$

Then it follows in addition with the fact that $v\left(x\right)$ is a quadratic form, that inequalities (1.4) are realized.

Sufficiency. Since inequalities (1.4) are realized, then Theorem 1.1 implies existence of the positively definite function $v\left(x\right)$ and $\delta > 0$ such that derivative of the function $v\left(x\right)$ by virtue of any of systems

$$\dot{x} = A_k x, \qquad k = 1, 2, \ldots, N,$$

does not exceed $-\delta \parallel x \parallel^2$. But then it is obvious that

$$w\left(x\right) \le -\delta \parallel x \parallel^2,$$

which implies that zero solution of the selector-linear differential inclusion (1.1) is asymptotically stable.

2. CONDITIONS FOR STABILITY OF LINEAR HOMOGENEOUS SWITCHED SYSTEMS

2.1. Preliminary Notes

Consider the system

$$\dot{x} = A(u)x, \tag{2.1}$$

where *x* is a vector of phase variables, *u* is a scalar piecewise constant function mapping a set of non-negative numbers $T^+ = \{t : 0 \le t < \infty\}$ to a finite set of integers:

$$u: \ T^+ \to \{1,\ldots,N\}, \tag{2.2}$$

and $A(u)$ is a $n \times n$ -matrix function defined by relations

$$A(i) = A_i, \qquad i = 1,\ldots,N.$$

Let

$$t_0 = 0 < t_1 < t_2 < \ldots \tag{2.3}$$

be a sequence of discontinuity points of the function u (t). Then on intervals

$$T_j = \{t : \ t_{j-1} \le t < t_j\}, \qquad j = 1,2,\ldots,$$

the function u (t) is constant. Assume the sequence (2.3) be such that there exists $\delta > 0$ for it that satisfies condition

$$\min_{j=1}^{\infty} \tau_j \ge \delta, \tag{2.4}$$

where τ_j denotes length of the interval T_j, i.e.

$$\tau_j = t_j - t_{j-1}, \qquad j = 1,2,\ldots. \tag{2.5}$$

Let U_δ denote a set of all functions $u(t)$ defined by conditions (2.2) and (2.4) for some fixed δ.

System (2.1) is named stable uniformly with respect to $u \in U_\delta$ if for any $\varepsilon > 0$ and $\delta > 0$ there exists $\eta > 0$, such that for any $u \in U_\delta$ all the solutions of system (2.1) starting in η-neighborhood of zero are in its ε-neighborhood with time.

Now necessary and sufficient conditions for uniform stability with respect to U_δ of system (2.1) must be obtained.

2.2. Necessary and Sufficient Conditions for Uniform Stability With Respect to U_δ of System 2.1.

Let A_i^t denote a linear operator corresponding to matrix $e^{A_i t}$, $\overline{J_\delta(M)}$ denote a closedness of set

$$J_\delta(M) = \bigcup_{i=1}^{N} \bigcup_{t \in [\delta, \infty)} A_i^t(M)$$

and

$$\rho(M,Q) = \min_{x \in M, y \in Q} \| x - y \|$$

denote a distance between sets *m* and *Q*.

Theorem 2.1: For the system (2.1) to be stable uniformly with respect to U_δ, it is necessary and sufficient a bounded closed set M_δ to exist for any $\delta > 0$ and contain a point zero as an interior point, such that

$$\overline{J_\delta(M_\delta)} \subset M_\delta.$$

Proof

Necessity. Let the system (2.1) be stable uniformly with respect to U_δ. Now it must be proved that in this case there exist sets satisfying conditions of the theorem.

Consider some $\delta > 0$ and an arbitrary bounded closed set M_1 containing point zero as its interior point. Let the following set be constructed:

$$M_2 = M_1 \cup \overline{J_\delta(M_1)}.$$

If $M_2 \subset M_1$, then the process ends. Otherwise, the following set is constructed

$$M_3 = M_2 \cup \overline{J_\delta\left(M_2\right)}.$$

If $M_3 \subset M_2$, then the construction process ends. Otherwise, it continues as it is described above. Continuing this process to infinity, the following system of sets is obtained:

$$M_1 \subset M_2 \subset \ldots.$$

Suppose

$$M_\delta = \overline{\left(\bigcup_{j=1}^{\infty} M_j\right)}.$$

First of all, let the boundedness of the set M_δ be proved. Suppose to the contrary that the set M_δ is unbounded. Then for an arbitrarily large $\Delta > 0$ there exists an index *k* such that a set $M_k \subset M_\delta$ contains a subset *Q* for which the condition *Q* implies $\| x \| > \Delta$. Let some point $x^* \in Q$ be fixed. By the definition of the set M_k, there exist a set of numbers

$$\tau_1, \ldots, \tau_k, \qquad \tau_j \geq \delta, \quad j = 1, \ldots, k,$$

defined by relations (2.5), a set of numbers

$$l_1, \ldots, l_k, \qquad l_j \in \left(1, \ldots, N\right), \quad j = 1, \ldots, k,$$

and a point $x_1 \in M_1$, such that

$$\left(\prod_{s=0}^{k-1} e^{A_{l_{k-s}} \tau_{k-s}}\right) x_1 = x^*.$$

Since Δ is arbitrary, the obtained equality means that if *u* runs over the set U_δ, then a set of system (2.1) solutions starting at $t = 0$ in set M_1 is not uniformly bounded. Then taking into account the boundedness of the set M_1 and linearity of system (2.1), it follows that this system can not be stable uniformly with respect to U_δ. This contradiction proves the boundedness of the set M_δ.

The set M_δ closed by its construction, and moreover it consist of point zero as its interior point, since $M_1 \subset M_\delta$. Thus, to complete the proof of this part of the theorem it only remains to show existence of the inclusion

$$\overline{J_\delta(M_\delta)} \subset M_\delta.$$

Suppose it does not exist. Then, for some operator $A_{l_k}^\tau$, $l_k \in (1,\ldots,N)$, $\tau \geq \delta$ there exists an interior point $x_0 \in M_\delta$ by virtue of its continuity, such that

$$A_{l_k}^\tau(x_0) \ni M_\delta.$$

But since point x_0 is an interior point of the set M_δ, for some *j* the point $x_0 \in M_j$. Taking into account that by definition

$$M_{j+1} = M_j \cup \overline{J_\delta(M_j)},$$

a chain of inclusions can be obtained

$$A_{l_k}^\tau(x_0) \in A_{l_k}^\tau(M_j) \subset M_{j+1} \subset M_\delta,$$

from which it follows that

$$A_{l_k}^\tau(x_0) \in M_\delta.$$

But this contradicts the assumption that the point $A_{l_k}^\tau(x_0)$ is not in the set M_δ, which proves existence of the required inclusion.

Taking into account the arbitrary choice of $\delta > 0$, it is clear that if the system (2.1) is stable uniformly with respect to U_δ. Then for any $\delta > 0$ there exists a set $\delta > 0$ satisfying all the requirements of the theorem.

Sufficiency. Let conditions of the theorem be satisfied. Then for an arbitrary $\delta > 0$ there exists a bounded closed set M_δ containing the point zero as its interior point and such that

$$\overline{J_\delta(M_\delta)} \subset M_\delta.$$

This means that all the solutions of the system (2.1) for any $u \in U_\delta$ starting in the M_δ are in the set

$$\overline{J_0(M_\delta)} = \overline{\bigcup_{i=1}^{N}\bigcup_{t=0}^{\infty} A_i^t(M_\delta)} = \overline{J_\delta(M_\delta)} \cup J'(M_\delta),$$

where $J'(M_\delta)$ denotes the closedness of set

$$\bigcup_{i=1}^{N}\bigcup_{t=0}^{\delta} A_i^t(M_\delta).$$

Let K_R denote a ball of radius R with a center at 0. Taking into account the uniform boundedness of norms of the operators A_i^t, $t \in [0,\delta]$. $i = 1, 2, N$ it follows that there exists an R such that

$$\overline{J_0(M_\delta)} \subset K_R.$$

Further, since the set M_δ contains the point zero as its interior point, there is an r such that the following inclusion is realized:

$$K_r \subset M_\delta.$$

From the fact that for any $u \in U_\delta$ all the solutions of the system (2.1) starting in M_δ are in $\overline{J_0(M_\delta)}$ it follows that all the solutions of this system starting in K_r are in the ball K_R.

Consider an arbitrary $\varepsilon > 0$ and μ so small that $\mu R < \varepsilon$. Since the system (2.1) is linear, the set

$$\mu M_\delta = \{y : y = \mu x, x \in M_\delta\}$$

has all the properties of the set M_δ. Setting $\eta = \mu r$ it follows that for any $\varepsilon > 0$ there exists $\eta > 0$ such that for any $u \in U_\delta$ all the solutions of the system (2.1) starting in η-neighborhood of zero are in its ε-neighborhood. But this implies that the system (2.1) is uniformly stable with respect to U_δ. Taking into account the arbitrariness of δ, it is clear that the theorem is proved.

Theorem 2.2: If the system (2.1) is stable uniformly with respect to U_δ, then for any $\delta > 0$ there exists a convex set M_δ satisfying all the requirements of Theorem 2.1. Moreover, for any $\mu > 0$, $\mu \neq 1$

$$\mu \partial M_\delta \cap \partial M_\delta = \varnothing$$

and for any $q \in [0,1]$

$$qM_\delta \subset M_\delta.$$

Here, ∂M_δ denotes the boundary of the set ∂M_δ.

Proof

Consider an arbitrary $\delta > 0$ and a convex bounded closed set M containing the point zero as its interior point. By conditions, the system (2.1) is stable uniformly with respect to Y_δ and, therefore, for the considered δ there exists a set $M_\delta^{'} \subset M$ satisfying all the requirements of Theorem 2.1.

Consider arbitrary $t \geq \delta$ and $i \in (1,\ldots,N)$. By Theorem 2.1

$$A_i^t\left(M_\delta^{'}\right) \subset M_\delta^{'}.$$

Given that $M_\delta^{'} \subset M$ it follows that

$$A_i^t\left(M_\delta^{'}\right) \subset M$$

and simultaneously,

$$A_i^t\left(M_\delta^{'}\right) \subset A_i^t\left(M\right).$$

But then

$$A_i^t\left(M_\delta^{'}\right) \subset M \cap A_i^t\left(M\right).$$

Applying an operator A_i^{-t}, which exists due to nondegeneracy of the operator A_i^t, to this inclusion the following is obtained:

$$M_\delta^{'} \subset M \cap A_i^{-t}\left(M\right).$$

This inclusion holds for any $t \geq \delta$ and for any $i \in (1,\ldots,N)$. This proves validity of inclusion

$$M_\delta^{'} \subset M_1 = \bigcap_{i=1}^{N}\bigcap_{t\geq\delta}\left(M \cap A_i^{-t}\left(M\right)\right).$$

The operator A_i^{-t} is linear and continuous, and therefore a set $A_i^{-t}\left(M\right)$ is closed and convex, as well as the set M. Further, since the intersection of any number of closed and convex sets is also a closed and convex set, which is applicable for the set M_1 as well. In addition, given that

$$M_\delta^{'} \subset M_1 \subset M,$$

It can be concluded that M_1 is bounded and contains point zero as its interior point.

Applying the described procedure to M_1, a bounded closed convex set M_2 such that

$$M_\delta^{'} \subset M_2 \subset M_1.$$

is obtained.

Continuing this process, the following sequence of embedded closed convex sets is obtained:

$$M \supset M_1 \supset \ldots \supset M_k \supset \ldots.$$

Let M_δ denote a limit of this sequence. It must be proved first M_δ to satisfy all the requirements of Theorem 2.1 for the considered δ.

By definition

$$M_\delta = \bigcap_{k=1}^{\infty} M_k,$$

which implies that M_δ is a bounded closed convex set due to the sets M_k.

Further, by its construction for any k $M_\delta^{'} \subset M_k$, and therefore $M_\delta^{'} \subset M_\delta$. Hence the zero point is an interior point of the set M_δ.

Now it must be proved that

$$\overline{J_\delta}\left(M_\delta\right) \subset M_\delta.$$

Indeed

$$M_{k+1} = \bigcap_{i=1}^{N} \bigcap_{t \geq \delta} \left(M_k \cap A_i^{-t}\left(M_k\right)\right).$$

Passing to the limit in this relation, it follows that

$$M_\delta = \bigcap_{i=1}^{N} \bigcap_{t \geq \delta} \left(M_\delta \cap A_i^{-t}\left(M_\delta\right)\right).$$

Then for any $i \in \left(1, \ldots, N\right)$ and $t \geq \delta$ there exists an inclusion

$$M_\delta \subset M_\delta \cap A_i^{-t}\left(M_\delta\right),$$

which implies that

$$M_\delta \subset A_i^{-t}\left(M_\delta\right).$$

Applying the operator A_i^t to the last inclusion it follows that

$$A_i^t(M_\delta) \subset M_\delta.$$

Taking into account the arbitrariness of *i* and *t* and the closedness of the M_δ it can be concluded that

$$\overline{J_\delta}(M_\delta) \subset M_\delta.$$

Thus, for an arbitrary δ a convex set M_δ is constructed which satisfies all the requirements of Theorem 2.1. Now, it must be shown that this set also satisfies the remaining requirements of Theorem 2.2.

Let us prove that for any $q \in [0,1]$

$$qM_\delta \subset M_\delta.$$

Consider an arbitrary point $x_0 \in \partial M_\delta$. Since the set M_δ is convex and the zero point is its interior point, then the segment $[0, x_0]$ containing points 0 and x_0 is also in the set M_δ. But for any $q \in [0,1]$

$$qx_0 \in [0,1],$$

which implies that

$$q\partial M_\delta \subset M_\delta.$$

Multiplication by a scalar takes the boundary of the initial set to the boundary of the finite one, and consequently

$$\partial(qM_\delta) = q\partial M_\delta \subset M_\delta.$$

This fact implies that

$$qM_\delta \subset M_\delta.$$

Let us show that for any $\mu > 0$, $\mu \neq 1$,

$$\partial M_\delta \cap \mu\partial M_\delta = \varnothing.$$

Suppose the contrary situation, i.e. suppose for some $\mu_0 \in (0,1)$ the sets ∂M_δ and $\mu\partial M_\delta$ have a nonempty intersection. Then there exists at least one point $x_0 \in \partial M_\delta$ such that

$$\mu_0 x_0 = x_0^{'} \in \partial M_\delta.$$

Since the point zero is an interior point of the set M_δ, there exists $\varepsilon > 0$ which a ball $\varepsilon > 0$ of radius ε with center at zero is in the set M_δ for. Let $\left[x, x_0\right]$ denote the rectilinear segment joining points x and x_0. Consider a set

$$Q = \bigcup_{x \in K_\varepsilon} \left[x, x_0\right].$$

Since M_δ is convex, an interval $\left[x, x_0\right]$ is in M_δ for any M_δ. Hence Q is also in M_δ. By construction, the point zero is an interior point of the set Q, and $x_0^{'} \in \left(0, x_0\right)$. Then $x_0^{'}$ is an interior point of the set Q, and therefore, of the set M_δ. But this fact contradicts the assumption that $x_0^{'} \in \partial M_\delta$. This contradiction proves the validity of the statement.

Thus, it is proved that for an arbitrary $\delta > 0$ it is possible to construct the set M_δ satisfying all the requirements of Theorem 2.2.

Let the system (2.1) be named asymptotically stable uniformly with respect to U_δ if it is stable uniformly with respect to U_δ and for any U_δ the solutions of the system (2.1) starting in the η-neighborhood of zero tend to zero as $t \to \infty$.

Theorem 2.3: For the system (2.1) to be stable uniformly at U_δ, it is necessary and sufficient for any U_δ to exist a bounded closed set M_δ, containing the point zero as its interior point, such that

1. M_δ is closed bounded set containing zero as its interior point;
2. For any $\mu \in \left[0, 1\right)$
3. $\mu M_\delta \subset M_\delta$ and $\mu \partial M_\delta \cap \partial M_\delta = \varnothing$;
4. For some $q \in \left(0, 1\right)$

$$\overline{J}_\delta \left(M_\delta\right) \subset q M_\delta.$$

Proof

Necessity. Due to initial condition system (2.1) is stable uniformly at U_δ. Then it follows from the Theorem 2.1 and Theorem 2.2 that for any $\delta > 0$ there exists a set M_δ that satisfies the requirements 1 and 2 of this theorem, which proves the necessity of those propositions.

Let us prove the necessity of condition 3. If

$$\overline{J}_\delta \left(M_\delta\right) \cap \partial M_\delta = \varnothing,$$

then due to isolation of sets M_δ and $\overline{J_\delta}\left(M_\delta\right)$ and the continuity of operators A_i^t, $i=1, 2, \ldots, N$, $t \geq 0$:

$$\rho\left(\overline{J_\delta}\left(M_.\right), \partial M_\delta\right) = \alpha > 0,$$

and consequently there exists $q \in \left(0,1\right)$ satisfying condition 3.

Let us choose an arbitrary $\delta > 0$ and some set $M_\delta^{'}$, satisfying conditions 1 and 2 and show how in general case it is possible to construct a set M_δ satisfying all the requirements of this theorem.

Firstly, let us show that if

$$\overline{J_\delta}\left(M_\delta^{'}\right) \cap \partial M_\delta^{'} \neq \varnothing,$$

then at least for one of the operator A_i^t, $i = 1, 2, \ldots, N$, $t \geq \delta$, for instance, A_1^{t-0} there exists a point $x_0 \in \partial M_\delta^{'}$, such that $A_1^{t_0}\left(x_0\right) \in \partial M_\delta^{'}$.

Indeed, let us denote a point contained in the intersection $\overline{J_\delta}\left(M_\delta^{'}\right) \cap \partial M_\delta^{'}$ as $x_0^{'}$, and show that this point is in set $J_\delta\left(M_\delta^{'}\right)$. Since $x_0^{'} \in \overline{J_\delta}\left(M_\delta^{'}\right)$, there are sequences $\left\{x_0^k\right\} \subset M_\delta^{'}$, $\left\{t_k\right\}$, $t_k \geq \delta$ at all k and a set of numbers $\left\{i_k\right\}$, $i_k \in \left(1, \ldots, N\right)$, such that

$$\lim_{k\to\infty} x_{i_k}\left(x_0^k, t_k\right) = x_0^{'}.$$

Here $x_j\left(\xi, \tau\right)$ denotes the point where the solution of the system

$$\dot{x} = A_j x,$$

started at the point ξ, $j = 1, 2, N$ at $t = 0$, will get at the moment τ. Without loss of generality it can be considered that $i_k = const$ as it is always possible to be contained by making a sequence. Suppose $i_k = 1$ at all *k*. Taken into consideration that the system

$$\dot{x} = A_1 x$$

is asymptotically stable, it is proved that the sequence $\left\{t_k\right\}$ can not increase unlimitly, because otherwise due to finiteness of sequence $\left\{x_0^k\right\}$:

$$\lim_{k\to\infty} x_1\left(x_0^k, t_k\right) = x_0^{'} = 0 \in \partial M_\delta^{'}.$$

But since point zero is an interior point of the set $M_\delta^{'}$, it can not be on its border.

Let us denote the limit of the sequence $\{t_k\}$ by t_0 and the limit of the sequence $\{x_0^k\}$ by x_0. The set $M_\delta^{'}$ is closed which means that $x_0 \in M_\delta^{'}$, then $t_0 \geq \delta$ since $t_k \geq \delta$ at any *k*. Considering the continuity of solutions of a linear system with constant coefficients at the initial data, we obtain

$$\lim_{k\to\infty} x_1\left(x_0^k, t_k\right) = x_1\left(x_0, t_0\right) = x_0^{'},$$

which means that $x_0^{'} \in J_\delta\left(M_\delta^{'}\right)$.

Since an operator $A_1^{t_0}$ is bounded and continuous, it maps the boundary of the set pre-image onto the boundary of set image. Hence the boundary $\partial M_\delta^{'}$ of set $M_\delta^{'}$ is mapped by $A_1^{t_0}$ onto the boundary $\partial A_1^{t_0}\left(M_\delta^{'}\right)$ of set $A_1^{t_0}\left(M_\delta^{'}\right)$. But due to the condition

$$x_0^{'} \in \partial M_\delta^{'},$$

and shown above

$$x_0^{'} \in A_1^{t_0}\left(M_\delta^{'}\right) \subset M_\delta^{'},$$

the following is true:

$$x_0^{'} \in \partial A_1^{t_0}\left(M_\delta^{'}\right).$$

Thus, after mapping $A_1^{t_0}$ it follows that the point $x_0^{'}$ belongs to the boundary of the set image and consequently, the point x_0 must belong to the boundary of the set pre-image, i. e. a set $\partial M_\delta^{'}$.

It follows from these considerations that the pre-images of point $x_0^{'}$ either belong to the set $\partial M_\delta^{'}$ or do not belong even to the set $M_\delta^{'}$.

Indeed, all operators A_i^t, $i = 1, 2,, N$, $t \geq \delta$ are continuous and bounded, so that, if there is a point $x_0 \in M_\delta^{'} \setminus \partial M_\delta^{'}$ which is a pre-image of the point $x_0^{'}$, then for this point there is a finite moment $t_0^{'} \geq \delta$ and some number $i \in \left(1, \ldots, N\right)$, such that

$$A_i^{t_0^{'}}\left(x_0\right) = x_0^{'}.$$

But this is infeasible, since

$$A_i^{t_0^{'}}\left(M_\delta^{'}\right) \subset M_\delta^{'},$$

and continuous bounded operator maps the boundary of the set translates into a boundary.

If the point $x_0 \in \partial M_\delta^{'}$, and if for some integer l there exist sets of numbers $i_1,\dots,i_l$ and $t_1,\dots,t_l$, $i_j \in (1,\dots,N)$, $t_j \geq \delta$, $j = 1, 2, \dots, l$, such that exist ratios

$$A_{i_j}^{t_j}\left(x_{j-1}\right) = x_j \in \partial M_\delta^{'},$$

then it can be considered that there are l pre-images of the point x_0

Let any point $x_0 \in M_\delta^{'}$ have no more than one image. Let us denote a set of points with exactly one image by $H_1 \subset \partial M_\delta^{'}$, and show that H_1 is closed.

Consider an arbitrary sequence $\{x_k\}$ from H_1. It is bounded, and therefore, it can assumed that $x_k \to x_0$ at $k \to \infty$. Since x_k has one image, then at any k point x_k can be associated with the number $i_k \in (1,\dots,N)$ and $t_k \geq \delta$, such that

$$A_{i_k}^{t_k}\left(x_k\right) = x_k^{'} \in \partial M_\delta^{'}.$$

But the parameter i is a finite number of values, and therefore, for some $i \in (1,\dots,N)$ from sequences $\{x_k\}$ and $\{t_k\}$ it is rational to extract subsequences $\{\xi_k\}$ and $\{\tau_k\}$, such that

$$A_i^{\tau_k}\left(\xi_k\right) = \xi_k^{'} \in \partial M_\delta^{'}.$$

Taken into consideration that the sequence $\{\tau_k\}$ is limited, so that it can be considered to be convergent to $\tau_0 \geq \delta$, and that the operator $A_i^t(x)$ is bounded and continuous, it follows that

$$A_i^{\tau_0}\left(x_0\right) = \lim_{k\to\infty} A_i^{\tau_k}\left(\xi_k\right) = \lim_{k\to\infty} \xi_k^{'} = x_0^{'} \in \partial M_\delta^{'}.$$

This means that the point x_0, that is an arbitrary limit point of set H_1, also has one image and, therefore, belongs to this set, which proves that it is closed.

If all points of set $\partial M_\delta^{'}$ have no more than one image, then

$$\rho\left(H_1, \overline{J_\delta}\left(M_\delta^{'}\right)\right) = h_1 > 0.$$

Suppose it is not true. Then, due to the sets H_1 and $\overline{J_\delta}\left(M_\delta^{'}\right)$ are closed they have at least one common point x_0. Since $x_0 \in H_1$, it has some point $x_0^{'} \in \partial M_\delta^{'}$ as an image. On the other hand, the point $x_0 \in \partial M_\delta^{'} \cap \overline{J_\delta}\left(M_\delta^{'}\right)$, and therefore, from argued above it follows that it has at least one preimage

$\xi_0 \in \partial M_\delta^{'}$. Thus, the point ξ_0 obviously has two images, which contradicts the assumption that the points of set $\partial M_\delta^{'}$ have no more than one image. This contradiction argues that the two closed sets H_1 and $\overline{J_\delta}\left(M_\delta^{'}\right)$ can not have common points, so that the distance between them can not be equal to zero. This argues the existence of a positive h_1 satisfied the required ratio.

By analogy with previous considerations, we can show that if all points of set $M_\delta^{'}$ have no more than *K* images, then

$$\rho\left(H_K, \overline{J_\delta}\left(M_\delta^{'}\right)\right) = h_K > 0,$$

where $H_K \subset \partial M_\delta^{'}$ denotes the set of points having exactly *K* images.

Let T_K be an open set containing set H_K, and such that

$$\rho\left(T_K, \overline{J_\delta}\left(M_\delta^{'}\right)\right) \geq 1/2\ h_K.$$

Let us show that set

$$M' = M_\delta^{'} \setminus T_K$$

does not consist of the points having *K* images.

Let the boundary of set M' be represented as following

$$\partial M' = M^* \cup M^{**},$$

where

$$M^* = \partial M' \cap \partial T_K,$$

$$M^{**} = \partial M' \cap \partial M_\delta^{'}.$$

Let $x_0 \in M^*$ be an arbitrary point. If x_0 has at least one image $x_0^{'} \in \partial M'$, then either $x_0^{'} \in M^*$, or $x_0^{'} \in M^{**}$. Consider these two alternatives.

Since

$$\rho\left(\partial T_K, \overline{J_\delta}\left(M'\right)\right) \geq 1/2\ h_K$$

due to the choice of the set T_K, then $x_0^{'}$ can not belong to M^*. But $x_0^{'}$ can also not belong to set M^{**}, since x_0 is an interior point of set $M_\delta^{'}$ and at any continuous mapping $A_i^t\left(x_0\right) = x_0^{'}$, $i = 1,2,\ldots,N$, $t \geq \delta$ it must stay an interior point of set $M_\delta^{'}$, i.e. $x_0^{'}$ can not belong to $\partial M_\delta^{'}$.

Therefore, if $x_0 \in M^*$, then x_0 does not have any image in M'.

Now let $x_0 \in M^{**}$. By the choice of set T_K the point x_0 does not have any image belonging to M^*, since

$$\rho\left(M^*, \overline{J_\delta}\left(M'\right)\right) \geq 1/2\ h_K.$$

But in set M^{**} point x_0 can not have K images, since otherwise it would belong to set H_K that is not in M' by construction.

Thus, all points of $\partial M'$ can not have more than *K – 1* images.

Reiterating these arguments for set M' gives the new closed set M'' by elimination of some open set T_{K-1} from M', and all the point of the new set have no more than *K – 2* images.

Keeping this process, no more than in *K* iterations we construct a closed bounded set

$$M_\delta = M_\delta^{'} \setminus \bigcup_{l=1}^{K} T_l,$$

whose points do not have any images.

It was shown above that if

$$\rho\left(\partial M_\delta, \overline{J_\delta}\left(M_\delta\right)\right) = 0,$$

then on the boundary of set M_δ there exist points having at least one image. But the set M_δ we constructed is known not to have such points, and therefore, the following inequality is true for this set:

$$\rho\left(\partial M_\delta, \overline{J_\delta}\left(M_\delta\right)\right) = h > 0.$$

Finally, it can be concluded that the constructed set M_δ satisfies condition 3.

Note that in each of *K* iterations the set T_l, $l = 1, 2, \ldots$, *K* can be selected in such a way a remaining set to satisfy condition 2. For instance, for set T_K it is sufficient to require for any $x_0 \in M^*$ an interval $\left[0, x_0\right]$ to belong to set M' and have no any other points in common with M^*. Thus, the sets T_l, $l = 1$, 2,…, *K* can be chosen so that the set M_δ constructed above satisfies all the requirements of this theorem.

To complete the proof of this part of the theorem it is necessary to show that if system (1) is asymptotically stable uniformly in U_δ, then for set $M_\delta^{'}$, satisfying the requirements of the Theorem 2.1 and Theorem 2.2, there exists a finite $M_\delta^{'}$, such that all the points of $\partial M_\delta^{'}$ have no more than *K* images.

Suppose there is no such *K*. Then from set $\partial M_\delta^{'}$ it is possible to extract a sequence $\left\{y_j\right\}$ with such a property that point y_j has less images than point y_{j+1}, $j = 1, 2, \ldots$. Since the set $M_\delta^{'}$ is bounded, then the sequence $\left\{y_j\right\}$, without loss of generality, can be considered to converge to some point x_0. Let us show that point x_0 can not have a finite number of images.

Let y_{jk} denote k^{th} image of point y_j, $j = 1, 2, \ldots$. Let $k = 1$. Then, by definition of the image, for each $j = 1, 2, \ldots$ there exist such $i_j \in \left(1, \ldots, N\right)$ and $t_j \geq \delta$, that

$$A_{i_j}^{t_j}\left(y_j\right) = y_{j1}.$$

Parameter i_j takes a finite number of values, and the sequence i_j, as mentioned above, is limited, therefore, it is possible to extract from $\left\{y_j\right\}$ a subsequence $\left\{y_j^1\right\}$, such that sequence $\left\{y_{j1}^1\right\}$ converges to some limit x_1, sequence $\left\{t_j^1\right\}$ converges to $\tau_1 \geq \delta$ and parameter i_j^1 has the same value $i \in \left(1, \ldots, N\right)$ for all $j = 1, 2, \ldots$ at once.

Since the operator $A_i^t\left(x\right)$ is bounded and continuous, then for any $\varepsilon > 0$ there exist $\gamma > 0$ and j_0, such that for all $j > j_0$

$$\| y_j^1 - x_0 \| < \gamma,$$

$$\rho\left(y_{j1}^1, A_i^{\tau_1}\left(x_0\right)\right) < \varepsilon.$$

But sequence $\left\{y_{j1}^1\right\}$ converges to a point x_1 by construction, which implies that

$$x_1 = A_i^{\tau_1}\left(x_0\right).$$

Point x_1 is a limit point of sequence $\left\{y_{j1}^1\right\}$ and belongs to a closed bounded set $\partial M_\delta^{'}$. This point is in set $\partial M_\delta^{'}$, which means that point x_1 is an image of point x_0.

Reiterating all these arguments for sequence $\left\{y_{j1}^1\right\}$, we construct a point $x_2 \in \partial M_\delta^{'}$ being the second image of point x_0.

This procedure can be reiterated any finite number of times, which proves the existence of infinite number of images for point x_0.

Thus, if the noted *K* does not exist, then there is a point $x_0 \in \partial M_\delta^{'}$, which has an unlimited number of images. But this means that for some $u \in U_\delta$ the solution of system (2.1), starting at point $x_0 \in \partial M_\delta^{'}$, can not asymptotically approach to zero, since $\partial M_\delta^{'}$ is separated from zero. This obviously contradicts the assumption of asymptotic stability of system (2.1) uniform in U_δ.

This contradiction shows that if system (2.1) is asymptotically stable uniformly in U_δ, then there exists such $K\infty$, that all the points from set $\partial M_\delta^{'}$ have no more than K images. But then, with the use of the method given above it is possible to extract from set $M_\delta^{'}$ a subsequence satisfying all the requirements of the theorem for a specified δ in a finite number of iterations. This note about the arbitrariness of $\delta > 0$ is a conclusion of the proof of the necessity in the conditions of theorem.

Sufficiency. Consider an arbitrary $\delta > 0$. By assumption, for the specified δ there exists a set M_δ satisfying the conditions 1 - 3 of the theorem. Without loss of generality the set M_δ can be considered to contain the unit ball centered at 0. Let $u_0 \in U_\delta$ be an arbitrary control and let us show that system (2.1) is asymptotically stable at $u = u_0$.

Let

$$0 = t_0 < t_1 < t_2 < \ldots$$

be a sequence of discontinuity points of function $u_0(t)$. Each interval $\left[t_{j-1}, t_j\right]$, $j = 1, 2, \ldots$ is assigned with numbers r_j and γ_j defined by the following relation

$$\frac{t_j - t_{j-1}}{r_j + 1} < \delta \leq \frac{t_j - t_{j-1}}{r_j} = \gamma_j.$$

Let $t > 0$ be an arbitrary moment. Then there exists such a number $k = 1, 2, \ldots$, that $t \in \left[t_k, t_{k+1}\right]$. If $t - t_k \geq \delta$, then in interval $\left[t_k, t\right]$ the numbers $r(t)$ and $\gamma(t)$ are defined by the inequalities

$$\frac{t - t_k}{r(t) + 1} < \delta \leq \frac{t - t_k}{r(t)} \leq \gamma(t).$$

It is easy to see that in this case at $u = u_0$ for any solution of system (2.1) starting at $\mathrm{t} = t_0$ in point $x_0 \in M_\delta$, the following estimation is true

$$\frac{\| x(t) \|}{\| x_0 \|} = \| \left(e^{A_{i_{k+1}}\gamma(t)}\right)^{r(t)} \left[\prod_{j=1}^{k} \left(e^{A_{i_{k-j+1}}\gamma_{k-j+1}}\right)^{r_{k-j+1}}\right] x_0 \| / \| x_0 \| \leq Lq^{r(t) + \sum_{j=1}^{k} r_j},$$

where values $i_j \in (1, \ldots, N)$, $j = 1, 2, \ldots$ are defined by choice of control u_0 and

$$L = \max_{x \in M_\delta} \| x \|.$$

If $t - t_k < \delta$, then the following estimation is true:

$$\frac{\| x(t) \|}{\| x_0 \|} = \frac{\| e^{A_{i_{k+1}}(t-t_k)} \left[\prod_{j=1}^{k} \left(e^{A_{i_{k-j+1}} \gamma_{k-j+1}} \right)^{r_{k-j+1}} \right] x_0 \|}{\| x_0 \|} \le CLq^{\sum_{j=1}^{k} r_j},$$

where

$$C = \max_{i \in (1,\ldots,N), t \in [0,\delta]} \| e^{A_i t} \|.$$

Two last estimations are obtained using linearity of system (2.1), condition 3 and the fact that if $x \in qM_\delta$, then $\| x \| \le qL$.

Combining these estimations, it is finally found that at $u - u_0$ for any solution $x(t)$ of system (1) starting at $t = t_0$ in point $x_0 \in M$ we have an estimation

$$\| x(t) \| \le D \| x_0 \| q^{\Gamma(t)},$$

where D is a sufficiently large constant and $\Gamma(t)$ denotes an integer increasing function, such that $\Gamma(t) \to \infty$ as $t \to \infty$. Since $q \in (0,1)$, the obtained estimation implies that $x(t) \to 0$ as $t \to \infty$.

Given the arbitrariness of $\delta > 0$ and $u_0 \in U_\delta$, under the conditions of the theorem system (2.1) is asymptotically stable uniformly in U_δ.

For the estimations obtained while proving proposition 1 it follows that if system (2.1) is asymptotically stable uniformly in U_δ, then for any $\delta > 0$ there exist quantities $a_l(\delta) > 0$ and $b_l(\delta) > 0$, $l = 1, 2, \ldots$, such that at $u \in U_\delta$ for any solution $x(t)$ of system (2.1) starting at $t = t_0$ in point x_0 the following estimation it true

$$\| x_0 \| a_1(\delta) e^{-b_1(\delta)(t-t_0)} \le \| x(t) \| \le \| x_0 \| a_2(\delta) e^{-b_2(\delta)(t-t_0)}.$$

Indeed, the lower-bound estimate follows from the boundedness below of all eigenvalues of matrices A_i, $i = 1, 2, \ldots, N$, and the upper-bound estimate follows from the fact proved above that $\| x(t) \|$ is bounded above by a geometric progression with a ratio less than one. But for such progression it is possible to fit the exponential a lower rate of decrease.

2.3. Application of Lyapunov's Functions

Using the method of Lyapunov functions, we obtain the new necessary and sufficient conditions for the stability of system (2.1) uniform in U_δ. Note that the propositions proved in this section actually follow from the Theorem 2.1, 2.2 and 2.3.

Theorem 2.4: For system (2.1) to be stable uniformly in U_δ it is necessary and sufficient for any $\delta > 0$ to exist a continuous function $v_\delta(x)$ satisfying the following requirements: $v_\delta(x)$ is positively definite function, $v_\delta(0) = 0$; for sufficiently small $c_1 > 0$ there exists $c_2 > 0$, such that $v_\delta(x) > c_2$ at $\| x \| > c_1$; for any $i \in (1,\ldots,N)$, x_0, t_0 and $t \geq t_0 + \delta$

$$v_\delta(x_0) \geq v_\delta\left(e^{A_i(t-t_0)}x_0\right).$$

Proof

Necessity. Let system (2.1) be stable uniformly in U_δ. Let $\delta > 0$ be arbitrary. According with Theorem 2.1 for the specified δ there exists a closed bounded set M_δ containing point zero as its interior point, and such that

$$\overline{J_\delta}(M_\delta) \subset M_\delta.$$

The boundary of this set is denoted as before by ∂M_δ.

Let $c(\mu)$ be a continuous steadily increasing on $[0,\infty)$ function $c(\mu)$, $c(0) = 0$ and let us construct function $v_\delta(x)$, assuming that

$$v_\delta(x) = c(\mu), \qquad x \in \mu\partial M_\delta.$$

We additionally require the set M_δ to satisfy the requirements of the Theorem 2.2. Then the constructed function is continuous positively definite and $v_\delta(0) = 0$, i.e. v_δ satisfies condition 1.

Now let us show that it satisfies condition 2. To this end let $c_1 > 0$ be small enough and μ_0 be such that

$$\mu_0 M_\delta \subset \bar{K}_{c_1}(0) = \{x \,\|\, x \,\|\leq c_1\}.$$

Assume

$$c_2 = c(\mu_0).$$

Then due to steadily increasing of function $c(\mu)$ at $\| x \| > c_1$ the following is true

$$v_\delta(x) > c_2,$$

which proves the proposition.

Finally, according with Theorem 2.1

$$\overline{J_\delta}(M_\delta) \subset M_\delta,$$

which in view of the monotonicity of function $c(\mu)$ shows that v_δ satisfies condition 3.

Given the arbitrariness of $\delta > 0$ we see the necessity of the proposal conditions.

Sufficiency. Now we fix an arbitrary $\delta > 0$ and let v_δ be a function satisfying the requirements of 2 - 3. Assume

$$M_\delta = \{x v_\delta(x) \le c_2\}.$$

Due to 2 we have $\| x \| \le c_1$ at $x \in M_\delta$, i.e. the set M_δ is bounded. Given the continuity and positive definite of function M_δ and a fact that $v_\delta(0) = 0$, it follows that M_δ is closed and contains a point zero as its interior point. Finally, condition 3 means that

$$\overline{J_\delta}(M_\delta) \subset M_\delta.$$

Thus, for any $\delta > 0$ there exists a set M_δ satisfying all the requirements of Theorem 2.1, which proves the uniform in U_δ stability of system (2.1).

Theorem 2.5: In order the system (2.1) to be asymptotically stable uniformly in U_δ it is necessary and sufficient for any $\delta > 0$ to exist a continuous function $v_\delta(x)$ satisfying the following requirements:

$v_\delta(x)$ is positively definite homogeneous function of degree $\alpha > 0$, $v_\delta(0) = 0$;

For sufficiently small $c_1 > 0$ there exists $c_2 > 0$, such that for $\| x \| > c_1$

$$v_\delta(x) > c_2;$$

Exists $h \in (0,1)$, such that for any $i \in (1, \ldots, N)$, x_0, t_0 and $t \ge t_0 + \delta$

$$v_\delta\left(e^{A_i(t-t_0)} x_0\right) \le h v_\delta(x_0).$$

Proof

Necessity. Let system (2.1) be asymptotically stable uniformly in U_δ and $\delta > 0$ be arbitrary. According with Theorem 2.3 for specified δ there exists $q \in (0,1)$ and set M_δ satisfying all its requirements.

Let $c(\mu)$ be continuous steadily increasing on $[0,\infty)$ homogeneous function of degree α, $c(0)=0$ and now we construct a function v_δ assumed that

$$v_\delta(x) = c(\mu), \qquad x \in \mu\partial M_\delta.$$

By assumption, system (2.1) is stable uniformly in U_δ and constructed here function v_δ does not significantly differ from the function constructed in the proof of Theorem 2.4, so that the function v_δ satisfies the requirements 1 and 2 of this theorem.

By the condition 3 of Theorem 2.3

$$\overline{J_\delta}(M_\delta) \subset qM_\delta.$$

This means that if at the moment t_0 point $x_0 \in \mu\partial M_\delta$ for some $\mu > 0$, then for all $i \in (1,\ldots,N)$ and $t \geq t_0 + \delta$ point $e^{A_i(t-t_0)}x_0$ belongs to set $q\mu M_\delta$. Given these ratios and steady monotony and homogeneity of function $c(\mu)$ it follows:

$$v_\delta\left(e^{A_i(t-t_0)}x_0\right) \leq c(q\mu) = q^\alpha c(\mu) = hv_\delta(x_0),$$

where

$$q^\alpha = h \in (0,1).$$

But this proves that condition 3 is true.

Given the arbitrariness of δ we see the necessity of the proposal conditions.

Sufficiency. Let us fix an arbitrary $\delta > 0$ and let v_δ be a function satisfying the requirements 1 - 3. Assume

$$M_\delta = \{x \mid v_\delta(x) \leq c_2\}.$$

The proof of this theorem is similar to the proof of Theorem 2.4, where it is shown that the set M_δ satisfies condition 1 of Theorem 2.3.

Consider set

$$M_\delta' = \{y \mid v_\delta(y) \leq hc_2\},$$

where $h \in (0,1)$ is taken from 3. Given the homogeneity of function v_δ and assumed that $q = h^{1/\alpha}$ set M_δ' can be represented as following:

$$M_{\delta}^{'} = \{y \mid v_{\delta}\left(\frac{y}{q}\right) \le c_2\}.$$

Then

$$M_{\delta}^{'} = qM_{\delta}.$$

By construction

$$M_{\delta}^{'} \subset M_{\delta},$$

and due to continuity of v_{δ} the boundaries of these sets can not have common points, since otherwise the function v_{δ} takes two different values at once in the points of the boundary crossing Thus, M_{δ} satisfies condition 2 of Theorem 2.3.

Finally, in view of arguments given above condition 3 means that if at $t = t_0$ point x_0 belongs to x_0, then for all $i \in (1,\ldots,N)$ and $t \ge t_0 + \delta$ point $e^{A_i(t-t_0)}x_0$ belongs to qM_{δ}. This means that

$$\overline{J_{\delta}}(M_{\delta}) \subset qM_{\delta}.$$

Thus, for any $\delta > 0$ there exists a set M_{δ} satisfying all the requirements of Theorem 2.3, which proves the uniform in U_{δ} asymptotic stability of system (2.1).

3. STABILITY OF LINEAR HOMOGENEOUS SYSTEMS WITH ARBITRARY SWITCHING

3.1. Preliminary Remarks

In this section the system defined by relations 2.1 – 2.5 is under consideration:

$$\dot{x} = A(u)x. \tag{3.1}$$

As before, let U_{δ} denote a set of all functions $u(t)$ satisfying conditions 2.2 - 2.4 for some fixed $\delta > 0$. Suppose

$$U = \bigcup_{\delta > 0} U_{\delta}.$$

The system (3.1) is stable uniformly with respect to U if for any $\varepsilon > 0$ there exists $\eta(\varepsilon) > 0$ such that for any $u \in U$ all solutions of the system (3.1) starting in η-neighborhood of zero stay in ε-neighborhood. If all these solutions tend to zero as for any $u \in U$, then the system 3.1 asymptotically stable uniformly with respect to *U*.

Let the necessary and sufficient conditions for uniform with respect to *U* stability of the system (3.1) be found.

3.2. Necessary and Sufficient Conditions for Uniformly With Respect to *U* Stability of System 3.1.

Let A_i^t denote a linear operator corresponding to the matrix $e^{A_i t}$ as above, and $\bar{J}(M)$ denote a closure of set

$$J(M) = \bigcup_{i=1}^{N} \bigcup_{t \geq 0} A_i^t(M),$$

where *M* is an arbitrary set.

Theorem 3.1: For the system (3.1) to be stable uniformly with respect to *U* it is necessary and sufficient the bounded closed set *M* to exist and contain point zero as an interior point, such that

$$\bar{J}(M) \subset M.$$

Proof

Necessity. Now let the procedure given at the beginning of Theorem 2.1 proof be used. Let a set M_δ be constructed for an arbitrarily chosen $\delta > 0$ and for a bounded closed set M_1 containing point zero as an interior point. By assumption, the system (3.1) is stable uniformly with respect to *U*, and consequently stable uniformly with respect to U_δ. Hence, by Theorem 2.1 it follows that the set M_δ is bounded.

Now for an arbitrary $\delta' \in (0, \delta)$ a set $M_{\delta'}$ is constructed on the basis of the set M_1. This set does not essentially differ from the set M_δ, and consequently it is also bounded. In addition, it is obvious that

$$M_\delta \subset M_{\delta'}.$$

Assume

$$M = \overline{\bigcup_{\delta \geq 0} M_\delta}$$

Then the boundedness of the set *M* must be proved.

On the contrary, let *M* be unbounded. Then for an arbitrarily large $\Delta > 0$ there exists $\delta_1 > 0$ such that a set $M_{\delta_1} \subset M$ contains the subset Q for which $\| x \| > \Delta$ by the condition $x \in Q$. Ley some point $x^* \in Q$ be fixed. By the definition of the set M_{δ_1}, there exist a control $u^* \in U_{\delta_1}$ and a point $x_1 \in M_1$ such that the solution of the system (3.1) starting at x_1 gets to the point x^* with time at $u = u^*$. But due to the arbitrariness of Δ it follows that for $u \in U$ the solutions of the system (3.1) starting in the bounded set M_1 are not uniformly bounded. Therefore, by the linearity of the system (3.1) it can be concluded that the system (3.1) is not stable uniformly with respect to *U*. This contradiction proves the boundedness of the set *M*.

Thus, the set M is bounded, closed by construction and contains point zero as its interior point, since $M_1 \subset M$. Therefore, in order to show that *M* satisfies all the requirements of the theorem, it must be proved that

$$\bar{J}(M) \subset M.$$

Assume the converse. Then for some operator A_i^τ, $i \in (1, \ldots, N)$, $\tau > 0$, due to its continuity there exists an interior point $x_0 \in M$ such that

$$A_i^\tau (x_0) \notin M.$$

But since x_0 is an interior point of *M*, then for some $\delta > 0$ the point $x_0 \in M_\delta \subset M$ where $\delta < \tau$ can be realized. Then, by Theorem 2.1 and the fact that $\delta < \tau$, there exists an inclusion

$$A_i^\tau (x_0) \in A_i^\tau (M_\delta) \subset M_\delta \subset M,$$

that contradicts the assumption that

$$A_i^\tau (x_0) \notin M.$$

This contradiction proves the required inclusion as well as the necessity of the theorem conditions.

Sufficiency. To contribute stability with respect to *U* for the system (3.1) the proof of sufficiency of the Theorem 3.1 conditions is actually the same as the proof of the sufficiency of the Theorem 2.1 conditions if it is assumed that $\delta = 0$.

Theorem 3.2: If the system (3.1) is stable uniformly with respect to *U*, then there exists a convex set *M* satisfying all the requirements of Theorem 3.1 and being such that for any $q \in [0,1)$ $qM \subset M$ and for any $\mu > 0$, $\mu \neq 1$

$$\mu \partial M \cap \partial M = \varnothing.$$

The proof of this assertion is carried out according to the scheme of Theorem 2.2 proof if δ is assumed to be equal to zero.

Theorem 3.3: For the system (3.1) to be asymptotically stable uniformly with respect to *U*, it is necessary and sufficient a set *M* to exist and satisfy the following conditions:

M is a bounded closed set containing point zero as an interior point;
For any $\mu \in [0,1)$

$$\mu M \subset M$$

and

$$\mu \partial M \cap \partial M = \varnothing;$$

There exists a strictly decreasing function $q(r)$, $q(0) = 1$, $q(r) \to 0$ as $r \to \infty$, such that for any $\delta \geq 0$

$$\overline{J_\delta}(M) \subset q(\delta) M.$$

Proof

Necessity. By assumption, the system (3.1) is stable uniformly with respect to *U*. Consequently, there exists a set M_0 satisfying all the requirements of Theorems 3.1 and 3.2. By Theorem 2.3, for any $\delta > 0$ there exists a set $M_\delta \subset M_0$ satisfying the requirements of the theorem for some $q = q_\delta$. Moreover, it is obvious that for an arbitrary $\delta' \in (0,\delta)$ the corresponding set $M_{\delta'} \subset M_0$ can be taken so that it is a superset of the set M_δ, in other words the following inclusion is realized:

$$M_\delta \subset M_{\delta'}.$$

Let *M* denote a closure of the set $\bigcup_{\delta \geq 0} M_\delta$. Let us show that *M* satisfies all the requirements of the theorem.

By construction, the set *M* is closed and belongs to the bounded set M_0, which implies that *M* is also bounded. Since for any $\delta > 0$ $M_\delta \subset M$, then *M* obviously contains point zero as its interior point. Thus, *M* satisfies requirement 1.

Let us show that for any $\mu \in (0,1)$

$\mu M \subset M.$

Indeed, if this is not so, then there exists an interior point $x_0 \in M$ such that $\mu x_0 \notin M$. But considering construction of the set *M* there exists $\delta > 0$ for the point x_0 such that $x_0 \in M_\delta$. Then

$$\mu x_0 \in \mu M_\delta \subset M_\delta \subset M \not\ni \mu x_0.$$

This contradiction proves the required inclusion.

Considering the method of the set M_δ construction given in the proof of Theorem 2.3 and the convexity of the set M_0, it follows that the sets M_δ can be taken so that for any point $x_0 \in \partial M$ the interval $\left[0, x_0\right)$ does not have points in common with the boundary of the set *M*. Thus, *M* satisfies the requirements 2.

Now the necessity of conditions 3 must be proved. Contrary to the statement, suppose that function $q\left(r\right)$ can not be strictly decreasing. Then such $\delta > 0$ and $\delta' \in \left(0, \delta\right)$ must exist that

$$q\left(\delta'\right) = q\left(\delta\right) = q.$$

Note that the set *M* obviously satisfies the requirements of Theorem 3.1, and therefore $q \leq 1$.

From the assumption that the function $q\left(r\right)$ can not be taken strictly decreasing and from the choice of δ and δ it follows that

$$\overline{J_{\delta'}}\left(M\right) \subset qM$$

and

$$\overline{J_\delta}\left(M\right) \subset qM.$$

Moreover, *q* can be reduced in none of these inclusions. But by virtue of the inclusion

$$\overline{J_\delta}\left(M\right) \subset \overline{J_{\delta'}}\left(M\right),$$

such a situation is obviously possible only if the sets $\partial\overline{J_{\delta'}}\left(M\right)$, $\partial\overline{J_\delta}\left(M\right)$ and ∂qM have at least one point in common. Let this point be denoted as x_0.

Repeating the arguments given in the proof of Theorem 2.3, it is possible to show that for the point x_0 there exist an index $i \in \left(1, \ldots, N\right)$, a time point $t_1 \geq \delta$ and a point $x_1 \in \partial M$ such that

$$A_i^{t_1}\left(x_1\right) = x_0.$$

Further, for any $t \geq \delta'$

$$A_i^t\left(\partial M\right) \subset \overline{J_{\delta'}}\left(M\right) \subset qM.$$

In particular, these inclusions are satisfied for $t \in \left[\delta', t_1\right)$. This implies that for the point x_1 there exist a time point $t_2 < t_1$ and a point $x_0 \in \partial qM$ such that

$$A_i^{t_2}\left(x_1\right) = x_2.$$

The points x_0 and x_2 belong to the set ∂qM. Consider the group property of the operator A_i^t and linearity of the system (3.1). Since the function $q\left(r\right)$ can not be taken strictly decreasing, then for $t^* = t_1 - t_2 > 0$ and for the points $x_0 / q = y_0 \in \partial M$ and $x_2 / q = y_1 \in \partial M$ the following relation is true:

$$A_i^{t^*}\left(y_1\right) = y_0.$$

Let us show that $y_1 \notin M_\delta$ for any $\delta > 0$. Indeed, if there exists $\delta^* > 0$ for which $y_1 \in M_{\delta^*}$, then for any $\tilde{\delta} \in \left(0, \delta^*\right)$

$$y_1 \in M_{\tilde{\delta}}.$$

by virtue of the inclusion

$$M_{\delta^*} \subset M_{\tilde{\delta}},$$

But by virtue of the set $M_{\tilde{\delta}}$ properties, for any $j \in \left(1, \ldots, N\right)$ and $\tilde{t} > \tilde{\delta}$

$$A_j^{\tilde{t}}\left(y_1\right) \in q_{\tilde{\delta}} M_{\tilde{\delta}}.$$

Since by Theorem 2.3 $q_{\tilde{\delta}} < 1$, then

$$q_{\tilde{\delta}} M_{\tilde{\delta}} \cap \partial M = \varnothing.$$

This implies that

$$A_j^{\tilde{t}}\left(y_1\right) \notin \partial M.$$

Considering the arbitrariness of *j* and $\tilde{t}$, these relations are true, in particular, for $j = i$ and $\tilde{t} = t^*$, if $\tilde{\delta}$ is sufficiently small. This implies that the following relation is true:

$$y_0 \notin \partial M.$$

The obtained contradiction proves that

$$y_0 \notin M_\delta$$

for any $\delta > 0$.

The next step is to show that from the statement just proved it follows that for any $t \in [0, t^*]$

$$A_i^t(y_1) \in \partial M$$

and for any $\delta > 0$

$$A_i^t(y_1) \notin M_\delta.$$

Indeed, assume the contrary, in other words that for some $\overline{t} \in (0, t^*)$ $y = A_i^{\overline{t}}(y_1)$ is an interior point of *M*. Then, as it is shown above, for any $t > 0$ $A_i^t(y)$ does not get on the boundary of the set *M*, which contradicts the inclusion $y_0 \in \partial M$. This contradiction implies that

$$A_i^t(y_1) \in \partial M$$

for any $t \in [0, t^*]$.

The validity of the second assertion is almost obvious, since if for some $\overline{t} \in (0, t^*)$ and $\delta > 0$ the relation

$$y = A_i^{\overline{t}}(y_1) \in M_\delta.$$

is true, then repeating for the point *y* all the arguments given for the point y_1, there appears the wrong conclusion that the point y_0 does not belong to the boundary of the set M.

So, since the function $q(r)$ can not be taken strictly decreasing, then there exist an index $i \in (1, \ldots, N)$, a time point $t^* > 0$ and points $y_0 \in \partial M$ and $y_1 \in \partial M$ such that

$$A_i^{t^*}(y_1) = y_0$$

and for all $t \in [0, t^*)$ and any $\delta > 0$

$$A_i^t(y_1) \in \partial M, \quad A_i^t(y_1) \notin M_\delta.$$

Assume that

$$t' = 1/2t^*, \quad y' = A_i^{t'}(y_1).$$

Considering the prove given above $y' \in \partial M$, and for any $\delta > 0$ $y' \notin M_\delta$. Due to Theorem 2.3 it is clear that the set M_δ is obtained from the set M_0 by eliminating some open set T_δ such that

$$\rho\left(T_\delta, \overline{J_\delta}(M_0)\right) > 0.$$

Since $y' \notin M_\delta$, then $y' \in T_\delta$. Therefore

$$\rho\left(y', \overline{J_\delta}(M_0)\right) > 0.$$

But

$$\overline{J_\delta}(M) \subset \overline{J_\delta}(M_0),$$

and moreover

$$\rho\left(y', \overline{J_\delta}(M)\right) > 0.$$

The last inequality is true for any $\delta > 0$ and, in particular, for $\delta = t'$. Taking this into account, it can be concluded that the following inequality is realized

$$A_i^{t'}(y_1) - y' > 0,$$

which contradicts the way of the choice of point y'.

The obtained contradiction proves that on the set ∂M the points y_0 and y_1 satisfying the inequality

$$A_i^{t^*}(y_1) = y_0.$$

for some $i \in (1,\ldots,N)$ and $t^* > 0$ actually do not exist.

It is clear that $q(0) = 1$, and the fact that $q(r)$ can be taken to tend to zero as $r \to \infty$ follows from the fact that diameter of the set $J_\delta(M)$ tends to zero as $\delta \to \infty$.

Thus, it is shown that the set *M* satisfies all the requirements of the theorem, which completes the proof of the necessity of its conditions.

Sufficiency. Suppose that for the system (3.1) there exists a set *M* satisfying the requirements 1 – 3 of the theorem. Then, for any $\delta > 0$ the set *M* obviously satisfies all the requirements of Theorem 2.3 for $q = q(\delta)$. This remark implies that for any $u = u_0 \in U$ the system (3.1) is asymptotically stable, since by definition of the set *U* for any control $u_0 \in U$ there exists $\delta > 0$ such that $u_0 \in U_\delta$.

Thus, if conditions 1 to 3 are satisfied, then the system (3.1) is stable uniformly with respect to *U* by Theorem 3.1. As noted above, all its solutions asymptotically approach to zero, which implies the asymptotic uniform with respect to *U* stability of the system (3.1).

3.3. Application of Lyapunov's Functions

Consider the new necessary and sufficient conditions for uniform with respect to *U* stability of the system (3.1) using the method of Lyapunov functions given in §2.

Theorem 3.4: For the system (3.1) to be stable uniformly with respect to U, it is necessary and sufficient a continuous function $v(x)$ to exist and satisfy the following requirements: $v(x)$ is a positively definite function, $v(0) = 0$; for sufficiently small $c_1 > 0$ there exists $c_2 > 0$ such that

$$v(x) > c_2,$$

For any x_0, t_0, $t \geq t_0$ and $i \in (1,\ldots,N)$

$$v(x_0) \geq v\left(e^{A_i(t-t_0)}x_0\right).$$

The proof of this theorem is carried out according to the scheme of the Theorem 2.4 proof with reference to Theorems 3.1 and 3.2.

Theorem 3.5: For the system (3.1) to be asymptotically stable uniformly with respect to U, it is necessary and sufficient a continuous function $v(x)$ to exist and satisfy the following requirements: $v(x)$ is a positively definite homogeneous function of degree $\alpha > 0$; for sufficiently small $c_1 > 0$ there exists $c_2 > 0$ such that

$$v(x) > c_2,$$

There exists a strictly decreasing function $h(r)$, $h(0)=1$, $h(r)\to 0$ as $r\to\infty$, such that for any x_0, t_0, $t\geq t_0$ and $i\in(1,\ldots,N)$

$$v\left(e^{A_i(t-t_0)}x_0\right)\leq h(t-t_0)v(x_0).$$

The proof is carried out according to the scheme of the Theorem 2.5 proof with reference to Theorem 3.3.

CONCLUSION

In this paper the stability problem of parametrically perturbed linear systems is considered. In §1 it is shown that the problem of the zero solution stability of a selector linear differential inclusion reduces to the problem of solvability of a algebraic inequalities system whose coefficients are uniquely defined by coefficients of the initial differential inclusion.

In §§2 and 3, the problem of stability of linear differential systems is considered in terms of the second Lyapunov's method, i.e. this reduces to the problem of a function with certain properties existence.

The set-theoretical approach to the solution of the considered problem is the new in this paper. Such an approach is not only of theoretical interest, but it also helps to formalize the necessary and sufficient conditions for the stability and asymptotic stability of zero solution of the considered system with the use of Lyapunov's functions and to simplify the proof of the introduced assertions validity.

REFERENCES

Alferov, G., Ivanov, G., Efimova, P., & Sharlay, A. (2017). Study on the Structure of Limit Invariant Sets of Stationary Control Systems with Nonlinearity of Hysteresis Type. ICNAAM-16, Greece, AIP Conference Proceeding 2017. doi:10.1063/1.4992264

Alferov, G.V., & Gorizontov, A.M. (1987). Resource management in flexible automated production. *Kibernetika i Vychislitel'naya Tekhnika, 1987*(73), 76-81.

Alferov, G. V., & Malafeyev, O. A. (1996) The robot control strategy in a domain with dynamical obstacles. Lecture Notes in Computer Science, 1093, 211-217.

Alferov, G. V., Malafeyev, O. A., & Maltseva, A. S. (2015). Programming the robot in tasks of inspection and interception. 2015 Internetional Confference on Mechanics - Deventh Polyakhov's Reading, 7106713. doi:10.1109/POLYAKHOV.2015.7106713

Ivanov, G., Alferov, G., Sharlay, A., & Efimova, P. (2017). Conditions of Asymptotic Stability for Linear Homogeneous Switched Systems. ICNAAM-16, Greece, AIP Conference Proceeding 2017. doi:10.1063/1.4992263

Ivanov, G. G., & Sharlay, A. S. (2015). On Stability of Linear Homogeneous Switched Systems. *2015 International Conference "Stability and Control Processes" in Memory of V. I. Zubov, SCP 2015*, 13-15. 10.1109/SCP.2015.7342039

Kulakov, F., Alferov, G., & Efimova, P. (2015). Method of remote control over space robots. 2015 Internetional Confference on Mechanics - Deventh Polyakhov's Reading, 7106742. doi:10.1109/POLYAKHOV.2015.7106742

Kulakov, F., Alferov, G. V., Efimova, P., Chernakova, S., & Shymanchuk, D. (2015) Modeling and control of robot manipulators with the constraints at the moving objects. *2015 International Conference on "Stability and Control Processes" in Memory of V.I. Zubov,* 102-105.

Kulakov, F., Sokolov, B., Shalyto, A., & Alferov, G. (2016). Robot Master Slave and Supervisory Control with Large Time Delays of Control Signals and Feedback. *Applied Mathematical Sciences, 10*(33-36), 1783–1796. doi:10.12988/ams.2016.6380

Malafeyev, O., Alferov, G., & Andreyeva, M. (2015) Group strategy of robots in game - theoretic model of interception with incomplete information. *2015 Internetional Conference on Mechanics - Seventh Polyakhov's reading*, 7106751. 10.1109/POLYAKHOV.2015.7106751

Meilakhs, A. M. (1975). O stabilizatsii upravlayemih system v usloviyah neopredelennosti. *Avtomatika I Telemechanika, 2*, 182-189.

Molchanov, A. P. & Pyatnitskii, E. S. (1986). Funkcii Lyapunova, opredelyaushie neobhodimiye I dostatochniye usloviya absolutnoi ustoichivosty nelineinih nestacionarnih system upravleniya. *Avtomatika I Telemechanika, 4*, 5-15.

Neverova, E. G., Malafeyev, O. A., Alferov, G. V., & Smirnova, T. E. (2015) Model of interaction between anticorruption authorities and corruption groups. *2015 International Conference on "Stability and Control Processes" in Memory of V.I. Zubov*, 488-490.

Philipov, A. F. (1985). *Differencialnye uravneniya s nerazrivnoy pravoy chastyu*. Moscow: Nauka.

Pichugin, Yu., Alferov, G., & Malafeyev, O. (2016). Parameters estimation in mechanism design. *Contemporary Engineering Sciences, 9*(1-4), 175–185. doi:10.12988/ces.2016.512319

Chapter 5
Stochastic Approximation-Based Transport Profiling for Big Data Movement Over Dedicated Connections

Daqing Yun
Harrisburg University of Science and Technology, USA

Chase Q. Wu
New Jersey Institute of Technology, USA

ABSTRACT

High-performance networks featuring advance bandwidth reservation have been developed and deployed to support big data transfer in extreme-scale scientific applications. The performance of such big data transfer largely depends on the transport protocols being used. For a given protocol in a given network environment, different parameter settings may lead to different performance, and oftentimes the default settings do not yield the best performance. It is, however, impractical to conduct an exhaustive search in the large parameter space of transport protocols for a set of suitable parameter values. This chapter proposes a stochastic approximation-based transport profiler, namely FastProf, to quickly determine the optimal operational zone of a protocol over dedicated connections. The proposed method is evaluated using both emulations based on real-life measurements and experiments over physical connections. The results show that FastProf significantly reduces the profiling overhead while achieving a comparable level of transport performance with the exhaustive search-based approach.

INTRODUCTION

Extreme-scale scientific applications are generating colossal amounts of datasets, now frequently termed as "big data", which must be transferred over long geographical distances for distributed processing and analysis. Such big data transfer requires stable and high-speed network connections, which, unfortunately, are not readily available in traditional shared IP networks, e.g., the Internet. High-performance

DOI: 10.4018/978-1-5225-5045-7.ch005

Networks (HPNs) such as ESnet (ESnet, 2017), Internet2 (Internet2, 2017), and Google's B4 (Jain et al., 2013) that provide on-demand dedicated high-speed network connections realized by technologies such as MPLS (Andersson & Swallow, 2003) and OpenFlow (McKeown et al., 2008) have emerged to support these data- and network-intensive applications. More recently, significant progress has been made for big data movement in various aspects including the deployment of 100Gbps networks with future 1Tbps capacity, the increase in end-host capabilities with multiple cores and buses, and the use of parallel file systems. However, such technology advancement and infrastructure investment have not led to corresponding improvement in big data transfer performance, especially at the application level. Maximizing the data transfer performance over such dedicated connections is still challenging, mainly because: i) the optimal operational zones of transport protocols are affected by the complex configurations and dynamics of the network segments, end-hosts, and protocol itself; ii) different parameter settings may lead to very different performance and oftentimes the default parameter setting does not yield the best performance; iii) due to the lack of accurate models for high-performance transport protocols such as UDT (Gu & Grossman, 2007), a widely used protocol in the HPN community (UDT, 2017a), and the complex dynamics of network environments, it is generally very difficult to model and derive the optimal parameter values analytically. For a given data transfer protocol, a careful selection of parameter values may result in a significant performance improvement over its default settings. As a motivating example, Figure 1 shows the instantaneous UDT throughput performance with different block sizes over a local back-to-back 10Gbps connection, where the performance is improved by three times on average through a simple adjustment made on the block size.

Figure 1. Instantaneous throughput measurements of UDT over a 10Gbps back-to-back connection with different block sizes

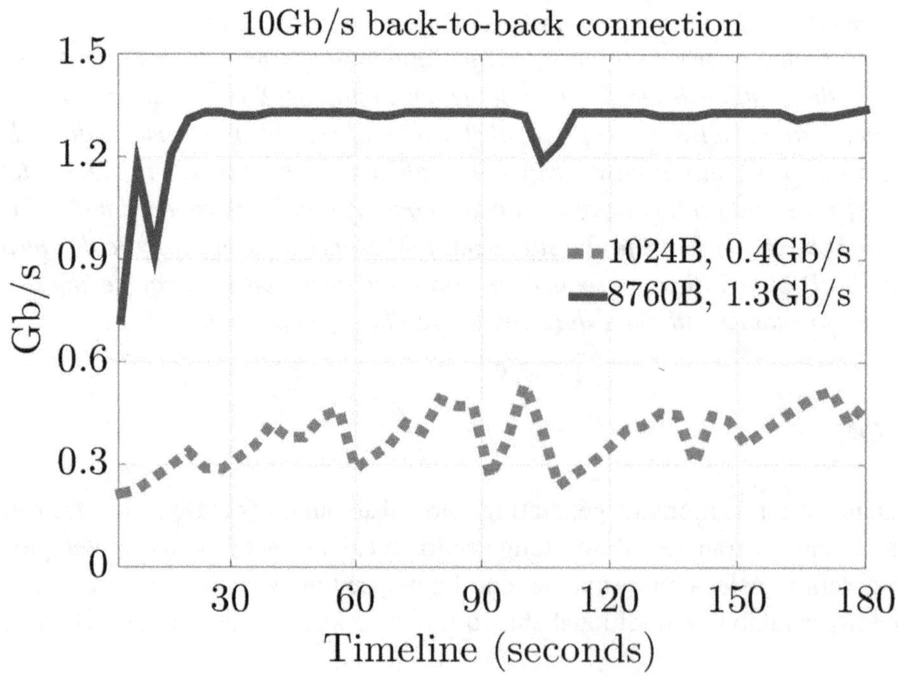

Transport profiling, which sweeps through the combinations of parameter settings such as socket options, application-specific parameters, and protocol-specific configurations, enables users to determine the "best" set of parameter values for the optimal data transfer performance. There exist several bandwidth estimation tools such as ESnet iperf3 (Iperf3, 2017), which uses continuous data transfer to estimate the achievable throughput along an end-to-end network path. It provides users with various functions and options for tuning TCP, UDP and SCTP. The Transport Profiler Generator (TPG) (Yun et al., 2015) complements ESnet iperf3 by providing throughput estimation for UDT, and additionally supports parallel data streams over multiple NIC-to-NIC connections. A survey of bandwidth estimation tools can be found in (Prasad et al., 2003). These bandwidth estimation tools could be utilized to conduct exhaustive transport profiling to optimize big data transfer performance. However, such exhaustive-based profiling method is prohibitively time consuming when there exists a large control parameter space, which is almost always the case in most big data transfer scenarios. In general, users may not be in favor of performing transport profiling if the profiling overhead is comparable with the time needed for their actual data transfer.

To improve the efficiency of transport profiling, we propose a stochastic approximation-based profiling method, namely FastProf, to help quickly determine the optimal operational zone of a given data transfer protocol/method in high-performance network environments. FastProf employs the Simultaneous Perturbation Stochastic Approximation (SPSA) algorithm (Spall, 1992) to accelerate the exploration of the parameter space, and is implemented by leveraging the capabilities of the existing toolkit, Transport Profile Generator (TPG) (Yun et al., 2015). We first present the theoretical framework of the SPSA-based profiling approach employed in FastProf, and then detail the design and implementation of FastProf. The performance superiority of FastProf is evaluated using both emulations with real-life performance measurements and experiments over physical connections with both short (2ms) and long (380ms) round-trip times (RTTs). Both emulation and experimental results show that FastProf significantly reduces the profiling overhead while achieving a comparable level of end-to-end throughput performance with the exhaustive search-based approach. FastProf makes it possible to conduct "on-line" profiling to support time-critical big data transfer, and provides an additional level of intelligence to existing profiling-oriented toolkits such as Iperf3 (Iperf3, 2017) and xddprof (XDD, 2017).

BACKGROUND

What Is Transport Profiling and Why Is Profiling Needed?

End-to-end data transfer is a complex process that involves various network segments and end-host components. Any of these components could become the bottleneck, and hence limit end-to-end data transfer performance observed by end users at the application level, where the parameter settings have a significant impact. Among these components and parameters, some can be accessed and controlled by the application, such as packet size, block size, socket buffer size, jumbo frame size, number of data transfer streams; while the others are mainly determined by the hardware configurations and network infrastructures, such as CPU frequency, memory size, memory bandwidth, bus speed, disk I/O speed, path MTU, round trip time (RTT), and link bandwidth. This work focuses on those parameters directly accessible and controllable by the user or application. In general, careful parameter tuning may help achieve a better performance, but it typically requires extensive domain knowledge in networking and systems that many science users lack. Moreover, even if they are able to manually conduct "fine tuning" in various aspects such as core affinities (Ahuja, Farrens, & Ghosal, 2012; Hanford et al., 2014) and IRQ balance/conflict (Leitao, 2009) at the system level, many application-level control parameters may still affect the end-to-end performance (mainly throughput). For illustration purposes, Figure 2 plots the UDT-based throughput measurements in response to different block sizes and different buffer sizes with different RTTs ranging from 0ms to 366ms on an emulated network testbed at Oak Ridge National Laboratory (ORNL). Similar to Figure 1 that plots the performance of UDT with different block sizes on a local testbed, Figure 2(a) shows that in a more complex network environment, different block sizes of UDT may result in very different throughput performances when the buffer size is fixed to be 256MB. In addition, Figure 2(b) shows that buffer sizes may have a more significant impact on the performance with a fixed block size of 89,559 bytes.

The throughput curves in Figure 2 generally follow a similar pattern, i.e., first increase and then stabilize as the block size increases (Figure 2(a)), and first increase and then decrease as the buffer size increases (Figure 2(b)). However, the parameter values yielding peak performance vary with different delays as the block size or the buffer size increases, which makes it difficult to derive an analytical model to determine the parameter settings for block size and buffer size to achieve satisfactory throughput performance in different environments.

A transport profile $TP_t(\langle h_s, h_r \rangle, e, \theta)$ is a control-response plot illustrating how a set θ of control parameters affect the performance of a given transport protocol t over a network connection or link e between a sender host h_s and a receiver host h_r. Such a transport profile indicates the qualitative behavior of each component involved in the data transfer process and provides an insight into maximizing the overall transport performance. A complete transport profile can be obtained by exhausting the combinations of the parameter values and collecting the corresponding performance measurements. Each data

Figure 2. UDT profiling over ORNL 10Gbps emulated connections with various RTTs. (a): throughput vs. block size; (b): throughput vs. buffer size.

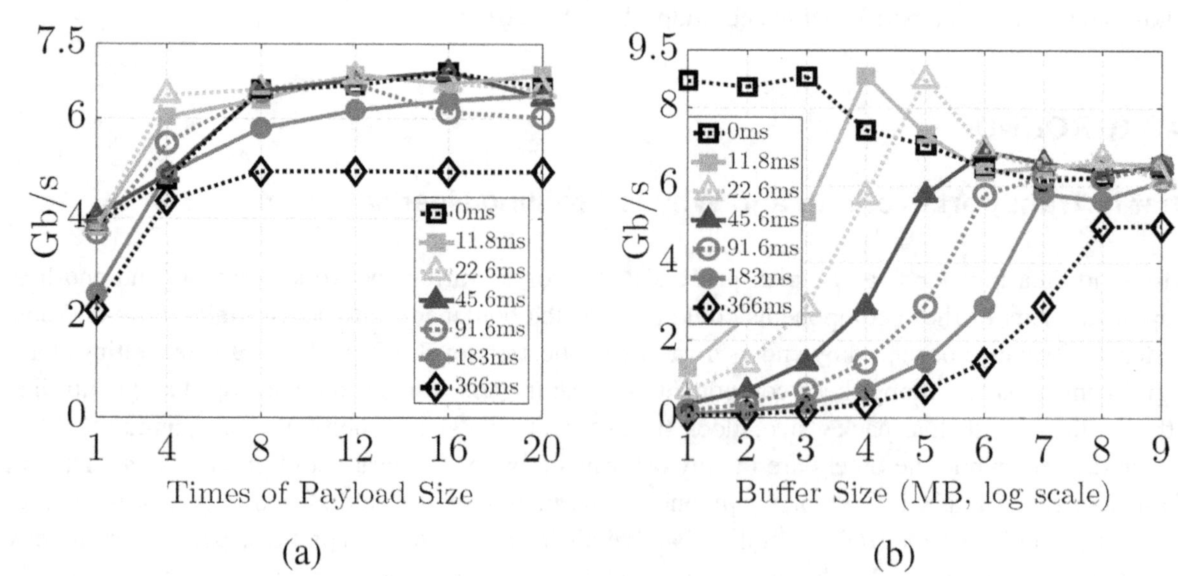

point in the profile is produced by a "one-time profiling" that sends a certain amount of data with a specific combination of parameter values θ during a certain time duration $[0, \Delta T]$ and measures the corresponding average throughput performance $G(\theta)$ as

$$G(\theta) = \frac{\int_0^{\Delta T} s(x, \theta)\, dx}{\Delta T}, \tag{1}$$

where $s(x, \theta)$ is the sending rate with respect to parameter values θ at time point x. During a specific profiling process, $\langle h_s, h_r \rangle$ and e are given, and θ is specified within an appropriate range of values.

Profiling Overhead

The goal of transport profiling is to find the parameter values θ^*, at which the end-to-end throughput performance $G(\theta^*)$ reaches its global maximum. An exhaustive search-based transport profiling can find the optima, but is very time consuming, and therefore is practically unsuitable to meet data transfer requirements of scientific applications in network environments that are subject to frequent changes. Even for HPNs where network connections are relatively stable in some aspects such as RTTs, different choices of protocols and corresponding parameter values for big data transfers over different connections (as determined by the sender-receiver pair specified in the user's data transfer request) may still largely affect the data transfer performance, which necessitates transport profiling.

As a numerical example, in the Transport Profile Generator (TPG) (Yun et al., 2015), the selected UDT (Gu & Grossman, 2007) transport method includes several commonly accessible parameters including packet size ($m \in \{m_1, m_2, \ldots, m_{N_m}\}$), block size ($l \in \{l_1, l_2, \ldots, l_{N_l}\}$), buffer size ($f \in \{f_1, f_2, \ldots, f_{N_f}\}$), and number of parallel data streams ($p \in \{p_1, p_2, \ldots, p_{N_p}\}$). If a one-time profiling takes Δt (typically on the order of several minutes) to finish, it takes a total of $\Delta t \cdot N_m \cdot N_l \cdot N_f \cdot N_p$ to generate a complete profile before the actual data transfer. Let us consider a simplified numerical example, where the packet size m and the number p of parallel data streams are both fixed (i.e., $N_m = 1$, $N_p = 1$), and only the block size varies from 1 to 25 times of the payload size (i.e., $N_l = 25$) and the buffer size varies from 1.0MB to 1.0GB at an interval of 2.0MB (i.e., $N_f = 513$). In this example, if a one-time profiling takes $\Delta t = 2$ minutes, the exhaustive search would take 25,650 minutes (around 18 days!) in total, and hence is impractical in real-life applications. As both the number of control parameters and the profiling resolution increase, the time to obtain a complete profile rapidly increases, making the exhaustive search-based approach practically infeasible.

FAST PROFILING BASED ON STOCHASTIC APPROXIMATION

To obviate the need of conducting an exhaustive profiling, we propose a fast profiling method based on the Simultaneous Perturbation Stochastic Approximation (SPSA) algorithm (Spall, 1992), referred to as FastProf, to quickly determine the "best" parameter values prior to the actual data transfer.

Rationale on the Use of Stochastic Approximation Method

Figure 3 shows the components involved in a typical data transfer process in a certain network environment to meet a user data transfer request. Since our work is mainly focused on transport profiling at the application layer rather than system tuning at lower layers, the whole data transfer process could be treated as a "black box" system, where the input is the set of control parameter values θ and the output is the corresponding throughput measurement $G(\theta)$. Based on this model, the SPSA algorithm is appropriate to be used for quickly determining the optimal parameter values. This is because the SPSA algorithm: i) does not require an explicit formula of $G(\theta)$ but only its "noise corrupted" measurements $y = G(\theta) + \xi$, which can be obtained by running a "one-time profiling" using existing tools such as Iperf3 (Iperf3, 2017) and TPG (Yun et al., 2015) with a set of specified parameter values; ii) does not require any additional information about system dynamics or input distribution; and iii) is of a low time complexity. These are highly desirable features as they not only account for the dynamics in the data transfer process and the randomness in the performance measurements, but also incur negligible computational overhead for practical deployment.

For a given data transfer over a given network connection, if jumbo frame is supported along the path, it is desirable to enable it to minimize per-packet overhead (Chase, Gallatin, & Yocum, 2001). Thus, the packet size m can be decided by exploring the Path MTU (PMTU) without profiling. Although multiple parallel data streams may improve data transfer performance (Allcock et al., 2005), they typically introduce inter-stream competition that may lead to complex transfer dynamics even over dedicated connections. Furthermore, since many high-performance transport methods including UDT are not best suited for environments with a high level of concurrency (Gu, Hong, & Grossman, 2004), we focus our work on one single data stream (i.e., $p = 1$). Like the Transport Profile Generator (TPG) (Yun et al., 2015), where UDT is used in the implementation as a use case, this chapter also uses UDT as an

Figure 3. The "black-box" data movement system

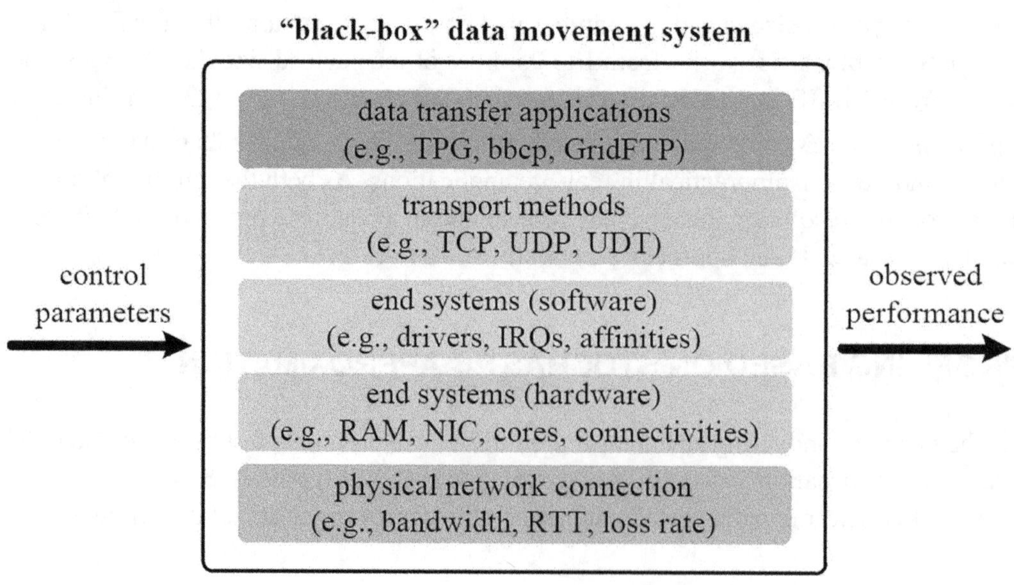

example transport method and considers block size l and buffer size f as the control parameters (or the input of the "black box" system in Figure 3), i.e., $\theta = \begin{bmatrix} l \\ f \end{bmatrix}$. For readers who are interested in transport profiling on the number of data streams, please refer to the profiling results presented in (Yun et al., 2017).

In the exhaustive search, a complete 2D table of performance measurements needs to be constructed by running $N_l \cdot N_f$ times of one-time profiling with $N_l \cdot N_f$ parameter combinations, as visualized in Figure 4, where the l-axis and f-axis represent block size and buffer size, respectively. Note that each data point in the graph is actually a 3-tuple $\langle l, f, G(l,f) \rangle$, where $G(l,f)$ is the observed throughput with respect to block size l and buffer size f. In FastProf, the SA-based profiling method attempts to explore a path of profiling data points (rather than all data points) in this 2D table to reach the global optimum within an "acceptable area".

Stochastic Approximation (SA) Methods

Based on the model shown in Figure 3, the average throughput performance is assumed as a function $G(\theta)$ of control parameters θ. The goal is then to find the control parameters θ^* that maximize $G(\theta)$ within the feasible parameter space Θ, i.e.,

$$\max_{\theta \in \Theta} G(\theta). \tag{2}$$

Following the Kiefer-Wolfowitz Stochastic Algorithm (KWSA) (Kiefer & Wolfowitz, 1952), the standard stochastic approximation form is given by

Figure 4. Visualized profiling process of FastProf

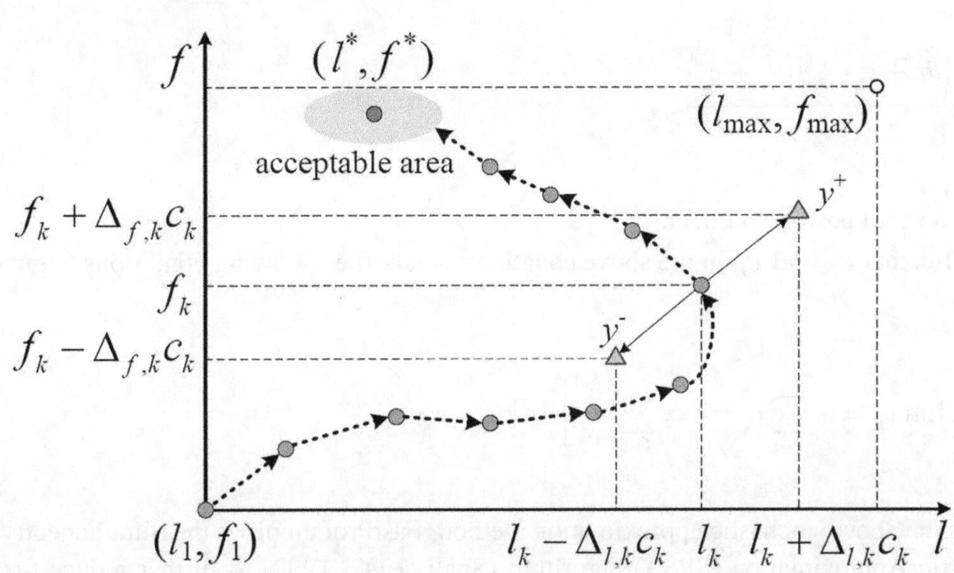

$$\hat{\theta}_{k+1} = \hat{\theta}_k + a_k \cdot \hat{g}_k(\hat{\theta}_k), \tag{3}$$

where $a_k > 0$ is a scalar gain coefficient, $g(\theta) \equiv \frac{\partial G(\theta)}{\partial \theta}$ is the gradient of $G(\theta)$, $\hat{\theta}_k$ is the set of control parameter values in the k-th iteration, and $\hat{g}(\hat{\theta}_k)$ is an approximation of $g(\theta_k)$.

The "noise corrupted" performance observation during transport profiling, denoted as $y(\theta)$, is available at any value of $\theta \in \Theta$, and given by

$$y(\theta) = G(\theta) + \xi, \tag{4}$$

where ξ is the noise incurred by the randomness in the network connection and the dynamics in the end-host systems. In fact, $y(\theta)$ is the observed average throughput performance of a one-time profiling with a specific θ during a specific time duration. Given $\theta = \begin{bmatrix} l \\ f \end{bmatrix}$ and its estimation denoted as $\hat{\theta} = \begin{bmatrix} \hat{l} \\ \hat{f} \end{bmatrix}$, the solution based on the classical KWSA method is a multi-variable recursive optimization procedure, defined as

$$\begin{bmatrix} \hat{l}_{k+1} \\ \hat{f}_{k+1} \end{bmatrix} = \begin{bmatrix} \hat{l}_k \\ \hat{f}_k \end{bmatrix} + a_k \begin{bmatrix} \hat{g}_{l_k}\left(\begin{bmatrix} \hat{l}_k \\ \hat{f}_k \end{bmatrix}\right) \\ \hat{g}_{f_k}\left(\begin{bmatrix} \hat{l}_k \\ \hat{f}_k \end{bmatrix}\right) \end{bmatrix}. \tag{5}$$

The gradient $g(\theta)$ of the function $G(\theta)$ is approximated by a "two-sided" finite difference given by

$$\hat{g}_k\left(\hat{\theta}_k\right) = \frac{y\left(\hat{\theta}_k + c_k\right) - y\left(\hat{\theta}_k - c_k\right)}{2c_k}, \tag{6}$$

where c_k is a small positive number.

The coefficients a_k and c_k in the above equations satisfy the following conditions to guarantee the convergence,

$$\lim_{k\to\infty} a_k = 0, \lim_{k\to\infty} c_k = 0, \sum_{k=1}^{\infty} a_k = \infty, \sum_{k=1}^{\infty}\left(\frac{a_k}{c_k}\right)^2 < \infty. \tag{7}$$

Based on the above stochastic approximation method, FastProf employs the Simultaneous Perturbation Stochastic Approximation (SPSA) algorithm (Spall, 1992, 1997) to further reduce the profiling overhead: instead of collecting observations along all dimensions of the gradient, it randomly perturbs the control parameter set in two separate directions and collects two corresponding performance measurements to approximate the gradient. The performance measurements at the *k*-th iteration based on the perturbations are then

$$\begin{cases} y_k^+ = y(\hat{\theta}_k + \Delta_k c_k) = y\left(\begin{bmatrix} \hat{l}_k + \Delta_{l,k} c_k \\ \hat{f}_k + \Delta_{f,k} c_k \end{bmatrix}\right) \\ y_k^- = y(\hat{\theta}_k - \Delta_k c_k) = y\left(\begin{bmatrix} \hat{l}_k - \Delta_{l,k} c_k \\ \hat{f}_k - \Delta_{f,k} c_k \end{bmatrix}\right) \end{cases}, \tag{8}$$

and the corresponding gradient approximation of $G(\hat{\theta})$ based on SPSA is then given by

$$\begin{cases} \hat{g}_{l_k}\left(\hat{\theta}_k\right) = \hat{g}_{l_k}\left(\begin{bmatrix} \hat{l}_k \\ \hat{f}_k \end{bmatrix}\right) = \dfrac{y_k^+ - y_k^-}{2\Delta_{l,k} c_k} \\ \hat{g}_{f_k}\left(\theta_k\right) = \hat{g}_{f_k}\left(\begin{bmatrix} \hat{l}_k \\ \hat{f}_k \end{bmatrix}\right) = \dfrac{y_k^+ - y_k^-}{2\Delta_{f,k} c_k} \end{cases}, \tag{9}$$

where the coefficient sequence $\{\Delta_{i,k}\}$ (i =1, 2, …, d, for d dimensional vector, and in this work $d = 2$, i.e., block size l and buffer size f) are independent and symmetrically distributed around 0 with finite inverse $E\left|\Delta_{i,k}^{-1}\right|$ over all parameter components i and time steps k. A simple and effective way to decide each component of $\{\Delta_{i,k}\}$ is to use symmetric Bernoulli ± 1 distribution with a probability of 0.5 for each outcome of either $+1$ or -1 (Spall, 1998).

Convergence of SPSA-Based FastProf

The convergence of SPSA-based FastProf is important as it affects the quality of the profiling results as well as the efficiency of the profiler. To explore the applicability of SPSA in the profiling optimization problem and investigate its convergence property, this section justifies the conditions that lead to the convergence in the context of FastProf.

As pointed out by Spall in (Spall, 2003, pp. 161), the conditions for convergence can hardly be all checked and verified in practice due to the lack of knowledge on $G(\theta)$. We provide the following arguments to justify the appropriateness of SPSA in solving the profiling optimization problem.

According to Theorem 7.1 in (Spall, 2003, pp. 186), if Conditions B.1″ – B.6″ hold and θ^* is a unique maximum of $G(\theta)$, then for SPSA, $\hat{\theta}_k$ almost surely converges to θ^* as $k \to \infty$.

The chosen coefficient sequences a_k and c_k in our work and the Bernoulli ± 1 distribution adopted to generate the simultaneous perturbations $\{\Delta_{i,k}\}$ could clearly validate Conditions B.1$''$ and B.6$''$ (see (Spall, 1998) for detail).

The main concern of Conditions B.2$''$ and B.3$''$ is to ensure that $\widehat{\theta}_k$ is close enough to θ^* such that $\widehat{\theta}_k$ has a natural tendency to converge to θ^*. These two conditions are intuitively valid in the profiling problem scenario because: i) the main requirement for these conditions, i.e., $\sup_{k\geq 0}\left\|\widehat{\theta}_k\right\| < \infty$, can be verified since the control parameter values of block size and buffer size are both finite positive integer numbers; ii) since the feasible regions of the control parameters of FastProf are finite and mapped to a limited range (e.g., [1.0, 25.0] in this work) of the iterative variables, $\widehat{\theta}_k$ (including the starting point) is sufficiently close to θ^*; iii) θ^* is not a single point but an "acceptable area" including a set of adjacent points (see Figure 4); iv) from a practical point of view, FastProf attempts to move $\widehat{\theta}_k$ to the nearest point within the feasible space in the case that it deviates from the feasible space, and then adjusts the step size accordingly to avoid such situations; and v) the extensive emulations and experiments conducted in this work confirm the closeness and natural tendency even if the starting point is randomly selected.

The way the simultaneous perturbations $\{\Delta_{i,k}\}$ are generated ensures that $\{\Delta_{i,k}\}$ is a mutually independent sequence, which is independent of $\hat{\theta}_0, \hat{\theta}_1, \ldots, \hat{\theta}_k$. The observed noise in a data transfer over high-speed dedicated connection is mainly caused by the dynamics of end-hosts and network segments. Since the "noise-corrupted" performance observation is the *average* throughput during time duration $[0, \Delta T]$ (see Equation 1), which captures both positive noise and negative noise, the long-term conditional expectation of the observed noise could be considered as zero, i.e., $E\left[\left(\xi^{(+)} - \xi^{(-)}\right) \middle| \left\{\hat{\theta}_0, \hat{\theta}_1, \ldots, \hat{\theta}_k\right\}, \Delta_k\right] = 0$. Since $\{\Delta_{i,k}\}$ is generated following the symmetric Bernoulli ± 1 distribution with a probability of 0.5 for each outcome of either $+1$ or -1, $E\left|\Delta_{i,k}^{-1}\right|$ is uniformly bounded. In addition, the observations $y\left(\hat{\theta}_k \pm \Delta_k c_k\right)$ are also bounded by the link capacity of the dedicated connection, so the ratio of measurement to perturbation $E\left[\left(\frac{y\left(\hat{\theta}_k \pm \Delta_k c_k\right)}{\Delta_{i,k}}\right)^2\right]$ is uniformly bounded over *i* and *k*. Hence, Condition B.4$''$ holds.

Condition B.5$''$ mainly concerns the smoothness of $G(\theta)$. It is difficult to verify whether $G(\theta)$ is three-times continuously differentiable and bounded since $G(\theta)$ is practically unknown, and even is not a continuous function since the control parameters are all discrete. However, the smoothness of $G(\theta)$ can be intuitively verified based on the nature of the problem because the throughput $G(\theta)$ as well as the gradient $g(\theta) \equiv \frac{\partial G(\theta)}{\partial \theta}$ are at least bounded by the connection capacity and the finite feasible region of the control parameter θ. More information including a formal proof of a "discrete" SPSA algorithm could be found in (Wang & Spall, 2011, 2013).

In addition to the above justification, it is worth pointing out that although the throughput performance should have a unique theoretical peak over the feasible control parameter space given a specific snapshot of the status of end-hosts and network environment, it has been observed in the experiments that different runs with identical parameter values may yield different throughputs, which makes the uniqueness of θ^* unverifiable. However, the observed performance $y = G(\theta) + \xi$ indeed shows a peak property over the feasible control parameter region in the experiments. As shown in Figure 2(a), increasing the block size improves the performance given the buffer is sufficiently large; otherwise, increasing the block size may decrease the performance especially when the block size is approaching the buffer size (Yun et al., 2015). In Figure 2(b), it is more clearly shown that larger buffer sizes may not always lead to better performance when all other parameters including block size remain the same, and this trend is consistent for different RTTs but with different turning points.

IMPLEMENTATION OF FASTPROF

An SPSA-Based Profiling Process

The SPSA-based profiling process is generally presented as follows.

- **Step 1:** Let $k = 0$, and select an either fixed or random starting point $\hat{\theta}_0 \in \Theta$ within the feasible space of block size l and buffer size f.
- **Step 2:** Check the termination conditions to continue or terminate the profiling process.
- **Step 3:** Calculate $a_k = \dfrac{a}{(A+k+1)^\alpha}$ and $c_k = \dfrac{c}{(k+1)^\gamma}$, where $\alpha = 0.602$, $\gamma = 0.101$ as suggested by (Spall, 1998), and $A = 0.0$ (or other values much less than the expected/allowed number of iterations). The step sizes *a* and *c* are determined empirically in FastProf based on the size of the entire search space.
- **Step 4:** Generate a pair of coefficients $\Delta_{l,k} \in \{+1, -1\}$ and $\Delta_{f,k} \in \{+1, -1\}$ following the symmetric Bernoulli ± 1 distribution with a probability of 0.5 for each outcome of either $+1$ or -1.
- **Step 5:** Run one-time profiling twice to collect two corresponding throughput observations y_k^+ and y_k^-, see Equation 8.
- **Step 6:** Compute the simultaneous perturbation approximation to the unknown gradient $g(\hat{\theta}_k)$ using Equation 9.
- **Step 7:** Use the standard stochastic approximation form (Equation 3) to update $\hat{\theta}_k$ to a new value $\hat{\theta}_{k+1}$, increase *k* by 1 ($k = k + 1$), and then go back to Step 2.

Profiling Precision

The two elements of the control parameter set used in the stochastic approximation model, denoted as $\theta' = \begin{bmatrix} l' \\ f' \end{bmatrix}$, are set as positive numbers within a reasonably selected range to ensure a comparable magnitude of each parameter. Rounding operations in calculating the actual values of the control parameters are necessarily performed in the case of fractional values of θ'.

The profiling unit of block size, denoted by μ_l, is defined as one payload size, and the block size l ($l \geq 1$) is defined as a multiple of the payload size. The profiling process transfers a data block of $\lambda_l(l') \cdot \mu_l$ bytes each time by calling the appsend()/apprecv() functions, which may in turn call the send()/recv() API functions of the underlying transport protocol multiple times to completely deliver an entire data block. If a UDP-based protocol such as SABUL (Gu & Grossman, 2003) or UDT (Gu & Grossman, 2007) is used, it is recommended to set the block size to be a multiple of the protocol's payload size if possible to avoid UDP automatic segmentation. For example, UDT's payload size is UDT data packet size (m) minus UDT header length (Gu & Grossman, 2003), i.e., the profiling unit of block size μ_l is given by

$$\mu_l = m - 16 = (\text{MTU} - 28) - 16 = \text{MTU} - 44, \tag{10}$$

where UDT header is of 16 bytes, and IP and UDP headers are of 20 and 8 bytes, respectively (hence 28 bytes in total). For example, if the jumbo frame is enabled and the MTU is set as 9,000 bytes, then μ_l is 8,956 bytes.

The profiling unit of buffer size, denoted by μ_f, is decided by the specific profiling precision chosen by the end user in any unit within a feasible profiling range, e.g., 1B, 512KB, 1.0MB, 2.0MB, or others.

Based on the above profiling units, the actual values of block size (l) and buffer size (f) for performance observations (i.e., the calculations of y^+ and y^- through one-time profilings) are calculated by

$$\begin{cases} l = round\left(\lambda_l\left(l'\right) \cdot \mu_l\right) \\ f = round\left(\lambda_f\left(f'\right) \cdot \mu_f\right) \end{cases}, \tag{11}$$

where λ_l and λ_f are scaling functions that may employ different profiling patterns. For example, with a function $\lambda_f\left(f'\right) = 2^{f'}$, the buffer size would exponentially increase as f' increases.

Termination Conditions

Many efforts (e.g., (Farrell, 1962; Wada & Fujisaki, 2013)) have been devoted to the termination conditions of SA-based methods since the KWSA algorithm was first proposed (Kiefer & Wolfowitz, 1952). In FastProf, the following three simple and practical conditions are adopted to guarantee the performance and the termination of a profiling process.

1. **Early Termination - the C Rule:** The best throughput performance y^* of a given data transfer method over a given network connection is unknown until a complete transport profile is obtained. FastProf defines the *performance gain ratio* of a one-time profiling as,

$$\rho = \frac{y}{y^*}, \tag{12}$$

where y is the observed throughput of a one-time profiling. Over a dedicated connection, the bandwidth B is considered as a known constant since it is reserved in advance through services such as OSCARS (OSCARS, 2017), thus $y^* = B$ is set in FastProf. When FastProf reaches an operational zone that produces a throughput y with a performance gain ratio no less than $\mathcal{C}$, i.e.,

$$\rho = \frac{y}{y^*} = \frac{y}{B} \geq \mathcal{C}, \tag{13}$$

FastProf terminates. Note that this condition may or may not be satisfied in a certain one-time profiling.

2. **Upper Bound - the $\mathcal{M}$ Rule:** FastProf terminates when the number of one-time profilings exceeds a threshold $\mathcal{M}$, which is typically set as $\mathcal{M} \ll (N_l \cdot N_f)$. If $\mathcal{M} = (N_l \cdot N_f)$, FastProf rolls back to the exhaustive search as in TPG (Yun et al., 2015).
3. **Impeded Progress - the $\mathcal{R}$ Rule:** If the number of consecutive iterations that do not produce any performance improvement compared with the best one observed so far exceeds an upper bound $\mathcal{R}$, FastProf terminates.

Preventing Local Optima

In addition to the aforementioned three termination conditions, we take a simple but "scalable" approach to move the profiling process out of a local optima region under the constraints of $\mathcal{M}$ and $\mathcal{R}$. Specifically, if no performance improvement is observed after a certain number of consecutive iterations that have reached a certain fraction of $\mathcal{R}$, we enlarge the profiling step sizes by certain factors $\tau_a > 1$ and $\tau_c > 1$, i.e., $a = a \cdot \tau_a$ and $c = c \cdot \tau_c$. This operation repeats until FastProf yields a better performance or one of the termination conditions aforementioned is met.

EMULATION-BASED PERFORMANCE EVALUATION USING REAL EXPERIMENTAL RESULTS

This section conducts profiling emulations using the profiling data collected on a real-life testbed to gain insights into the behaviors of FastProf and also compare FastProf with other search algorithms including random walk (Lawler & Limic, 2010) and Tabu search (Glover, 1986).

Data Collection

A complete profile (a 2D table) as shown in Figure 4 is built by exhausting $N_l \cdot N_f = 12{,}825$ combinations of block size and buffer size, whose profiling resolutions are set to be one payload size and 2MB, respectively. These results are collected by running TPG tests over a 10Gbps 380ms connection between a sender host at Argonne National Laboratory (ANL) and a receiver host at University of Chicago (UChicago). The profiling emulations in this section are conducted based on this complete profile.

Scaling Functions and Parameter Settings

In the profiling emulations, the actual control parameter values ($\theta = \begin{bmatrix} l \\ f \end{bmatrix}$) are calculated using Equation 14 based on the iterative parameter values ($\theta' = \begin{bmatrix} l' \\ f' \end{bmatrix}$),

$$\begin{cases} l = round\left(\lambda_l\left(l'\right)\cdot\mu_l\right) = round\left(\left(\left(l' - l'_{\min}\right)\cdot\dfrac{\left(l_{\max} - l_{\min}\right)}{\left(l'_{\max} - l'_{\min}\right)} + l_{\min}\right)\cdot\mu_l\right) \\ f = round\left(\lambda_f\left(f'\right)\cdot\mu_f\right) = round\left(\left(\left(f' - f'_{\min}\right)\cdot\dfrac{\left(f_{\max} - f_{\min}\right)}{\left(f'_{\max} - f'_{\min}\right)} + f_{\min}\right)\cdot\mu_f\right) \end{cases}. \tag{14}$$

In particular, the profiling range for block size is from 1 payload to 25 payloads (i.e., $l_{\min} = 1$, $l_{\max} = 25$, and $\mu_l = 8956$ bytes), and the profiling range for buffer size is from 1MB to 1024MB (i.e., $f_{\min} = 1$, $f_{\max} = 1024$, and $\mu_f = 1{,}048{,}576$ bytes). The iterative variables for both block size and buffer size are chosen to be from 1.0 to 25.0 (i.e., $l'_{\min} = 1.0$, $l'_{\max} = 25.0$, $f'_{\min} = 1.0$, and $f'_{\max} = 25.0$) to ensure that they are of comparable and consistent magnitudes.

Performance Evaluation Results

Four parameters are considered in these profiling emulations: i) the starting point (SP) (either fixed or randomly selected); ii) the $\mathcal{A}$ rule (either disabled or enabled, and if $\mathcal{A}$ is enabled, FastProf only searches the feasible space where the buffer size is larger than Bandwidth Delay Product (BDP), i.e., $f \geq BDP$; otherwise, it searches the entire feasible space); iii) the value of $\mathcal{R}$ (integers from 1 to 50); iv) the value of $\mathcal{C}$ (selected from {0.95, 0.90, 0.80}); and v) the value of $\mathcal{M}$ (either ∞ or $2\mathcal{R}$). Four types of performance metrics are measured in all 5,000 runs of each test case in the emulations: i) the average throughput achieved by the profiling; ii) the percentage among all 5,000 runs that leads to a desired performance; iii) the average profiling time as indicated by the average number of one-time profilings; and iv) the longest profiling time as indicated by the maximum number of one-time profilings. The subfigures in each figure in this section correspond to these four types of performance metrics labeled by $\mathcal{A}$, $\mathcal{M}$, $\mathcal{C}$, and starting point (SP), respectively.

1. **Overall Performance:** The overall performance results of FastProf plotted in Figure 5 show that FastProf is able to find a set of control parameter values result in satisfactory performance in a short period of time, where the starting point is randomly selected from the feasible region of control parameters, and the $\mathcal{A}$ rule is disabled.

The actual throughput obtained by FastProf is considered as the most important performance metric. Figure 5(a) shows that a satisfactory throughput performance can be achieved as long as a reasonably large $\mathcal{R}$ is specified (e.g., > 10 in the test cases shown in Figure 5). As $\mathcal{R}$ increases, the average

Figure 5. Overall performance of FastProf

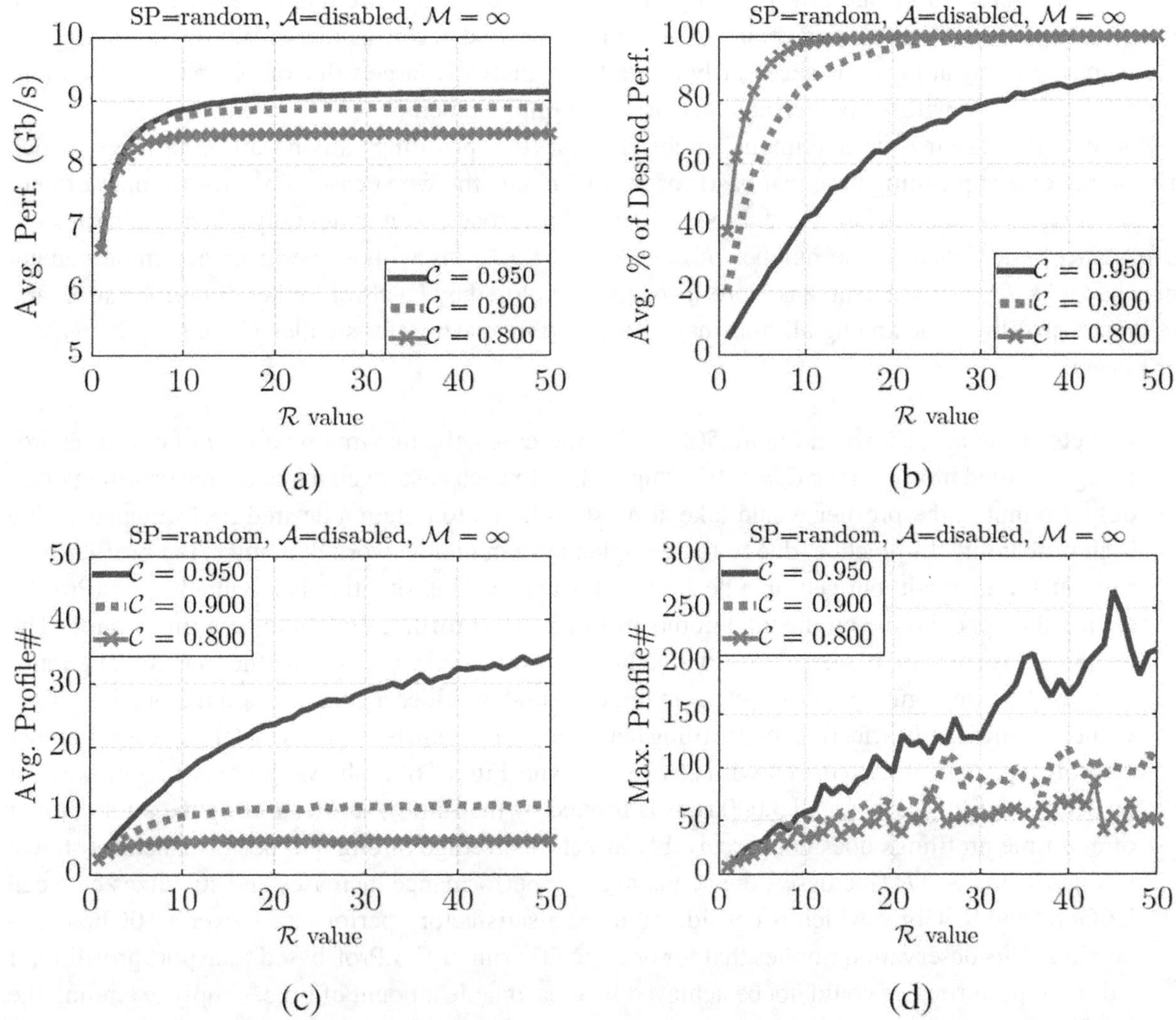

throughput performance explored by FastProf first increases and then stabilizes at the desired level as specified by the value of $\mathcal{C}$, which is very near to the connection capacity.

Figure 5(b) shows the percentage of cases that yield a user-desired performance among all 5,000 runs. As $\mathcal{R}$ increases, the percentage significantly increases up to 100% for $\mathcal{C} \in \{0.80, 0.90\}$. When $\mathcal{C} = 0.95$, the percentage does not reach 100% as $\mathcal{R}$ approaches 50, but FastProf still achieves slightly higher throughput performance near or above 9.0Gbps as $\mathcal{R}$ increases, as shown in Figure 5(a).

The profiling speed is measured by calculating the average number of one-time profilings consumed by the profiler among all 5,000 runs. As expected, this average number generally increases as $\mathcal{R}$ increases, as shown in Figure 5(c). For a relatively smaller $\mathcal{C}$ value (e.g., 0.80 and 0.90 in these test cases), it does not always increase as $\mathcal{R}$ increases since the percentage of yielding a desired performance reaches 100% quickly and FastProf terminates the profiling process without consuming more profiling time; while for a larger $\mathcal{C}$ value (e.g., 0.95 in these test cases), it takes longer to obtain a higher percentage for a desired performance. As $\mathcal{R}$ increases, there is a higher probability to obtain a desired performance by conducting more one-time profilings, but the actual *average* throughput performance does not increase as significantly as the percentage of obtaining a desired performance. This observation implies that an appropriately large $\mathcal{R}$ is necessarily needed, but an overly large value of $\mathcal{R}$ may lead to a longer profiling process without perceivable performance improvement.

Figure 5(d) measures the maximum number of one-time profilings among all 5,000 runs, which reflects the longest profiling time that FastProf may take, i.e., the worst case. Since the number of one-time profilings is not limited (since $\mathcal{M} = \infty$), the profiling process terminates only when either a desired performance is achieved or the number of consecutive iterations without performance improvement reaches $\mathcal{R}$. As $\mathcal{R}$ increases, there is more space to be explored by FastProf for better performance, and the longest profiling time among all runs may either stop increasing (for smaller $\mathcal{C}$) or keep increasing (for larger).

2. **Effects of $\mathcal{M}$:** As shown in Figure 5(d), as $\mathcal{R}$ increases, the maximum number of one-time profilings resulted from a larger $\mathcal{C} = 0.95$ is up to 250. In such case, even if a one-time profiling takes only 2 minutes, the profiler would take at most 10 hours to obtain a desired performance with a high probability. In practice, due to the complex system and network dynamics, the profiler may run much longer without getting a desired performance. This situation is obviated in FastProf by setting the upper bound on the total number of one-time profilings ($\mathcal{M}$) to be a finite value. The results in Figure 6 are based on $\mathcal{C} = 0.95$, the upper bound $\mathcal{M} = 2\mathcal{R}$, and the same other parameters as those in Figure 5. Setting such an upper bound provides a guarantee on the profiling time without significantly affecting the profiling performance. Figure 6(c) shows that the average number of profilings is slightly reduced with $\mathcal{M} = 2\mathcal{R}$, and Figure 6(d) shows the maximal number of one-time profilings among all 5,000 runs is limited by the finite $\mathcal{M}$. Restricting the total number of one-time profilings does not perceivably affect the average throughput performance, as shown in Figure 6(a). As $\mathcal{R}$ increases, the actual average performance increases and stabilizes between 8.0Gbps and 9.5Gbps, which is considered to be a satisfactory performance over a 10Gbps connection. This observation implies that for one specific run of FastProf-based transport profiling, if a desired performance could not be achieved in a reasonable amount of time, simply extending the profiling process may not improve the performance. Therefore, this upper bound $\mathcal{M}$ should be set for FastProf based on the expectation of the tolerable amount of profiling time. In the experiments conducted in next section, this value is set to be $\mathcal{M} = 2\mathcal{R}$. Note that since the user-desired performance tends to be achieved before the number of profilings reaches or exceeds $\mathcal{M}$ when $\mathcal{C}$ is relatively small (e.g., 0.80), the performance results plotted in Figure 6 are based on a relatively larger value $\mathcal{C} = 0.95$ to show the effects of $\mathcal{M}$ on the profiling performance. The measurements in Figure 6 suggest that setting $\mathcal{M}$ to be a finite value is more useful and even critical for an aggressive $\mathcal{C}$.
3. **Effects of $\mathcal{A}$:** Figure 7 shows the comparisons between the test cases where $\mathcal{A}$ is disabled and the test cases where $\mathcal{A}$ is enabled. Note that when $\mathcal{A}$ is enabled in Figure 7, the profiling range for buffer size is from $f_{min} = 475$ MB to $f_{max} = 1,024$ MB since a delay of 380ms over a connection of 10Gbps bandwidth has a bandwidth-delay product (BDP) of 475MB. Since the ranges of iterative variables l' and f' are both from 1.0 to 25.0, the buffer size range (in unit of MB) mapped to the range of iterative variable f' is actually $[475, 1024]$, i.e., $[475, 1024]$ is linearly and uniformly

Figure 6. Effects of $\mathcal{M}$ on profiling performance

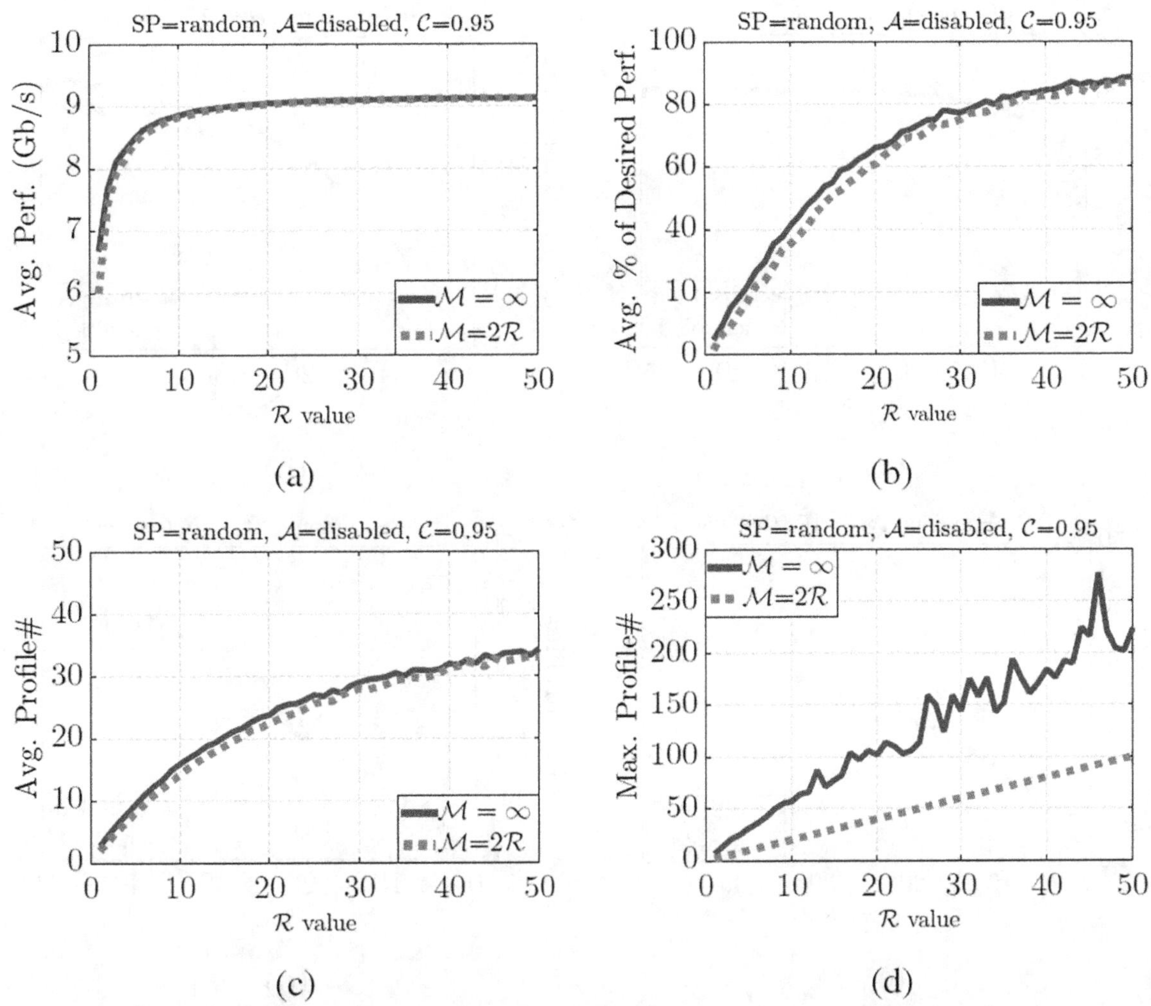

mapped to $[1.0, 25.0]$. Figure 7 shows the effects of cutting the search space by nearly half and searching only the half with larger buffer sizes. Firstly, for $\mathcal{R} < 10$, the average performance when $\mathcal{A}$ is enabled is slightly higher than that when $\mathcal{A}$ is disabled, as shown in Figure 7(a). This is because the profiling time (indicated by the number of profilings) is mainly limited by the value of $\mathcal{R}$ when it is relatively small. The profiling process terminates before FastProf further explores the space for performance improvement, and thus the performance mainly depends on the initial starting point. As $\mathcal{R}$ increases, the differences made by $\mathcal{A}$ become less obvious, and the average throughput performance converges and stabilizes at around 9Gbps. Secondly, when $\mathcal{C}$ is a large value that can be hardly achieved (e.g., $\mathcal{C} = 0.95$), as $\mathcal{R}$ increases, enabling $\mathcal{A}$ results in a slightly lower percentage of obtaining a desired performance and a slightly longer profiling time in comparison with the case where $\mathcal{A}$ is disabled (Figure 7(b) and Figure 7(c)). Figure 7(d) shows that enabling $\mathcal{A}$ does not make significant difference on the longest profiling time, which implies that

Figure 7. Effects of $\mathcal{A}$ on profiling performance

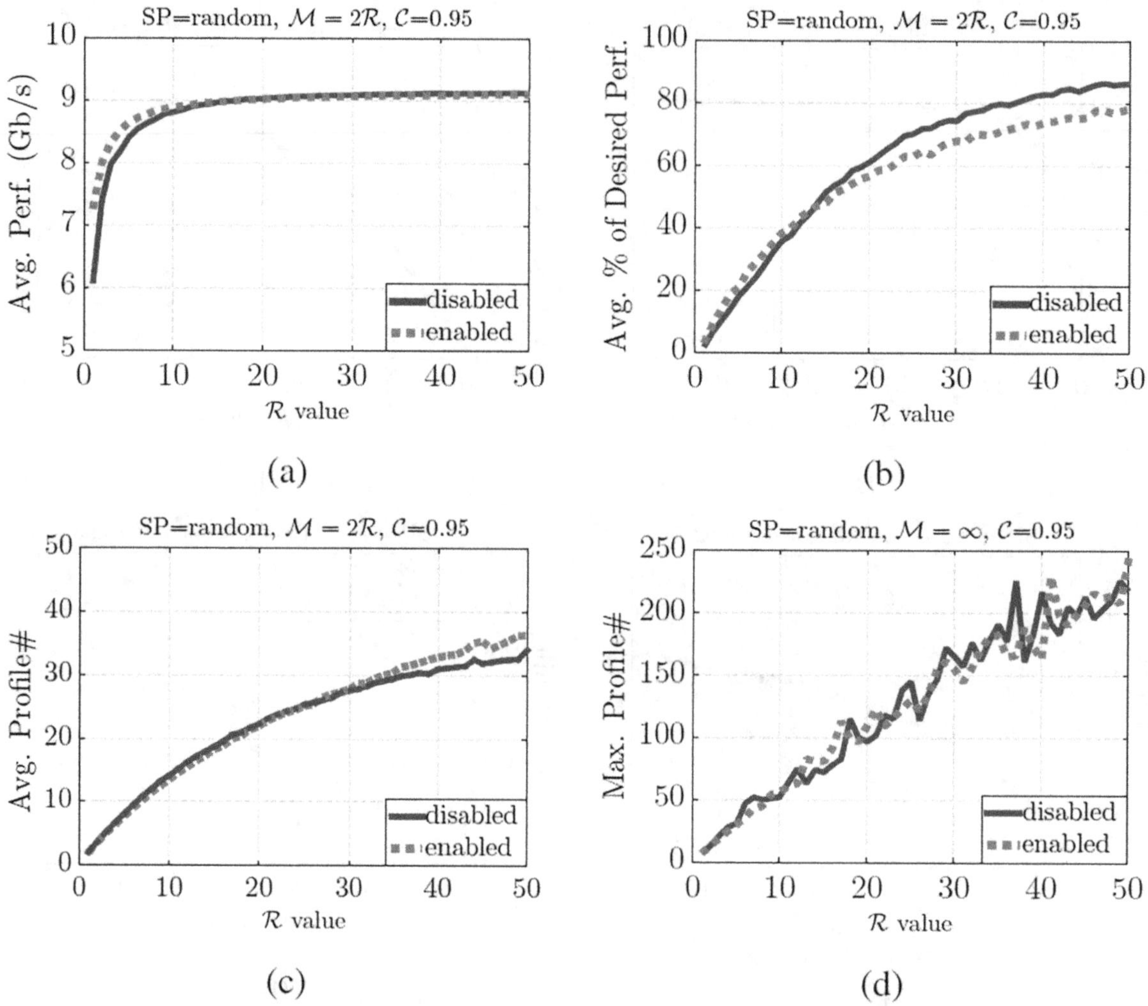

the profiling overhead of FastProf is not sensitive to the size of the parameter space. Note that in Figure 6(d), $\mathcal{M}$ is set to be ∞ rather than $2\mathcal{R}$ to eliminate the effects of $\mathcal{M}$ and show only the effects of $\mathcal{A}$ on the profiling time.

4. **Effects of Starting Point (SP):** To study the effects of the starting point on the profiling performance, the emulations are re-run with a fixed starting point $(l_1, f_1) = (1, 1)$, i.e., the "left-bottom" corner that represents one payload size for block size and 1MB for buffer size (see Equation 14 and Figure 4). Other parameter settings remain the same as the emulations previously presented. Figure 8 compares the results with either random or fixed starting points, and shows that a random starting point, in comparison with a fixed starting point, results in: i) either slightly higher (when $\mathcal{R}$ is small) or identical (when $\mathcal{R}$ is large) average performance, as shown in Figure 8(a); ii) a consistently higher average probability (percentage) of obtaining a desired performance, as shown in Figure 8(b); iii) a consistently shorter average profiling time, as shown in Figure 8(c); and iv) a more stable status for the worst case, as shown in Figure 8(d).

Figure 8. Effects of starting point (SP) on profiling performance

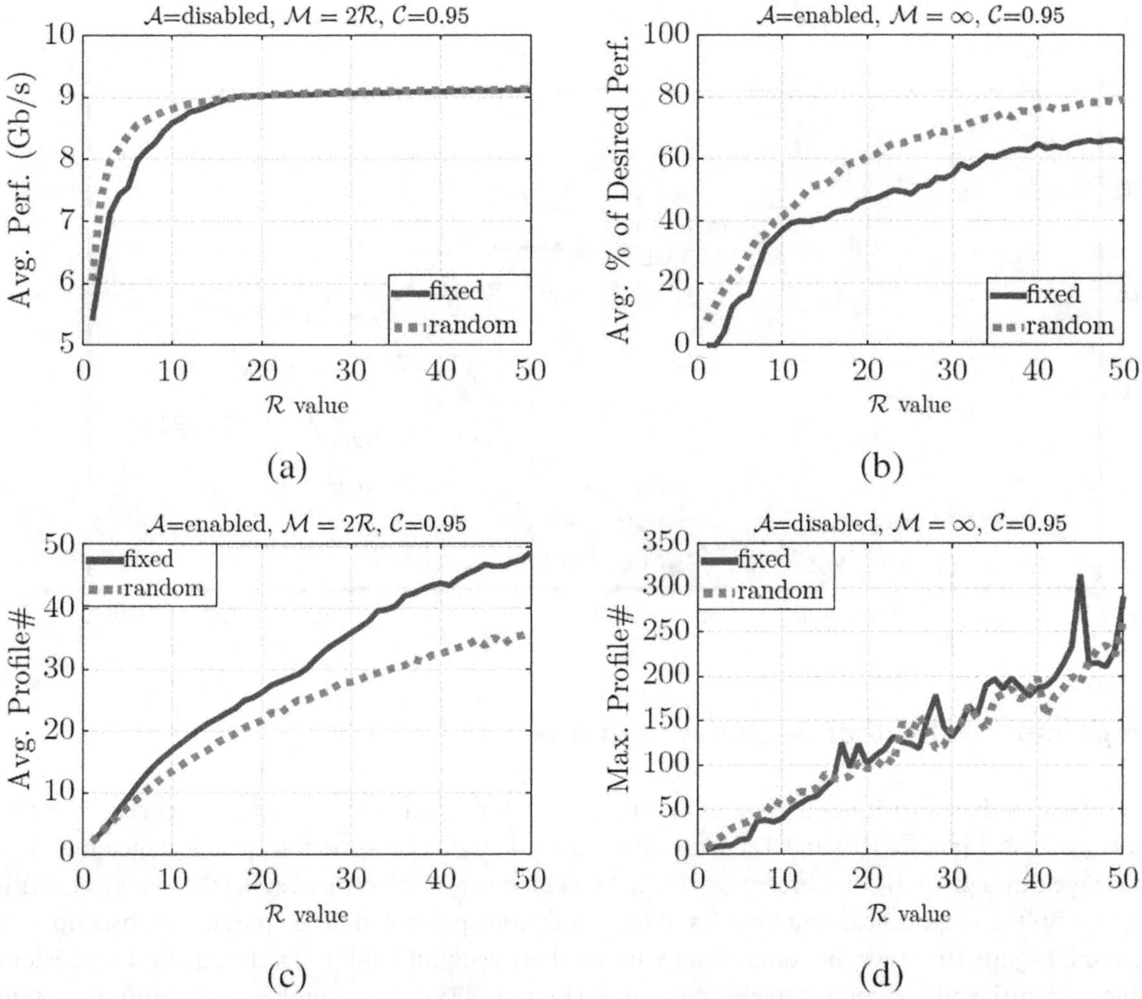

Trace of the Profiling Process

A profiling trace is conducted in this subsection to show the detailed profiling process of FastProf. In this tracing, $\mathcal{R} = 35$, $\mathcal{C} = 0.95$, $\mathcal{A}$ is disabled, and all pairs of parameter values (l, f) are recorded as the profiling process moves forward. Figure 9 plots the control parameter values of the data points profiled by FastProf, where the (green) *circles* represent the values of $\hat{\theta}_k$ in each iteration, the (cyan) *triangles* indicates the perturbations $\hat{\theta}_k \pm \Delta_k$, i.e., the values used by FastProf to measure y_k^+ and y_k^- in each iteration, and the (red) *circle* (i.e., the last data point on the path) is the control parameter values that produce a desired performance, i.e., the one in the "acceptable area" (user-desired) as shown in Figure 4. In this tracing experiment, $a = 25$, $c = 9.5$, and the starting point is fixed at (1, 1). Figure 9 shows that FastProf is able to explore a path in the feasible parameter space from the fixed starting point to an acceptable area.

Figure 9. Profiling process trace with $\mathcal{R} = 35$, $\mathcal{C} = 0.95$, and disabled $\mathcal{A}$.

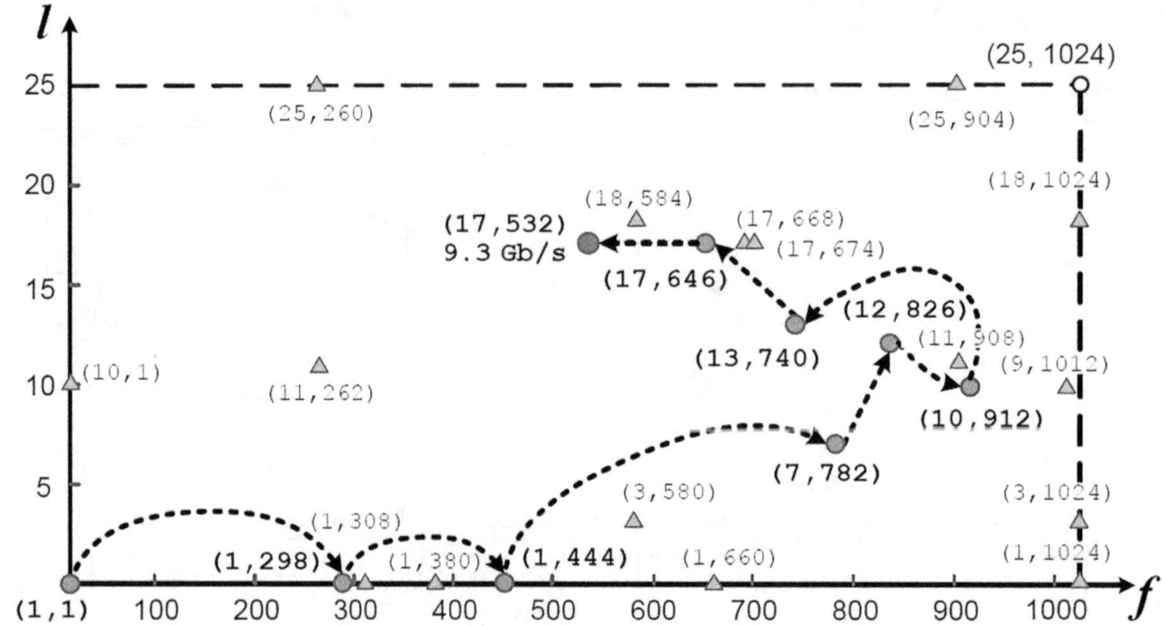

Comparison With Other Search Algorithms

This section conducts profiling emulations to compare FastProf with other two existing heuristics, random walk (Lawler & Limic, 2010) and Tabu search (Glover, 1986). The same four performance metrics, i) the average throughput; ii) the percentage that achieves a desired performance; iii) the average profiling time; and iv) the longest profiling time, as in the emulations presented in the previous subsections, are measured. Figure 10 shows, in comparison with random walk and Tabu search, FastProf consistently produces significantly better average performance (Figure 10(a)), has a higher probability to obtain a user-desired performance (Figure 10(b)), takes much shorter profiling time (Figure 10(c)), and has much better worst cases (Figure 10(d)) among all 5,000 runs.

EXPERIMENT-BASED PERFORMANCE EVALUATION OVER PHYSICAL CONNECTIONS

This section conducts experiments over two physical connections with 2ms and 380ms RTTs in real-life network environments.

Experimental Results on ANL Testbed

This subsection presents experimental results of running FastProf over 10Gbps physical connections from Argonne National Laboratory to University of Chicago with different delays (2ms and 380ms), where, $y^* = 10\,\text{Gbps}$ (i.e., the connection capacity), $a = 30.0$, $c = 9.5$, $\mathcal{R} = 30$, and $\mathcal{M} = 2\mathcal{R} = 60$,

Figure 10. Comparison of profiling effectiveness and efficiency between FastProf, random walk, and Tabu search

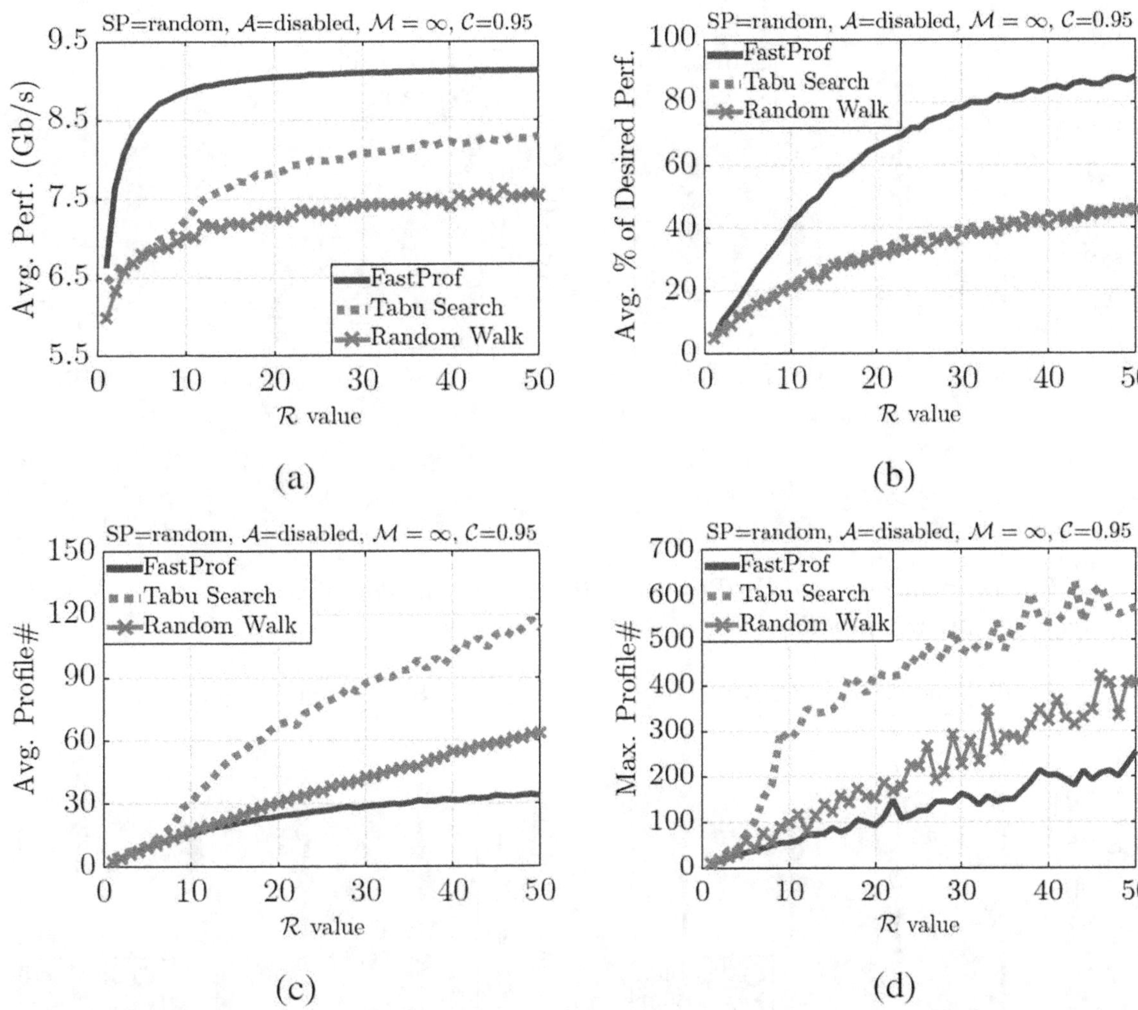

and the profiling precision is set to be one byte (i.e., the highest precision). Note that rounding operations are necessarily performed in these experiments to ensure that an integer number of bytes are set for both block size and buffer size.

Each experimental test is run for 10 times, and the average throughput performance and profiling time are measured together with their standard deviations. Figure 11 (2ms) and Figure 12 (380ms) plot the results for both disabled $\mathcal{A}$ and enabled $\mathcal{A}$ cases, which show that FastProf is able to consistently find a satisfactory set of control parameter values in a short profiling period for both short and long delays. For a short RTT of 2ms, Figure 11 shows that the throughput explored by FastProf is quite stable (> 8.0Gbps). When $\mathcal{C} \in \{0.80, 0.85\}$, FastProf achieves a user-desired performance in all 10 runs with less than 30 one-time profilings on average. When $\mathcal{C} \in \{0.90, 0.95\}$, the desired performance can only be occasionally achieved. Consequently, the early termination rule (i.e., the $\mathcal{C}$ rule) is less frequently triggered and most of the runs are terminated by the $\mathcal{R}$ rule or the $\mathcal{M}$ rule. Since FastProf attempts to

Figure 11. Experimental results of FastProf on a 10Gbps 2ms connection. (a): average throughput performance; (b): average profiling time.

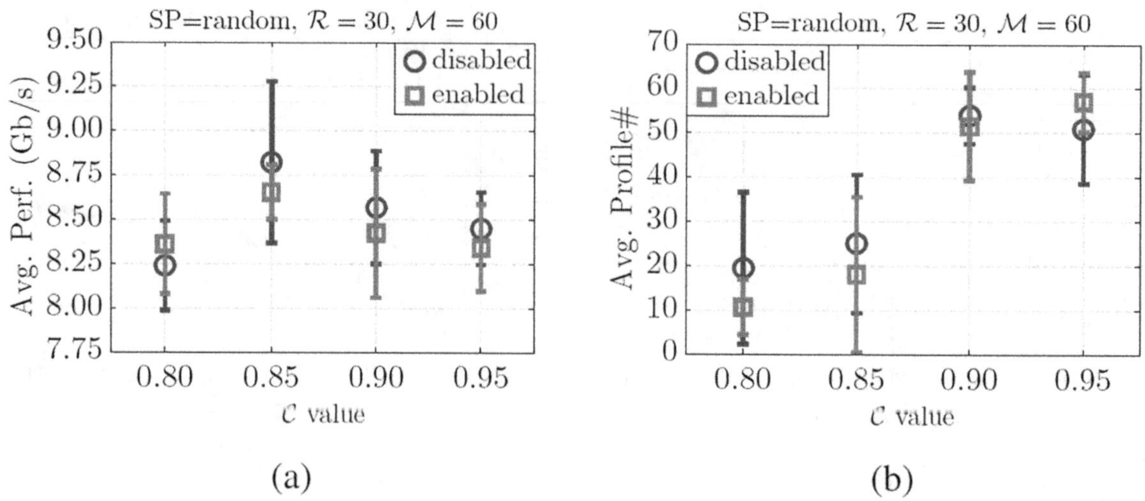

Figure 12. Experimental results of FastProf on a 10Gbps 380ms connection. (a): average throughput performance; (b): average profiling time.

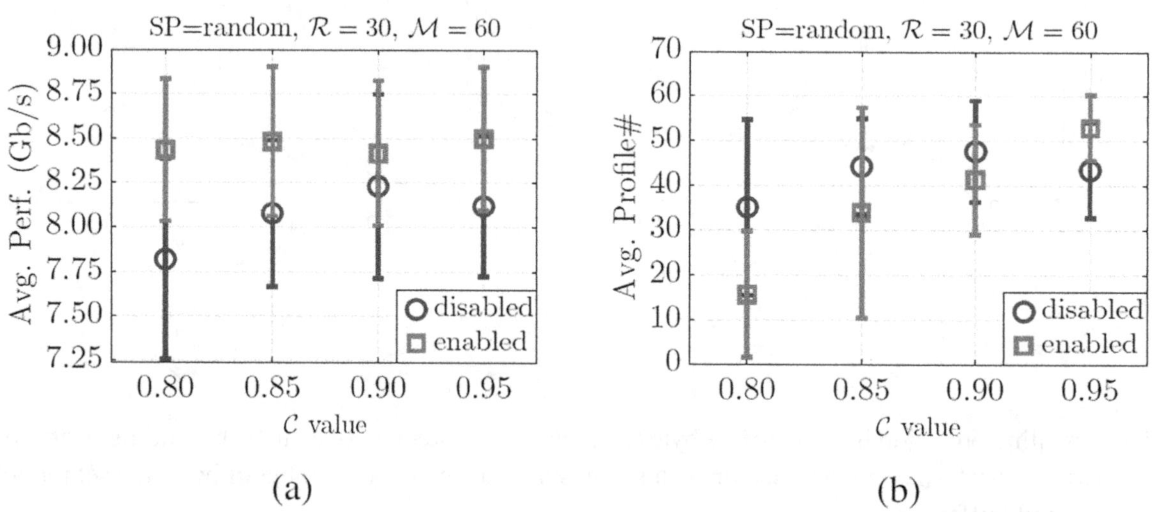

achieve a higher level of performance (due to the larger values of C), it takes a longer time (Figure 11(b)) and the performance is maintained at the same level (Figure 11(a)). For a longer RTT of 380ms, Figure 12 shows that FastProf is able to discover an appropriate set of values for block size and buffer size that result in an average performance between 7.5Gbps and 8.5Gbps. The performance achieved by FastProf is consistent with the complete transport profile manually collected for the emulations as presented in previous section. It is also observed that enabling the $\mathcal{A}$ rule does not necessarily improve the performance, which implies again (in addition to the results shown in Figure 7) that FastProf is not sensitive to the size of the search space.

FUTURE RESEARCH DIRECTIONS

To further improve the performance and accuracy of transport profiling, the proposed SA-based fast profiling method is worth further investigation in various aspects, including gradient approximation averaging, step size adaption, intelligent termination, formal convergence proof, etc. End-to-end big data transfer over dedicated connections is a complex task that involves both network segments and end-hosts, and the parameter selection for each may affect the application-level performance observed by end users. Motivated by FastProf, we are also interested in exploring the possibility of applying such SA-based approaches to storage profiling on end-hosts in our future research. Since FastProf makes it possible to conduct "on-line" profiling for big data transfer, it is also of our future interest to integrate it into existing big data movement solutions such as (Yun & Wu, 2015) to improve their time efficiency.

CONCLUSION

We proposed a stochastic approximation-based transport profiler, FastProf, to accelerate the profiling process for improving big data transfer performance in high-performance networks. FastProf was implemented based on the existing Transport Profile Generator (TPG) toolkit. Extensive profiling emulations in comparison with other search algorithms, and profiling experiments over physical network connections with various delays are both conducted to evaluate the performance superiority of FastProf. The emulation and experimental results showed that FastProf significantly reduces the profiling time while achieving a comparable level of end-to-end throughput performance, which enables "on-line" profiling to achieve optimal big data transfer performance in high-performance networks.

REFERENCES

Ahuja, V., Farrens, M., & Ghosal, D. (2012). Cache-aware Affinitization on Commodity Multicores for High-speed Network Flows. In *Proceedings of the ACM/IEEE Symposium on Architectures for Networking and Communications Systems* (pp. 39-48). Austin, TX: IEEE. 10.1145/2396556.2396564

Allcock, W., Bresnahan, J., Kettimuthu, R., Link, M., Dumitrescu, C., Raicu, I., & Foster, I. (2005). The Globus Striped GridFTP Framework and Server. In *Proceedings of the ACM/IEEE Conference on Supercomputing* (pp. 54-65). Seattle, WA: IEEE. 10.1109/SC.2005.72

Andersson, L., & Swallow, G. (2003). *The Multiprotocol Label Switching (MPLS) Working Group decision on MPLS signaling protocols*. IETF RFC 3468.

Chase, J., Gallatin, A., & Yocum, K. (2001). End System Optimizations for High-speed TCP. *IEEE Communications Magazine, 39*(4), 68–74. doi:10.1109/35.917506

Esnet. (2017). *Energy Sciences Network*. Retrieved August 28, 2017, from http://www.es.net/

Farrell, R. (1962). Bounded Length Confidence Intervals for the Zero of a Regression Function. *Annals of Mathematical Statistics, 33*(1), 237–247. doi:10.1214/aoms/1177704727

Glover, F. (1986). Future Paths for Integer Programming and Links to Artificial Intelligence. *Computers & Operations Research, 13*(5), 533–549. doi:10.1016/0305-0548(86)90048-1

Gu, Y., & Grossman, R. (2003). SABUL: A Transport Protocol for Grid Computing. *Journal of Grid Computing, 1*(4), 377–386. doi:10.1023/B:GRID.0000037553.18581.3b

Gu, Y., & Grossman, R. (2007). UDT: UDP-based Data Transfer for High-speed Wide Area Networks. *Computer Networks, 51*(7), 1777–1799. doi:10.1016/j.comnet.2006.11.009

Gu, Y., Hong, X., & Grossman, R. (2004). Experiences in Design and Implementation of a High Performance Transport Protocol. In *Proceedings of the ACM/IEEE Supercomputing Conference* (pp. 22-35). Pittsburgh, PA: IEEE. doi: 10.1109/SC.2004.24

Hanford, N., Ahuja, V., Farrens, M., Ghosal, D., Balman, M., Pouyoul, E., & Tierney, B. (2014). Analysis of the Effect of Core Affinity on High-Throughput Flows. In *Proceedings of the International Workshop on Network-Aware Data Management* (pp. 9-15). New Orleans, LA: Academic Press. 10.1109/NDM.2014.10

Internet2. (2017). *Internet2 Official Website*. Retrieved August 28, 2017, from http://www.internet2.edu

Iperf3. (2017). *ESnet iperf3*. Retrieved August 28, 2017, from https://github.com/esnet/iperf

Jain, S., Kumar, A., Mandal, S., Ong, J., Poutievski, L., Singh, A., ... Vahdat, A. (2013). B4: Experience with a Globally-Deployed Software Defined WAN. In *Proceedings of the ACM SIGCOMM Conference* (pp. 3-14). Hong Kong, China: ACM. 10.1145/2486001.2486019

Kiefer, J., & Wolfowitz, J. (1952). Stochastic Estimation of the Maximum of A Regression Function. *Annals of Mathematical Statistics, 23*(3), 462–466. doi:10.1214/aoms/1177729392

Lawler, G., & Limic, V. (2010). *Random Walk: A Modern Introduction.* New York, NY: Cambridge University Press. doi:10.1017/CBO9780511750854

Leitao, B. (2009). Tuning 10Gb Network Cards on Linux. In *Proceedings of the Linux Symposium* (pp. 169-184). Ottawa, Canada: Academic Press.

McKeown, N., Anderson, T., Balakrishnan, H., Parulkar, G., Peterson, L., Rexford, J., ... Turner, J. (2008). OpenFlow: Enabling Innovation in Campus Networks. *Computer Communication Review, 38*(2), 69–74. doi:10.1145/1355734.1355746

OSCARS. (2017). *On-Demand Secure Circuits and Advance Reservation System.* Retrieved August 28, 2017, from http://www.es.net/oscars

Prasad, R., Murray, M., Dovrolis, C., & Claffy, K. (2003). Bandwidth estimation: Metrics, measurement techniques, and tools. *IEEE Network, 17*(6), 27–35. doi:10.1109/MNET.2003.1248658

Spall, J. (1992). Multivariate Stochastic Approximation Using a Simultaneous Perturbation Gradient Approximation. *IEEE Transactions on Automatic Control, 37*(3), 332–341. doi:10.1109/9.119632

Spall, J. (1998). Implementation of the Simultaneous Perturbation Algorithm for Stochastic Optimization. *IEEE Transactions on Aerospace and Electronic Systems, 34*(3), 817–823. doi:10.1109/7.705889

Spall, J. (2003). *Introduction to Stochastic Search and Optimization: Estimation, Simulation, and Control*. Hoboken, NJ: John Wiley & Sons, Inc. doi:10.1002/0471722138

UDT. (2017a). *A List of Projects/Software/Companies that use UDT*. Retrieved September 1, 2017, from http://udt.sourceforge.net/poweredby.html

UDT. (2017b). *UDT: Breaking the Data Transfer Bottleneck*. Retrieved August 28, 2017, from http://udt.sourceforge.net/

Wada, T., & Fujisaki, Y. (2013). A Stopping Rule for Simultaneous Perturbation Stochastic Approximation. In *Proceedings of the European Control Conference* (pp. 644-649). Zurich, Switzerland: Academic Press.

Wang, Q., & Spall, J. (2011). Discrete Simultaneous Perturbation Stochastic Approximation on Loss Function with Noisy Measurements. In *Proceedings of the American Control Conference* (pp. 4520-4525). San Francisco, CA: Academic Press. 10.1109/ACC.2011.5991407

Wang, Q., & Spall, J. (2013). Rate of Convergence Analysis of Discrete Simultaneous Perturbation Stochastic Approximation Algorithm. In *Proceedings of the American Control Conference* (pp. 4771-4776). Washington, DC: Academic Press. 10.1109/ACC.2013.6580576

XDD. (2017). *XDD: The eXtreme dd toolset*. Retrieved August 28, 2017, from https://github.com/bws/xdd

Yun, D., & Wu, C. (2015). An Integrated Transport Solution to Big Data Movement in High-performance Networks. In *Proceedings of the IEEE International Conference on Network Protocols* (pp. 460-462). San Francisco, CA: IEEE. 10.1109/ICNP.2015.55

Yun, D., Wu, C., Rao, N., Liu, Q., Kettimuthu, R., & Jung, E.-S. (2017). Data Transfer Advisor with Transport Profiling Optimization. In *Proceedings of the IEEE Conference on Local Computer Networks* (pp. 269-277). Singapore: IEEE. doi: 10.1109/LCN.2017.23

Yun, D., Wu, C., Rao, N., Settlemyer, B., Lothian, J., Kettimuthu, R., & Vishwanath, V. (2015). Profiling Transport Performance for Big Data Transfer over Dedicated Channels. In *Proceedings of the International Conference on Computing, Networking and Communications* (pp. 858-862). Anaheim, CA: Academic Press. 10.1109/ICCNC.2015.7069458

KEY TERMS AND DEFINITIONS

Big Data Transfer: The transfer of datasets that are typically on the order of hundreds of gigabytes, terabytes, or larger.

Dedicated Connection: A network connection that is reserved in advance and established using tunneling technologies such as MPLS to support end-to-end big data transfer in high-performance networks (HPN).

High-Performance Networks: A high-speed computer network with the capability of advanced bandwidth reservation.

SPSA: Simultaneous perturbation stochastic approximation, an efficient stochastic algorithm for solving optimization problems in complex dynamic systems with multiple unknown parameters.

TPG: Transport profile generator, an iperf-like network toolkit for end-to-end data transfer performance test using UDT protocol.

Transport Profile: A transport profile is a control-response plot illustrating how a set of control parameters affect the performance of a given transport protocol/method over a network connection or link between a given sender host and a given receiver host. Such a transport profile indicates the qualitative behavior of each component involved in the data transfer process and provides an insight into maximizing the overall transport performance.

Transport Profiling: The process of obtaining a transport profile for a given data transfer protocol/method over a given network connection.

UDT: UDP-based data transfer protocol, a reliable high-speed data transfer protocol that is built upon UDP and is typically used to carry out data transfer over wide-area high-speed network connections.

Chapter 6
Stochastic Methods for Estimation and Problem Solving in Engineering:
Stochastic Methods of Decision Making in Aviation

Tetiana Shmelova
National Aviation University, Ukraine

ABSTRACT

In this chapter, the author presents stochastic methods in aviation. The stochastic methods are presented as methods of decision making (DM) of operators of air navigation systems (ANS) in risk and uncertainly. The ANS is presented as a socio-technical system (STS). Analysis influences the factors of professional and non-professional activities on DM of STS's operators. The author made an analysis of the International Civil Aviation Organization (ICAO) documents on risk assessment. To determine the quantitative characteristics of risk levels, models for DM by the operator of the aviation system under risk and uncertainty have been developed. The author demonstrates some interesting advantages offered by the new methodology of DM in STS and forecasting the behavior of the operator in an emergency situation (ES) for using models of DM in risk and uncertainty.

SAFETY REQUIREMENTS IN AVIATION

In aviation, significant attention is paid to Safety Management. Safety is the state in which the risk of harm to persons or of property damage is reduced to and maintained at or below, an acceptable level through a continuing process of hazard identification and risk management (International Civil Aviation Organization [ICAO], 2013). In determining an acceptable level of safety, it is necessary to consider such factors as the level of risk that applies the cost/benefits of improvements to the system, and public expectations on the safety of the aviation industry. To determine an acceptable level of risk and balance

DOI: 10.4018/978-1-5225-5045-7.ch006

between flight performance and safety requirements, it is necessary to have quantitative characteristics of DM under risk conditions.

Safety is a dynamic characteristic of aviation with the help of which risk factors for flight safety should steadily decrease. It is important to note that adoption of efficiency indices of ensuring flight safety is frequently influenced by internal and international standards and also by cultural features (ICAO, 2004). While risk factors for flight safety and operating errors are under control, such opened and dynamic system as civil aviation may be controlled providing the necessary balance between flight performance and safety requirements for passengers and their baggage. Aviation systems cannot be wholly free from dangerous factors and connected with them risks, while, the elimination of aviation events and serious incidents continues to be the final goal of human activity in the sphere of aviation safety. Neither human activity nor systems created by it guarantee a total absence of operating errors and their consequence (ICAO, 2013).

International civil aviation organization constantly develops and improves proactive, based on the risks evaluation, methods, directed on the further decrease in a number of aviation events in the world. Also, ICAO encourages aviation communities to recognize the importance of adherence to the single global approach for safety improvement and monitoring (ICAO, 2013). A modern approach, founded on the characteristics (performance-based approach – PBA), based on the next three principles monitoring (ICAO, 2009):

- The main accent on desired/necessary results;
- Decision making, oriented on desired/necessary results;
- Using facts and data while DM.

Herein the principle "using facts and data while decision making" admits that tasks shall comply with the widely known in Western management criteria SMART (*specific, measurable, achievable, relevant and timebound*) (ICAO, 2009). Such level of accuracy of tasks determination may be achieved only using the way of consistent and structural description of inhomogeneous components of aviation - aviation enterprises, aviation personnel, aviation infrastructure, techniques, rules and information that is used for creating of conditions and applying of aerospace by aircraft users.

In such a way, safety is a dynamic characteristic of aviation with the help of which risk factors for flight safety should steadily decrease. It is important to note that adoption of efficiency indices of ensuring flight safety is frequently influenced by internal and international standards and also by cultural features (ICAO, 2004). While risk factors for flight safety and operating errors are under control, such opened and dynamic system as civil aviation may be controlled providing the necessary balance between flight performance and safety requirements for passengers and their baggage. By taking a systematic approach to corporate decision-making and risk management, accidental losses are reduced (ICAO, 2013).

Nowadays a new approach for flight safety provision is forming in global practice. The ICAO's term «Safety Management System» (SMS) integrates operations and technical systems with the management of financial and human resources to ensure aviation safety or the safety of the public. Given statement implies binding into one, in a single system different objects and subjects of aviation activity. The subjects of aviation activity are designers and manufacturers of aviation equipment, airline operators, and organizations engaged in aircraft maintenance and repair, and also handling companies. The air traffic services providers, aviation educational institutions, research and design aviation organizations, aviation authorities are included to them. Each of subjects of aviation activity is used to provide manufacturing processes and functioning of civil aviation the appropriate objects: aircraft, their components and equipment, industrial equipment, ground vehicles, aerodrome and on-route equipment (radio- and electrotechnical facilities), engineering technical buildings, other movable and immovable assets, etc. And it is a very complex task to join the named elements of aviation. Since the period of «organizational era» (the 1990s – present), flight safety has been considered systematically, taking into account organizational, human and technical factors. Also at that date, a notion «aviation incident because of organizational reasons» appeared in aviation, taking into account impact of organizational culture and politics on the control systems efficiency for flight safety risk factors. The «Swiss Cheese Model» by James T. Reason shows that significant role in the incidents causation belongs to organizational and management factors (ICAO, 2013).

In Safety Management Manual was written safety requirements in aviation for estimation of "*Risk*" (ICAO, 2013). *Risks* are often expressed as probabilities; however, the concept of risk involves more than probabilities. To illustrate this with a hypothetical example, let us assume that the probability of the supporting cable of a 100-passenger cable car failing and allowing the cable car to fall was assessed as being the same as the probability of a 12-passenger elevator failing and allowing the elevator to fall. While the probabilities of the events occurring may be the same, the potential consequences of the cable car accident are much more severe. The risk is therefore two-dimensional. Evaluation of the acceptability of a given risk associated with a particular hazard must always take into account both the likelihood of occurrence of the hazard and the severity of its potential consequences. Therefore, it is proposed to assess the Risk taking into account the probabilities, outcomes, risk levels at each DM's stage (Kharchenko, Shmelova & Sikirda, 2012; 2016).

Air Navigation System as a Socio-Technical System

Statistical data show that human errors account for up 80% of all causes of aviation accidents (Leychenko, Malishevskiy & Mikhalic, 2006). The existing approaches to checking separate aspects (psychophysiological, behavioral, ergonomic, professional, etc.) do not give the proper consideration to the functional state of H-O in the conditions of the dynamic change of external and internal factors. The environmental conditions determine the reaction of an H-O, while the reaction of the latter, in its turn, changes the environmental conditions themselves. Representation of the ANS in the form of a socio-technical system (STS) first of all makes possible to take into account the influence of social, cultural environment of people who make decisions. Culture surrounds people and affects their values, convictions, and behavior, which they share along with other members of different social groups. Culture serves to bind us together as members of groups and to provide clues as to how to behave in both normal and unusual situations. The psychologist Hofstede suggests that culture is a "collective programming of the mind" (ICAO, 2002). Thus fatal mistakes can be committed by normal, healthy, highly motivated and well-equipped personnel. Russian scientists have used lately the term "departure of conscience" when they analyze the causes of aviation events conditioned by the insufficient development of the appropriate cultural values in a person that makes decisions (Leychenko, Malishevskiy & Mikhalic, 2006). One of the possible approaches to the solution of these problems is formalization and mathematical presentation of the ANS operators' activities in the form of a complex STS on the base of the systemic analysis. Taking into account in the act of DM by a human-operator (H-O) within ANS, besides the separate professional factors (knowledge, habits, skills, experience) also the factors of non-professional nature (individual psychological, psycho-physiological and socio-psychological) (Kharchenko, Shmelova &

Sikirda, 2011, 2012, 2016; Shmelova, & Sikirda, 2012, 2013a, 2013b, 2015, 2016, 2017) enables to predict the H-O's actions on the basis of modelling the foresight "large-scale" outcomes of individual actions with the aid of the reflexive theory (Lefebvre, 2001, 2008; Lefebvre, & Adams-Webber, 2002).

Currently, one of the main strategic problems of mankind on the path to sustainable development is the safety and stability of technogeneous production. Technogeneous production is a complex system that contains interrelated technical, economic and social objects. It has a multilevel hierarchical structure and a high level of risk (Keating, 2001). Recent results show that there are frequent and common emergency such as disaster, accidents, crashes in hydraulic engineering, chemical and military industries, gas and oil pipelines, nuclear power plants and transport (Keating, 2001; Bertsch, 2007). Aviation systems with its complex interrelation between a man and technologies have been evolved towards complex STS. The interfaces between people and the technologies that comprise these systems are highly interactive, interdependent and affected by similar environmental events. The STS also tend to have two common features: high technologies and high-risk activities. As such, they require much less direct operation due to the fact the technology replaces the human operator. On the other hand, require much more remote operator's supervision due to the modern tendency to supervise the technology by distance. The systems' work is not transparent due to increased difficulty to know exactly what the technology is being used. The systems are also highly hazardous and of high-risk, and have greater potential for catastrophic consequences (i.e. accidents) (ICAO, 2004).

As a result of the previous studies the factors which affect the DM by an H-O ANS have been determined, namely: level of knowledge, skills, abilities, preceding experience as well as the factors of non-professional nature (psycho-physiological, individual-psychological, social-psychological factors). The systemic analysis which has been carried out as well as the formalization of the factors which affect DM by H-O (individual-psychological, psycho-physiological and social-psychological) in the conditions of the progress of a flight situation from normal to catastrophic obtained (Kharchenko, Shmelova & Sikirda, 2011, 2012, 2016; Shmelova, & Sikirda, 2012, 2013a, 2013b, 2015, 2016, 2017):

- The models of preferences by an H-O under the influence of social-psychological factors;
- The models of preferences by a human-operator depending on the significance of individual-psychological factors in the conditions of development flight situations from normal to catastrophic ones;
- The models of diagnostics of psycho-physiological factors at the score of monitoring the emotional state of H-O.

To determine the quantitative characteristics of risk levels, models for DM by the operator of the aviation system under in risk and uncertainty have been developed. For the formalization of the behavioral activity of H-O ANS in flight situations, those models seem to be suitable which present the process of appearance of separate preconditions and their development into the causal chain of events in the form of proper diagrams of causal-consequential relations. Nowadays the most widely spread are the diagrams in the form of different graphs (or current states and transitions), trees of events as well as functional networks of stochastic structure.

The purposes of research's are: decomposition of the process of DM by H-O ANS, systemic analysis, and formalization of the influence of the factors on the DM within ANS treated as complex STS;

working-out of stochastic models DM by H-O in socio-technical ANS (DM under Risk and DM under Uncertainty); calculation the scenarios of the flight situation development using GERT's and Markov's networks.

Decision Making by Human-Operator in Flight Emergencies Under Risk

The environmental conditions determine the reaction of H-O, while the reaction of the latter, in its turn, changes the environmental conditions themselves. The systemic analysis which has been carried out as well as the formalization of the factors which affect DM by H-O (individual-psychological, psycho-physiological and social-psychological) in the conditions of the progress of a flight situation from normal to catastrophic, models in this situations as the models of DM in Risk and Uncertainly, the models of preferences by an H-O under the influence of social-psychological, individual-psychological factors and the models of diagnostics of psycho-physiological factors (Kharchenko, Shmelova & Sikirda, 2011, 2012, 2016; Shmelova, & Sikirda, 2012, 2013a, 2013b, 2015, 2016, 2017).

Regarding the Air Navigation System as a complex STS the research based on the methodology of analysis of H-O's DM has been carried out. The impact of individual-psychological and socio-psychological factors on the professional activities of H-O during the flight situation development from normal to catastrophic has been studied. On the basis of the reflexive theory of bipolar choice, the expected risks of DM of the ANS's operator have been studied and the influence of external environment, previous experience and intention of the H-O has been identified. On the basis of the methods for analysis of DM by the H-O of ANS using the graph, stochastic, GERT, Markov's networks, reflexive theory the methodology for analysis of flight situation development under influence DM by H-O ANS in flight emergencies (FE) has been developed (Table 1).

The structural analysis of developing FE and DM by aircraft (AC) crew and air traffic controller (ATC) in FE with the aid of decision tree enabled to obtain such results: graphical-analytical models of FE development and DM by a H-O (controller, pilot) in FE; stochastic models type GERT network (Graphical Evaluation and Review Technique), decision trees and Markov chains; reflexive models of bipolar choice in FE under the influence of external environment, previous experience and intentional choice by H-O:

- The deterministic models of DM in Certainty;
- The stochastic models of DM in Risk;
- The stochastic models of DM in Uncertainty.

Decision making by H-O in FE under certainty for using the network analysis of the actions of an AC crew and an ATC in FE with the aid of the network planning methods gave a chance to obtain:

- The structural-time table of the actions taken by H-O (controller, pilot) in FE;
- The network graph of taking the actions by an H-O (controller, pilot) in the FE;
- The critical time of taking the actions by an H-O (controller, pilot) in the FE.

For example, determined models for an H-O (controller) were obtained in accordance with the adopted technologies of controller's work ASSIST (Acknowledge, Separate, Silence, Inform, Support, Time) are presented on Figure 1, which were obtained in accordance with the adopted technologies of controller's

Table 1. Methodology for analysis of flight situation development under influence DM by H-O ANS in FE

Nº	Phase of Analysis	Models	Process	Result	Parameters
1	Modeling of DM by H-O in FE under Certainty	Determined models for an H-O (controller, pilot) with deterministic run-time procedure	Technology of work of H-O (controller, pilot) in FE Algorithm of work of H-O (controller, pilot) in FE	Structural-hourly table of the actions taken by H-O (controller, pilot) in FE	t_i T_{cr}
		Determined models for an H-O (controller, pilot) with probabilistic run-time procedure	Determination of time t_i (t_i'), required for the performance of *i*-procedure according to the algorithm of work of H-O in FE	Network graph of taking the actions by H-O (controller, pilot) in the FE	$t_{i'}$; T_{cr} T_{mid} T_{min} T_{max}
2	Modeling of DM by H-O in FE under Risk	Stochastic models type Decision Trees of DM by H-O in FE	The structural analysis of DM by H-O (controller, pilot) in FE	The optimal solution is found by the criterion of an expected value with the principle of risk	A_{dopt}
3	Modeling of DM by H-O in FE under Fuzzy logic	Fuzzy logic to determine quantitative estimates of potential loss	Determination of membership functions in fuzzy logic	Potential loss value in consumption of fuel, incident, breakage, accident, catastrophe	$\{g_r\}$
4	Modeling of DM by H-O in FE in Neural Network	Neural Network Model to determine potential alternative of the flight completion	Determination of weight coefficients of neural network and effectiveness of flight completion	Effectiveness of flight completion at the aerodrome (landing field). Weight coefficients of neural network	Y_G Y_{Gaer} Y_{Glf} W
5	Modeling of DM by H-O in FE under Uncertainty	Matrix of the possible results of the DM by H-O in FE	The analysis of DM by H-O (controller, pilot) in FE using Minimax, Laplace, Savage and Hurwicz criterions	The optimal solution is found by the criterion of Minimax, Laplace, Savage and Hurwicz	A_{dopt}
6	Modeling of FE developing	Stochastic models type GERT's network (Graphical Evaluation and Review Technique) of DM by H-O (controller, pilot) in FE and FE developing	Analysis of the flight situation development from normal to catastrophic and conversely	Mathematical expectation of flight situation development time t_{ij}; variance of flight situation development time t_{ij}; probability of flight situation development p_{ij}	t_{ij} $M[t_{ij}]$ δ^2 $[t_{ij}]$ $p_{ij'}$; p_{ji} p_{ii}
7		Stochastic models type Markov Chains of DM by H-O (controller, pilot) in FE	Analysis of the flight situation development from normal to catastrophic and conversely under choice towards the positive or negative pole	Transmission coefficient of (*i*, *j*)-arc in positive choice Transmission coefficient of (*i*, *j*)-arc in negative choice Value of risk during transition between flight situations. Probabilities of states of flight situation	W_{ij} (*A*) W_{ij} (*B*) R_{ij} R_{ji} $p_{tij}(A)$ $p_{tij}(B)$
8		Reflexive models of bipolar choice in FE under the influence of external environment, previous experience and intentional choice by H-O	Modeling of behavioral activity of H-O in FE under the influence of external environment, previous experience and intention in impact of individual-psychological and socio-psychological factors on the professional activities of H-O during the flight situation development	Expected risk of the DM by H-O with taking into account the criterion of the expected value minimization (choice towards the positive pole A). Expected risk of the DM by H-O with taking into account his model of preferences (choice towards the negative pole B).Mixed choice made by H-O. Concept of a rational individual's behaviour. System of the individual's preferences	R_A R_B R_{AB} γ ρ
9		Graphical-analytical models of FE development	Graphical-analytical modeling of flight situation development with taking into account individual qualities of H-O	The optimal action of H-O in FE	Y_G
10		Graphical-analytical models of DM by H-O (controller, pilot) in FE	Graphical-analytical modeling of H-O DM in FE under influence socio-factors	Scenarios of the flight situation development	S

Figure 1. The network graph of carrying out the actions by an air traffic controller in the ANS

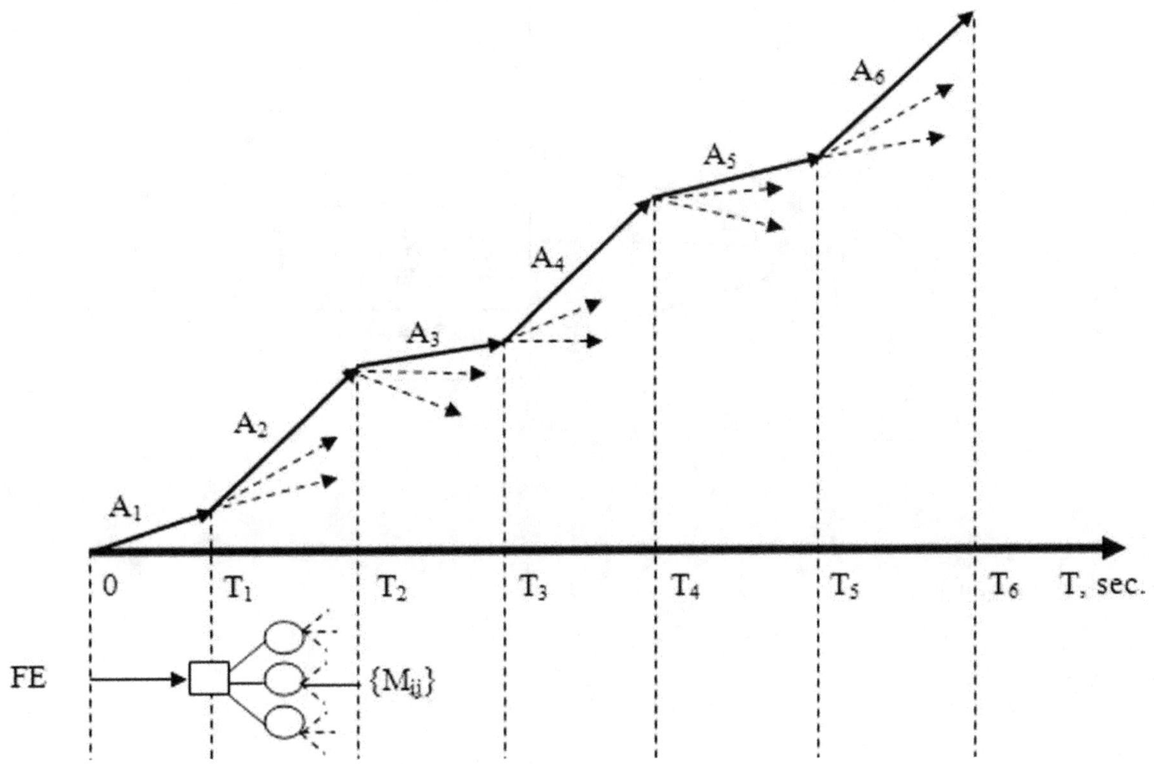

work in the FE (Kharchenko, Shmelova & Sikirda, 2012; 2016). This is a site of SKYbrary Aviation Safety for reference education aviation safety knowledge.

The obtained critical time for performing the operations by a controller in the FE namely: engine failure on take-off, depressurization of flying vehicle, hydraulic system faults, failure of the electric power supply system, etc. as well as the critical time of the crew actions in case of an engine failure on take-off and approach to land in the adverse meteorological conditions has been obtained. In the real situation, the influence of the external environment is taken with using Stochastic DM models have been developed.

DM under risk R - is known the law of the probability distribution of the random variable, such as the task in Risk. So, Risk - is the expected loss or profit, the probability of the event. Risk can be defined in a variety of different ways. In the ISO definition of risk is the «effect of uncertainty on objectives». In this definition, uncertainties include events (which may or may not happen) and uncertainties caused by ambiguity or a lack of information (ISO, 2009; ISO, 2002). It also includes both negative and positive impacts on objectives. In Safety Management Manual - *safety risk* - estimated probability and severity of consequences or outcomes danger. In Pilot's Handbook of Aeronautical Knowledge (Pilot's Handbook of Aeronautical Knowledge, FAA, 2008) qualitative risk estimation for using risk matrix for almost any operation by assigning likelihood and consequence. As one can see, this falls in the high risk area matrix of DM. Therefore, it is proposed to assess the Risk taking into account the probabilities ($P_j =\{ P_{1j}, P_{2j}, \dots P_{ij}, \dots, P_{mj} \}$), outcomes (Uj of event, expected cost (or profit) $U_j = \{ U_{1j}, U_{2j}, \dots U_{ij}, \dots, U_{mj}\}$ - outcomes of alternative actions $A = \{A_1, A_2, \dots A_i, \dots, A_m\}$ and probability p_j of the event *i*), risk levels at each decision-making stage fuzzy-logic ($R = \mu(x)$) too etc. (Figure 2).

Figure 2. Quantitative risk estimation

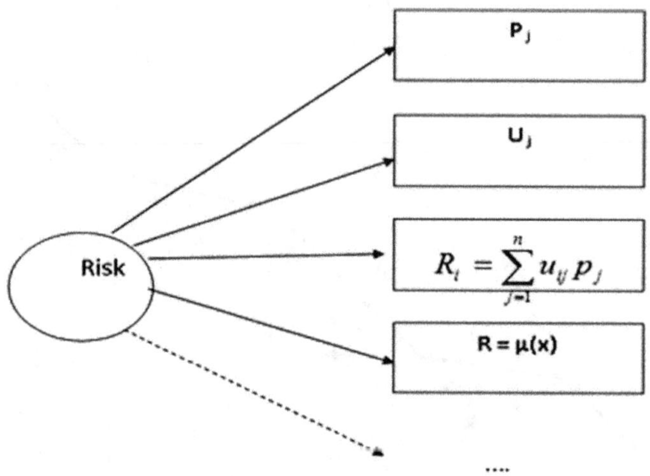

Risk R defined as:

$$R_i = \sum_{j=1}^{n} u_{ij} p_j \tag{1}$$

p_j – probability *j*;

u_{ij} - expected cost (or profit) - outcomes of alternative actions $A = \{A_1, A_2, \ldots A_i, \ldots, A_m\}$

The main method of DM in risk is analysis and solution using the graph in the form of a decision tree. For probabilistic data, decision trees comparing the expected cost (or profit) for the different alternatives $A = \{A_1, A_2, \ldots A_i, \ldots, A_m\}$ are the basis for reaching a decision. Decision tree - representation of the process of DM using GRAPH theory. GRAPH - a graphical representation of a mathematical model. Decision tree - a graph that has one root. Decision tree has tops - tops - solutions (square nodes) and tops - casual tops (round nodes) - probability of expected events (Figure 3)

Decision-making paradigm (steps):

1. Structural analysis of the situation - definition stages solutions.
2. Definition of alternatives at each stage.
3. Definition of probabilities for each alternative.
4. Definition of outcomes for each alternative.
5. Building a decision tree.

Calculation of the optimal solution using the criterion of Expected value and the Dynamic programming method

The structural analysis of developing FE and DM by AC crew and ATC in FE with the aid of decision tree enabled to obtain such results:

Figure 3 Structure of Decision tree; tops - solutions (square nodes) – 1, 4, 5 tops, casual tops - probability (round nodes) – 2, 3, 6

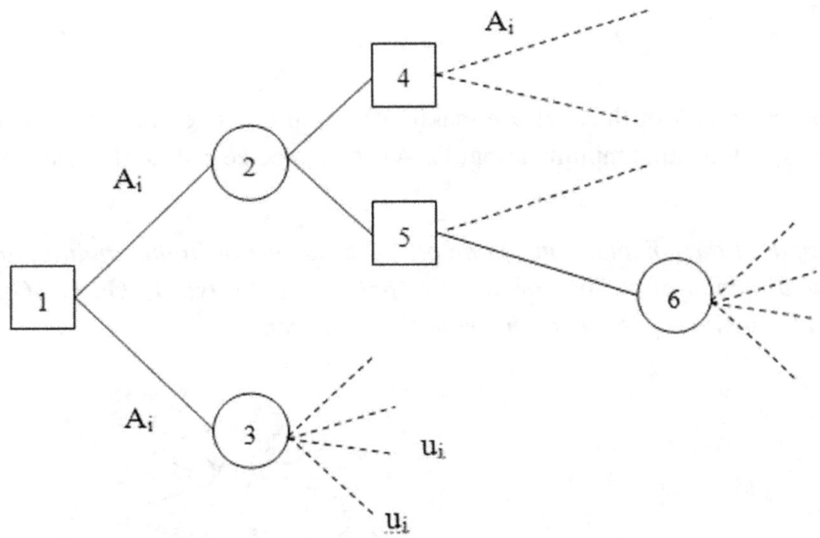

- Graphical-analytical models of FE development and DM by an H-O (controller, pilot) in FE;
- Stochastic models type GERT network (Graphical Evaluation and Review Technique), decision trees and Markov chains;
- Reflexive models of bipolar choice in FE under the influence of external environment, previous experience and intentional choice by H-O.

With the aid of the bipolar reflexive model of the behavioral activity of H-O in the extreme situations (Lefebvre, 2001, 2008; Lefebvre, & Adams-Webber, 2002), W-functions of the positive and negative choice have been obtained. The model represents the subject (H-O) who is on the verge of choosing one of the alternatives: A (positive pole) and B (negative pole) is described by the function:

$$X = f(x_1, x_2, x_3), \tag{2}$$

where X – is a probability with which the H-O is ready to choose the positive pole A in the reality; x_1 – is a pressure of the external environment on the H-O in the direction of the positive alternative at the moment of the choice, $x_1 \in [0, 1]$; x_2 – is a pressure of the previous H-O's experience in the direction of the positive alternative at the moment of the choice, $x_2 \in [0, 1]$; x_3 – is an intentional choice (intention) of H-O in the direction of the positive alternative at the moment of the choice, $x_3 \in [0, 1]$.

The expected risks R_A, R_B of making a decision in the ANS under the influence of the external environment x_1, the previous H-O's experience x_2 and the intentional choice of an H-O x_3 have been obtained. The expected risk in the process of DM of a H-O is equal:

$$R_{DM} = \begin{cases} R_A = \min\left\{R_{ij}\right\} \\ R_B = \left\{\gamma, \rho\right\} \\ R_{AB} = \left\{X(x_1, x_2, x_3), \gamma, \rho\right\} \end{cases}, \tag{3}$$

where R_A – is an expected risk of the decision making for a human-operator with taking into account the criterion of the expected value minimization; R_B – is an expected risk of the decision making for a

Figure 4. The example of the computation of one of the scenarios of flight situation development: A, B – selection in the direction of positive or negative pole respectively; G_1, G_2, G_3, G_4, G_5 – normal, complicated, difficult, emergency, catastrophic situations respectively.

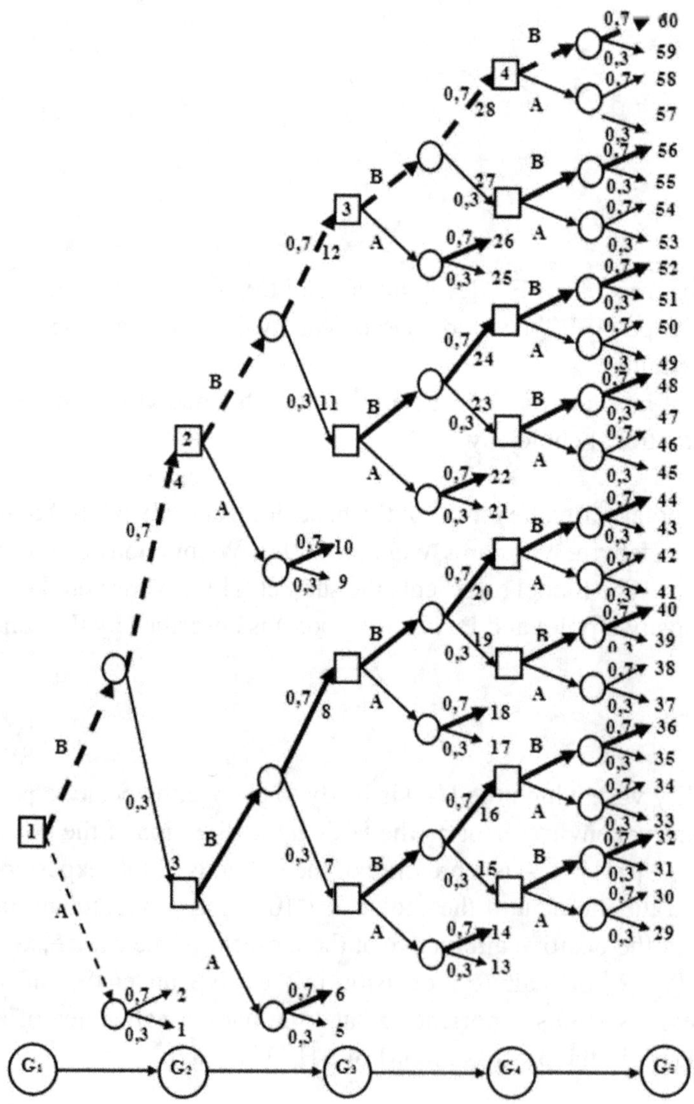

human-operator with taking into account his model of preferences; R_{ij} – is an expected risk for making A_{ij}-decision; γ – is a concept of a rational individual's behavior; ρ – is a system of an individual's preferences in a concrete situation of the choice; R_{AB} – is a mixed choice made by a H-O.

The computation of one of the scenarios of the flight situation development is presented in Figure 4 (for example – approach performed in bad weather conditions).

The results of the computation of the expected risks in the course of the transition between flight situations and the criterion of the expected value by means of the dynamic programming method are presented in Table 2.

The example of the computation of the expected risks in the course of the transition between flight situations is presented in Figure 5. The selection in the direction of the negative pole in compliance with the $S_{1\text{-}2\text{-}3\text{-}4B}$ scenario leads to the maximum expected risk R=1028 conventional units (c.u.). The choice in the direction of the positive pole when the FE occurs at the first stage of DM by H-O ANS (for example, a flight to a reserve aerodrome in the difficult meteorological conditions) has a risk which is 60,5 times lesser: R=17 c.u.

The next purpose - to calculate the scenarios of the flight situation development under influence socio-factors. The value of risk during a transition between normal, complicated, difficult, emergency

Table 2. The results of computation of the scenarios of flight situations development

Scenarios, S	Probabilities, p	Consequences, U	Expected risks, R, c.u.
S_{4B}	0,7	60	597
	0,3	59	
$S_{3\text{-}4B}$	0,7	28	874
	0,3	27	
$S_{2\text{-}3\text{-}4B}$	0,7	12	991
	0,3	11	
$S_{1\text{-}2\text{-}3\text{-}4B}$	0,7	4	1028
	0,3	3	
S_{1A}	0,7	2	17
	0,3	1	

Figure 5. Markov Chains of the development of flight situations: G_1, G_2, G_3, G_4, G_5 – normal, complicated, difficult, emergency, catastrophic situations respectively; R_{ij} – value of risk during transition between flight situations

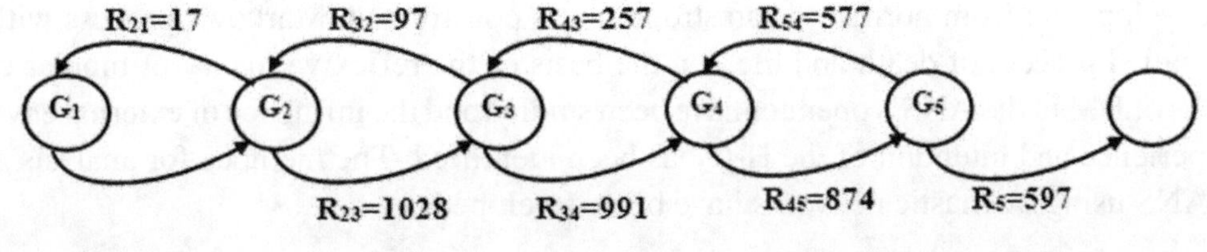

Figure 6. Stochastic models type GERT's network of DM by H-O (controller, pilot) in flight emergencies and developing flight emergencies: G_1 – normal situation; G_2 – complicated situation; G_3 – difficult situation; G_4 – emergency situation; G_5 – catastrophic situation; W_{ij} – W-function, transmission coefficient of (i, j)-arc

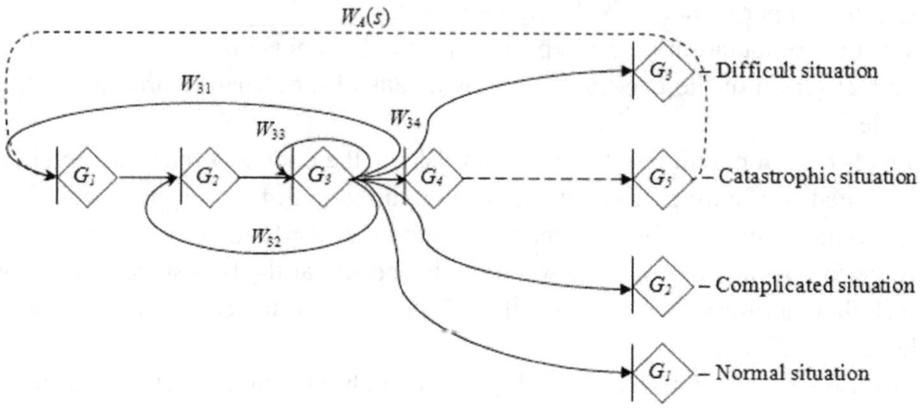

and catastrophic flight situations respectively has determined. The computation of one of the scenarios of the difficult situation development using GERT's network is presented in Figure 6.

Thus according to results of stochastic network analysis of the flight situation development from normal to catastrophic the following values have been obtained:

- Mathematical expectation of flight situation development time t_{ij} – $M[t_{ij}]$;
- The variance of flight situation development time t_{ij} – δ^2 [t_{ij}];
- The probability of flight situation development $p_{ij} - p_{ij}, p_{ji}, p_{ii}$.

In stochastic networks of the flight situation development of GERT type the tops are represented by stages of the situation (normal, complicated, difficult, emergency or catastrophic), and the arcs are represented by a process of transition between stages of the situation. The algorithm of stochastic network analysis was developed (Kharchenko, Shmelova & Sikirda, 2012; 2016). Thus according to results of stochastic network analysis of the flight situation development from normal to catastrophic the following values have been obtained: the mathematical expectation of flight situation's development time tij; variance of flight situation's development time tij; the probability of flight situation's development pij. For example, let's analyze catastrophic situation development under hazardous weather conditions using the decision tree and stochastic network GERT (Figure 7).

Based on the *W*-functions of positive and negative of H-O choice the Markov's network of flight situations' development from normal to catastrophic was constructed. Markov's process with discrete states W_{ij} is called process of death and life. On the basis of the reflexive theory of bipolar choice the expected risks of DM of the ANS's operator have been studied and the influence of external environment, previous experience and intention of the H-O has been identified. The methods for analysis of DM by the H-O in ANS using stochastic networks have been developed.

Figure 7. The example of the computation of one of the scenarios of flight situation development: A, B – selection in the direction of positive or negative pole respectively; G1, G2, G3, G4 and G5 – normal, complicated, difficult, emergency, catastrophic situations respectively.

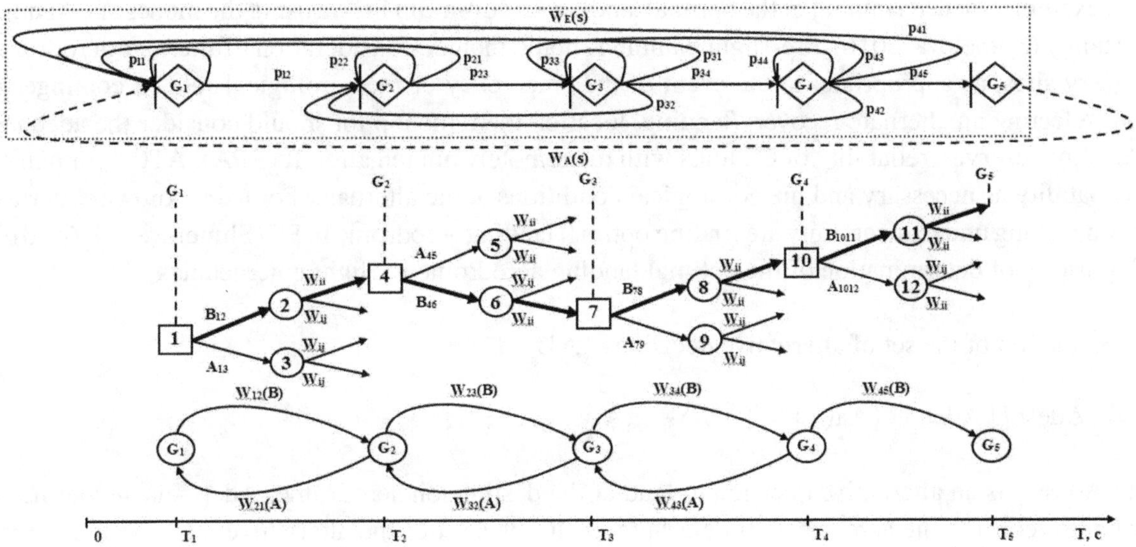

Decision Making by Human-Operator in Flight Emergencies Under Uncertainty

When we don't know the law of the probability distribution of the random variable, we have DM in uncertainty. The analysis of FE developing and DM of an AC crew and air traffic controller has been made with the aid of Minimax, Laplace, Savage and Hurwicz criterions which enabled to obtain the DM by H-O models in FE under uncertainty. The matrix of the possible results of the H-O's decision-making in the ANS in the flight emergencies are presented in Table 3. The consequences of the flight situation development uij have been obtained in conformity to the reflexive theory. It is advisable to employ the models of the development of flight situations as components of the Decision Making Support System which enables a H-O ANS to evaluate numerically the possible versions of FE developing and choose timely the strategy of behaviour with the minimum level of the potential loss in the conditions of the insufficient and uncertain available information.

For optimization the solution of problems are developed models of determination the optimal landing site in case of an extraordinary situation, search for optimal flight routes (Unmanned Aerial Vehicles) UAV's with the module «ASSSIST» (Acknowledge, Separate, Synergetic ((Coordinated, Cooperation,

Table 3. Matrix of the possible results of the decision making by H-O ANS in FE

Alternative Decisions	Factors		
Positive/negative pole	Pressure of the external environment on H-O	Pressure of the previous experience of H-O	H-O's intentional choice (intention)
A – choice in the direction of the positive pole	u_{11}	u_{12}	u_{13}
B – choice in the direction of the negative pole	u_{21}	u_{22}	u_{23}

Consolidation)). The investigation into the processes of modeling the DM by UAV's operator in the normal and unusual situations enabled to build the following models: DM under Certainty, DM under Risk and DM under Uncertainty.

For example, for determining of the optimal landing aerodrome in FE we using the model of DM under Uncertainty (Shmelova, 2016). Pre-flight planning should include consideration to alternate aerodromes / recovery sites, as appropriate, in the event of the emergency or meteorological-related contingency. Before selecting an alternate recovery/landing, location the remote pilot should consider the adequacy fuel/energy, reserves, reliability of C2 links with the remotely piloted aircraft (RPA), ATC communication capability as necessary and meteorological conditions at the alternate. For using known criteria of decision making under uncertainty are finding optimal landing aerodrome in FE (Shmelova, 2016; 2017).

Algorithm of determination of the optimal landing aerodrome in flight emergencies:

1. Formation of the set of alternative decisions {A}:

$\{A\} = \{Adest\ U\ Adep\ U\ \{Aalt\}\} = \{A_1, A_2, \ldots A_i, \ldots, A_n\}$,

where Adest – is an alternative decision to land at the destination aerodrome; Adep – is an alternative decision to return to the aerodrome of departure; Aalt – is a set of the alternative decision to alternate aerodromes / recovery sites.

2. Formation of the set of factors {λ}, influencing the choice of alternate aerodromes / recovery sites:

$\{\lambda\} = \lambda_1, \lambda_2, \ldots, \lambda_j, \ldots, \lambda_m$,

where $\lambda_1, \lambda_2, \ldots, \lambda_j$ is set of factors (fuel level, remoteness of the aerodromes, technical characteristics of runways of destination aerodromes, meteorological conditions, a reliability of C2 links with the RPA, etc.).

3. Formation of the set of possible results of DM under the influence of specified factors in FE, that were determined by the method of expert estimates by rating scales according to the regulations:

Table 4. Matrix of decisions in the task «choosing of the optimal landing aerodrome/recovery site

Alternative Decisions		**Factors Influencing the DM {λ}**					
{A}		Adequacy fuel /energy on RPA	Remoteness of the aerodromes/ recovery sites	Technical characteristics of runways of aerodromes/recovery sites	Meteorological conditions at aerodromes/recovery sites	Reliability of C2 links with the RPA	ATC communication capability
A_1	A_{dest}	u_{11}	u_{12}	…	u_{1j}	…	u_{1n}
A_2	A_{dep}	u_{21}	u_{22}	…	u_{2j}	…	u_{2n}
A_1	A_{alt}	u_{11}	u_{12}	…	u_{1j}	…	u_{1n}
A_n	A_{alt}	u_{m1}	u_{m2}	…	u_{mj}	…	u_{mn}

$\{U\} = U_{11}, U_{12}, \ldots, U_{ij}, \ldots, U_{nm}.$

4. Formation of the decision matrix

Table 4 summarises the decision matrix - alternative decisions, factors and results of DM under the influence of factors.

Choosing of the optimal aerodrome, in the case of forced landing is carried out by the methods of DM under uncertainty. Selection of criterion of DM under uncertainty (Wald, Laplace, Hurwitz, Savage) is conducted according to the type of flight. Wald criterion (min/max) is based on the principle of "conservative attitude", and is applied if it is necessary to find a guaranteed solution in case of primary flight:

$$A^* = \max_{A_i}\left\{\min_{\lambda_j} u_{ij}(A_i, \lambda_j)\right\}. \tag{4}$$

Laplace criterion is based on the principle of "insufficient reason" and applied is in the case of regular flight:

$$A^* = \max_{A_i}\left\{\frac{1}{m}\sum_{j=1}^{n} u_{ij}(A_i, \lambda_j)\right\}. \tag{5}$$

Hurwitz criterion uses the coefficient of a pessimism-optimism α ($0 \leq \alpha \leq 1$) and is applied in the case of flight once in 2 weeks:

$$A^* = \max_{A_i}\left\{\begin{matrix}\alpha \max_{\lambda_j} u_{ij}(A_i, \lambda_j) + \\ +(1-\alpha)\min_{\lambda_j} u_{ij}(A_i, \lambda_j)\end{matrix}\right\}. \tag{6}$$

The optimal solution for the Savage criterion can be found using a matrix of "regret". In the case of win, the elements of the "regret" matrix $r_{ij}(A_i, \lambda_j)$ are defined as the difference between the maximum value uij in the row and other values in the row. Then, with the help of the "regret" matrix according to the min/max principle the minimum deviations are determined:

$$r_{ij}(A_i, \lambda_j) = \underset{A_i}{\Delta} = \max_{\lambda_k} u_{ij}(A_i, \lambda_j) - u_{ij}(A_i, \lambda_j), \tag{7}$$

$$A^* = \min_{\lambda_j}\max_{A_i} r_{ij}(A_i, \lambda_j).$$

Thus the person, who makes a decision, expresses with the help of matrix "regret" if he can't make the best decision in the condition λj. Making this decision the person, who makes a decision, has a guarantee that in the worst conditions the obtained income would be not lower than the found income. If the flight of UAV is a scheduled one, Laplace, and Hurwitz ($0 \leq \alpha \leq 0.5$) criteria are used for decision making. If the flight of UAV is performed for the first time, Wald, Savage, and Hurwitz ($0.5 \leq \alpha \leq 1$) criterions are used for DM.

Applications of Decision Making by Human-Operator in Aviation

Applications of DM by H-O of Unmanned Aerial Vehicles and aircraft.

- **Example 1:** Within the pre-flight planning, alternate aerodromes/places for return procedure should be defined. These aerodromes/places will be used in the cases of an emergency situation or urgent situation that could be caused by inappropriate meteorological conditions. Stages to find optimal aerodrome/place of landing will be considered:
- Re-flight preparation piloted aircraft crew is one of the conditions of flight safety provision. According to regulatory documents, alternate aerodromes are selected by using of following parameters:
- The minimum pilot-in-command level of aircraft;
- Meteorological conditions on alternate aerodrome;
- The amount of fuel on board of aircraft;
- Distance from the alternate aerodrome.

With the aim to optimize the pre-flight preparation the automated systems of pre-flight preparation information and systems of decision making support were developed. The application of the models in these systems depends on the type of flight. An estimation of factors that influence on the selection of optimal landing aerodrome is realized with the help of method of expert estimation, two layers strictly directed network of two layers perceptron. The usage of RPA should be regulated by the same regulatory documents as for ordinary aircraft. One of the tasks that solve RPA pilot in the process of pre-flight planning is a selection of alternate aerodromes/places for return procedure in the case of an emergency situation that is caused by meteorological conditions. During pre-flight preparation adequate fuel/energy capacity should be predicted. This operation is necessary for a possible excursion from a planned place of landing/return. In this case, RPA can fly to the alternate aerodrome/place safely and perform approach and landing. An RPA pilot should include following factors: adequacy of fuel/energy reserve; distance to the alternate aerodrome/place; reliability of C2 lines for connection with RPA; the possibility of communication with ATC units; meteorological conditions on the alternate aerodrome/place. The multifunctional model of selection of alternate aerodrome/place of RPA is proposed for the economical effectiveness of flight realization of RPA which is used in the remote system of decision making support of UAV's operator. The decision of selection task of alternate aerodrome/place in the case of an emergency landing (difficult meteorological conditions, etc) by the method of decision making under uncertainty be means of the criteria of decision making under uncertainty: Wald, Laplace, Savage, Hurwicz (Shmelova, 2016).

Output data:

- A calculated route of direction;
- An aerodrome of departure (ADep) and its characteristics;
- An aerodrome of destination (ADest) and his characteristics;

- A list of alternate aerodromes (AA) according to the calculated route;
- A type of RPA and its tactical and technical characteristics (TTC);
- Current flight situation (difficulties in conditions of flight realization, difficult situation)
- Factors that influence on air traffic;
- Adequacy of fuel/energy reserve;
- Distance to the alternate aerodrome/place;
- Reliability of C2 lines for connection with RPA;
- Possibility of communication with ATC units;
- Meteorological conditions on the alternate aerodrome/place.

A solution of the finding of optimal landing aerodrome/place (alternate aerodrome/place, (AAP)) for return operation in the case of an emergency situation that is caused by meteorological conditions.

Formation of multiplication of alternative decisions $\{A\}$ from ADep, ADest, AAP:

$$\{A\} = \{AADest \text{ U } AADep \text{ U } \{AAAP\}\} = \{A_1, A_2, ...A_i, ..., A_n\},$$

where AADest – alternate decision about landing ADest; AADep - alternate decision about return to ADep; AAAP - multiplication of alternates APP.

Formation of multiplication of factors $\{\lambda\}$, that influence on the selection of AAP in the case of H-O decision making in conditions of forced landing of RPA (emergency landing, etc):

$$\{\lambda\} = \lambda_1, \lambda_2 ..., \lambda_j, ..., \lambda_m,$$

where λ_1 - availability of fuel/energy on board of RPA; λ_2 - distance of RPA from ADep, ADest, AAP; λ_3 - TTC, AAP, ADep, ADest; λ_4 - meteorological conditions on ADep, ADest, AAP; λ_5 – reliability of C2 lines for connection with RPA; λ_6 - possibility of communication with ATC units; λ_7 - subjective factor.

Formation of multiplication of possible consequences $\{U\}$ that influence on the selection of AAP in the case of H-O decision making in conditions of forced landing of RPA (emergency landing, etc):

$$\{U\} = U_{11}, U_{12}, ..., U_{ij}, ..., U_{nm},$$

where U_{ij} - is defined be means of the method of expert estimation according to estimation scale and data from the regulatory documentation.

Formation of decision matrix M = ||Mi|| (Table 5) and selection of criteria of DM under uncertainty in the case of an emergency situation that is caused by meteorological conditions: Wald (min max) criterion – the flight is performed rarely; Laplace criterion – the flight is performed often. The matrix of possible results of DM that relates to the selection of AAP for return operation in the case of an emergency situation or uncertain situation that is caused by meteorological conditions is represented in Table 5. During usage of Wald criterion, each action is estimated from the best to the worth states. This criterion uses estimated function that corresponds to the position of extreme caution. According to Wald criterion an optimal decision is defined for the rule, provides guaranteed result and completely excludes a risk:

$$A^* = \max\min \{u_{ij}\} = \max \{1, 2, 3, 2, 1\} = 3 = A_3$$

Table 5. The matrix of possible results of DM that relates to selection of AAP

Alternate Decisions		**Factors that Influence on DM {λ}**						**Solution**	
{A}		Availability of fuel/ energy onboard of RPA	Distance of RPA from ADep, ADest, AAP	TTC, AAP, ADep, ADest	Meteorological conditions on ADep, ADest, AAP	Reliability of C2 lines for connection with RPA	Possibility of communication with ATC units	Criterion	
		λ1	λ2	λ3	λ4	λ5	λ6	Wald criterion	Laplace criterion
A1	AADest	1	1	9	6	5	7	1	5
A2	AADep	8	9	9	2	8	8	2	**7.3**
A3	AAAP1	5	6	3	3	6	3	**3**	4.3
A4	AAAP2	6	5	2	5	5	2	2	4.7
A5	AAAP3	5	7	7	4	4	1	1	4.6

According to Laplace criterion

$$A_i^* = \max 0_i = \left\{ \frac{1}{n} \sum_{I=1}^{N} u\left(a_i, \lambda_j\right) \right\} = \max \{5,\ 7.3,\ 4.3,\ 4.7,\ 4.6\} = 7.3 = A_2$$

So, pre-flight planning for RPA is the main procedure to choose alternate aerodrome/place of landing in the case of an emergency situation that is caused by meteorological conditions. There are set of factors that should be calculated before the flight to perform safe approach and landing such as:

- Adequacy of fuel/energy reserve;
- Distance to the alternate aerodrome/place;
- Reliability of C2 lines for connection with RPA;
- The possibility of communication with ATC units;
- Meteorological conditions on the alternate aerodrome/place.

The model safe and economical effective landing during an emergency situation that is caused by meteorological conditions. It is possible due to the fact that four criteria were implemented to achieve the reliability of model: Wald, Laplace, Savage, Hurwicz. It is clear from calculations that according to Wald criterion a return to the place of alternative landing is better if flights are performed rarely. If flights are realized often return to the aerodrome of departure is the better solution according to Laplace criterion. The program is a useful and complex tool to find the optimal solution of any problem. Using this computer software you can compute profits and losses of your enterprise but in this project, this software is used for purposes to find an optimal place of landing of RPA. Other words, you will be able to find the best or the worst decision according to the decision of one or a multiple number of persons. Also, you can choose any desired number of assessment criteria. Four different methods of calculations help to optimize the solution.

Example 2. Chose the optimal aerodrome for landing due to bad weather conditions on the approach. Initial data: Route: Kharkiv - Kiev (Borispol). Flight: regular; for the first time; after 2 weeks. Make the optimal choice of landing aerodrome based on classical DM criteria: Vald, Laplace, Hurwicz, Savage.

Alternative actions (Alternative aerodromes) -A = $\{A_1, A_2, \dots A_i, \dots, A_m\}$:

- A_1 - Borispol;
- A_2 - Kharkiv;
- A_3 -Juliani;
- A_4 -Cherkassy
- A_5 – Hostomel

States of nature (Factors) $\lambda = \{\lambda_1, \lambda_2, \dots \lambda_j, \dots, \lambda_n\}$:

- λ_1 - Availability fuel on board.
- λ_2 - Remoteness.
- λ_3 - TC RWY.
- λ_4 - Weather conditions.
- λ_5 - The lighting system approach.
- λ_6 - System approach.
- λ_7 - The navigation tools approach.
- λ_8 - Characteristics of the apron, taxiways.
- λ_9 - Air navigation charges.

Outcomes of payoff matrix uij obtained for using Expert Judgment Method in Table 6.

Wald criterion gives a guaranteed result - the best solution of the worst alternatives (A_3). This criterion is used in cases when a rare decision, the flight is performed the first time. The optimal solution by Laplace determined if you frequently make a decision or if all factors are taken equally. This criterion is used in cases where a decision is often, a regular flight. When using the Laplace criterion for finding optimal solutions - select alternate aerodrome optimal strategy is to choose the alternative A_1 - Borispol airport. The optimal solution by Hurwicz criterion - A_3. Using the criterion of Savage, which minimizes

Table 6. The payoff matrix of possible results of DM

Alternative Aerodromes		**Factors {λ}**									**Solution**			
{A}		λ_1	λ_2	λ_3	λ_4	λ_5	λ_6	λ_7	λ_8	λ_9	Wald criterion	Laplace criterion	Hurwicz criterion α=0,5	Savage criterion
A_1	Borispol	9	10	10	1	10	10	10	10	10	1	**8, 9**	**9**	9
A_2	Kharkiv	3	1	7	8	8	7	8	7	9	1	6, 4	8	8
A_3	Cherkassy	8	9	8	10	6	7	6	7	8	**6**	7, 67	2	**2**
A_4	Juliani	5	7	6	6	7	6	7	7	5	5	6, 2	2	**2**
A_5	Hostomel	6	8	9	4	9	6	9	7	5	4	7	5	5

the loss of decision-makers, it is determined that the optimal choices of alternate aerodrome strategies will be A_3 and A_2.

FUTURE RESEARCH DIRECTIONS

It is our belief, that the Models of DM can be used in fields other than aviation. The Algorithms and Methodology for analysis of situation development under influence DM by H-O can be useful in any technogenous production. Technogenous production is a complex system that contains interrelated technical, economic and social objects. Technogenous production has a multilevel hierarchical structure and a high level of risk. Recently, is a frequent and common emergency such as disaster, accident, crash on hydraulic engineering, chemical and military industries, gas and oil pipelines, nuclear power plants and transport. The main strategic problems of mankind on the path to sustainable development is the safety and stability of technogenous production in complex system.

What's Next?

On next steps we are planning consider stochastic models in DM for other different complex STS, such as hydraulic engineering, chemical and military industries, gas and oil pipelines, nuclear power plants and transport etc. It is necessary to analyze the factors influencing the DM of operators in these systems in order to predict the development of the technoogenic catastrophe and prevent it using models in DM of H-O in emergency situation. We may to prepare these models for air crash investigators to analyze air crashes.

CONCLUSION

Further research should be directed to the solution of practical problems of actions UAV's operator in case of emergencies, software creation. Models of FE development and of DM by UAV's in FE will allow predicting the H-O's actions with the aid of the informational-analytic and diagnostics complex for research H-O behavior in the extreme situation. The ANS is presented as a complex socio-technical system. The influence on the DM by H-O of the ANS of the professional factors (knowledge, skills, abilities, experience) as well as the factors of non-professional nature (individual-psychological, psycho-physiological and socio-psychological) has been defined. The network analysis of the actions of an AC crew and air traffic controller in the flight emergencies has been made with the aid of the network planning methods; also the logic determined models DM by H-O ANS in FE have been developed. The numerical indices of an expected risk in the stochastic models DM by H-O ANS under Risk and Uncertainty have been defined.

The scenarios of developing a flight situation in case of selecting either the positive or negative pole under the pressure of the external environment, the preceding experience of a human-operator and the intentional selection (intention) in accordance with the reflexive theory have been obtained.

REFERENCES

Bertsch, V., Treitz, M., Geldermann, J., & Rentz, O. (2007). Sensitivity analyses in multi-attribute decision support for off-site nuclear emergency and recovery management. *International Journal of Energy Sector Management*, *1*(4), 342–365. doi:10.1108/17506220710836075

Federal Aviation Administration. (2008). Aeronautical Decision Making. In Pilot's Handbook of Aeronautical Knowledge. FAA-H-8083-25. 17.1–17.32. FAA.

Flueler, T. (2006). Decision Making for Complex Socio-Technical Systems: Robustness from Lessons Learned in Long-Term Radioactive Waste Governance. In *Environment & Policy*. Springer.

International Civil Aviation Organization. (2004). *Cross-Cultural Factors in Aviation Safety: Human Factors Digest Nº 16*. Circ. ICAO 302-AN/175. Montreal, Canada: Author. doi:10.1007/1-4020-3529-2

International Civil Aviation Organization. (2002). Human Factors Guidelines for Safety Audits Manual. Doc. ICAO 9806-AN/763. Canada, Montreal: Author.

International Civil Aviation Organization. (2009). *Manual on Global Performance of the Air Navigation System. Doc. 9883*. Montreal, Canada: Author.

International Civil Aviation Organization. (2013). Safety Management Manual (SMM) (3rd ed.). Doc. ICAO 9859-AN 474. Montreal: Author.

International Civil Aviation Organization. (2013). *State of Global Aviation Safety*. Montreal, Canada: Author.

ISO 31000 (2009) / ISO Guide 73:2002

Keating, C., Fernandez, A. A., Jacobs, D. A., & Kauffmann, P. (2001). A methodology for analysis of complex sociotechnical processes. *Business Process Management Journal*, *7*(1), 33–50. doi:10.1108/14637150110383926

Kharchenko, V., Shmelova, T., & Sikirda, Y. (2012). *Decision-Making of Operator in Air Navigation System*. Kirovograd: KFA of NAU.

Kharchenko, V., Shmelova, T., & Sikirda, Y. (2016). *Decision-Making in Socio-Technical Systems*. Kyiv: NAU.

Kharchenko, V., Shmelova, T., & Sikirda, Y. (2011). Graphanalytical Models of Decision-Making by Air Navigation System's Human-Operator. *Proceedings of the National Aviation University*, *1*, 5–17.

Kharchenko, V., Shmelova, T., & Sikirda, Y. (2011). Methodology for Analysis of Decision Making in Air Navigation System. *Proceedings of the National Aviation University*, *3*, 85–94.

Kharchenko, V., Shmelova, T., & Sikirda, Y. (2012). Modeling of Behavioral Activity of Air Navigation System's Human-Operator in Flight. *Proceedings of the National Aviation University*, *2*, 5–17.

Kharchenko, V., Shmelova, T., & Sikirda, Y. (2012). Methodology for analysis of flight situation development using GERT's and Markov's networks: In *Proceedings of the 5 World Congress Aviation in the XXIst century.*

Kharchenko, V., Shmelova, T., & Zubrytskyi, V. (2017). Decision making in air navigation system as a socio-technical system. *Proceedings of the International Conference "AVIA 2017"*.

Lefebvre, V. A. (2001). Algebra of Conscience (2nd ed.). Dordrecht, The Netherlands: Kluwer Publishers.

Lefebvre, B., & Adams-Webber, J. (2002, January-June). Functions of Fast Reflexion in Bipolar Choice. *Reflexion Processes and Control, 1*, 29–40.

Lefebvre, B. (2008, January-June). Intentional-Reflective Model of the Agent. *Reflexion Processes and Control, 1*, 69–78.

Leychenko, S., Malishevskiy, A., & Mikhalic, N. (2006). *Human Factors in Aviation: monograph in two books*. Kirovograd: YMEKS.

Shmelova, T., & Sikirda, Y. (2012). Analysis of Human-Operator's Decision-Making in Air Navigation System. *Radioelectronic and Computer Systems*, *7*(59), 319–324.

Shmelova, T., Sikirda, Y., & Jafarzade, T. (2013). Models of Flight Situations Development while Decision-Making by Air Navigation System's Human-Operator. *Information Processing Systems*, *8*(115), 136–142.

Shmelova, T., Sikirda, Y., & Sunduchkov, K. (2013). Socio-technical Analysis of Air Navigation System. *Science and technology of the Air Forces of the Armed Forces of Ukraine*, *4*(13), 34–39.

Shmelova, T., Bondarev, D., & Znakovska, Y. (2016). Modelling of the Decision Making by UAV's Operator in Emergency Situations. In *Proceedings of the IEEE 4th International Conference on Methods and Systems of Navigation and Motion Control (MSNMC-2016)*. Kyev: IEEE.

Shmelova, T., Mirskova, M., & Krepak, D. (2017). Decision making in emergency situation of air navigation system's operator. *Proceedings of the International conference "AVIA -2017"*.

Shmelova, T., & Konovalova, A. (2017). Decision making of ATCO in emergency situation "aircraft decompression". *Proceedings of the International conference "AVIA -2017"*.

Shmelova, T., & Shulimov, O. (2017). Method of determination of optimal place of landing for remotely piloted aircraft. *International Conference "AVIA 2017"*.

Shmelova, T., & Sikirda, Y. (2016). Calculation the scenarios of the flight situation development using GERT's and Markov's networks. *Proceedings of the Conference "Management of high-speed moving objects training operators of complex systems"*.

Shmelova, T., Dolhov, D., & Muliar, P. (2015). Mathematical models of operators activities in aeronavigation sociotechnical system. In *Proceedings of the Conference "Techniques and technology. Science yesterday, today, tomorrow"*. Warsaw.

SKYbrary Aviation Safety. (n.d.). Retrieved from http://www.skybrary.aero/index.php/Main_Page

Chapter 7
Stochastic Methods in Microsystems Engineering

Irina Stanciu
University Politehnica of Bucharest, Romania

ABSTRACT

The main objective of this chapter is to present a stochastic modeling and simulation methodology for estimating the variation of functional parameters of MEMS devices as a result of fabrication tolerances. The approach used for achieving this objective is to implement in the simulation process the variations of the geometrical parameters caused by the fabrication dispersion as random variables and to model the influence of these variations on the performance of the simulated device. The proposed method is demonstrated using two test structures: a micro-com resonator and a passive micromixer. In each example there are underlined important characteristics of the proposed simulation method: the ability to predict variation ranges of functional parameters, the ability to improve the design in function of the fabrication process, and the possibility of obtaining more precise results than the traditional deterministic methods.

INTRODUCTION

Microsystems Engineering is a growing field, that can be described as the ensemble of techniques, technologies, and practices involved in the realization of structures and devices with dimensions of the order of micrometers (Banks. D., 2006). Microsystems Engineering draws its roots from the microelectronics industry, but as the field evolved, the technological framework has been adapted and modified, in order to allow the development of new innovative devices, with unprecedented functionalities.

Since the arising of the micro-engineering domain, thousands of researchers have spent a lot of time developing new components design or exploring basic functionalities, such as piezoresistive, capacitive and vibration sensing schemes, in the case of micro-sensor devices or electrostatic driving in the case of micro-actuators. Microsystem devices, also known as MEMS ("Micro Electro Mechanical Systems") devices that have been built to date are highly specialized to a particular application, and are realized in custom technologies that may not be widely available. The progress in microsystem development has

DOI: 10.4018/978-1-5225-5045-7.ch007

been limited by the absence of standard commercial components. Most of today's devices are application specific, which requires that unique components must be developed for each application with their own specific requirements. In the present, it does not exist a standardized cost-efficient technology for these devices (Gianchandani Y. B., 2008, Fair R. B., 2007).

Standardization of fabrication technologies implies definitions of standards on manufacturing tolerances, especially for the new polymer micro-manufacturing techniques, that have not yet been fully established, while most of the data is 'in-house' determined (Qin Y., 2010). Design for manufacturability and fault tolerance are expected to emerge as compulsory design considerations (Chakrabarty K. & Zeng J., 2014). Due to the multiple energy domains employed and mixed technology used for developing MEMS devices, they exhibit unique failure mechanisms and defects as compared to microelectronic systems.

In the recent years, fault tolerance analyses and design for manufacturability methodologies for MEMS have become a major concern in the scientific community as well as in the industry. However, from this perspective, the research has focused mainly on the development of experimental techniques for reliability testing of MEMS devices. The fabrication tolerances lead to variations of the geometrical parameters of the fabricated device, which leads to variations of the performance parameters. These variations in performance need to be quantified and controlled. Most of the techniques developed up to date for monitoring the effects of the deviations of geometrical parameters on the behaviour of microsystems are based on experimental methods (Chakrabarty K., 2014, Frangi A., 2008). The experimental testing of the device is not always convenient, since many applications like bio-medicine, pharmaceutics or environmental monitoring require the development of sterile or disposable devices and physical testing would imply the device contamination. In other applications like biophysics or biochemistry, functionalised surfaces are needed and the experimental testing would imply the damaging of these surfaces. Another approach has been to integrate testing mechanisms in the device (like optical sensors, capacitance measurement electrodes, impedance sensors, etc). This approach brings along two main disadvantages: the design complexity and implicitly the cost of the device are increased and on the other hand, there are needed supplementary fabrication steps, which are not always compatible with the materials used for manufacturing the MEMS device.

Given the unique challenges presented by the experimental testing of MEMS devices, in this chapter, it is proposed a stochastic simulation based methodology for estimating the influence of geometrical parametric deviations generated by the microfabrication tolerances on the general performance of the device (Stanciu I., 2017). This methodology is developed by following the general flow of uncertainty modelling methods for studying the influence of random factors on the behaviour of engineering systems, but by adapting it to the design specifics of microsystems. This simulation method brings important benefits such as the possibility of determining values of certain parameters that cannot be measured experimentally and the development of precise and detailed behavioural models of MEMS devices without contaminating or damaging the device.

METHODOLOGY

The influence of the fabrication tolerances on the performance of a MEMS device is estimated by using a combined Monte Carlo – Finite Element Method simulation procedure. This method can be applied for any microfabrication process for which a probabilistic model of the geometric tolerances is known or it can be determined. One iteration step of the Monte Carlo algorithm consists in a finite element simulation of the investigated device. A statistical analysis is performed on the results of the Monte Carlo simulation. This way it is possible to predict the percentage of devices that could fail due to parametric faults; hence it helps to determine if the design and the microfabrication process are suited for the desired application.

Monte Carlo Simulation Method

Monte Carlo methods are computational algorithms that take as input random objects (random variables, random vectors, stochastic processes) and are used for computing numerical quantities (expectations, probabilities) in simulating physical and mathematical systems (Frenkel D., 2004, Kroese D.P., 2011). These methods are used for simulating systems with significant uncertainty in inputs or for simulating systems with many coupled degrees of freedom for which either it is not possible or it is too time or resources consuming to compute an exact result with a deterministic algorithm (Rubinstein R.Y. & Kroese D.P., 2008). Monte Carlo simulation methods have applicability in almost all scientific domains: physical sciences, engineering, telecommunications, reliability and maintenance, traffic control, computational biology, finance and business, game theory, applied statistics, robotics, etc. The Monte Carlo method analyzes uncertainty propagation [Ayyub B.M.& Klir G.J., 2006], in order to determine how the dispersion of the input generated by random variation, lack of knowledge, or error affects the sensitivity, performance, or reliability of the system that is being modelled.

In Monte Carlo based methods, the input parameters (x_1, x_2, …, x_k) are random variables defined by certain probability distributions. They are introduced in the model of the simulated system, which is described analytically using a function f of parameters (x_1, x_2, …, x_k). The output parameters are analyzed using statistical methods. A schematic of the basic principle of a Monte Carlo simulation is presented in figure 1.

The result of a Monte Carlo simulation is a probabilistic entity that has to be analyzed using statistical methods such as probability plots, data summarization or hypothesis testing.

For applying a Monte Carlo simulation for a MEMS device, the geometrical are implemented as random variables that follow given probability distributions. The choice of the most appropriate probability model for a parameter has to be done by closely studying the utilized fabrication process. This can be done either experimentally or by process simulation.

A Finite Element Simulation is implemented in the same manner as for typical deterministic simulations. This FEM simulation is performed repeatedly for a given number of times, taking each time as input a randomly generated geometry. Each repetition of the FEM simulation is in fact an iteration of the Monte Carlo Algorithm. The results of the repeated numerical simulations are collected in a data set and they are analyzed statistically. This way, the variation of the output due to the variation of the input geometry can be estimated (Helton J. C. & Davis F.J., 2003, Moens D. & Vandepitte D., 2005). Probabilistic laws of variation of the output, confidence limits and sensitivities can be deducted. This results offer a more realistic view of the behaviour of the analyzed MEMS device.

Stanciu I. (2017) presented the steps of the probabilistic method of estimation of functional parameters variation of microfluidic devices as result of fabrication dispersion:

1. Design the geometry of the MEMS device
2. Parameterize the geometry
3. Build the partial differential equation model of the MEMS device

Figure 1. Schematic of the basic principle of Monte Carlo simulation (Stanciu I., 2017)

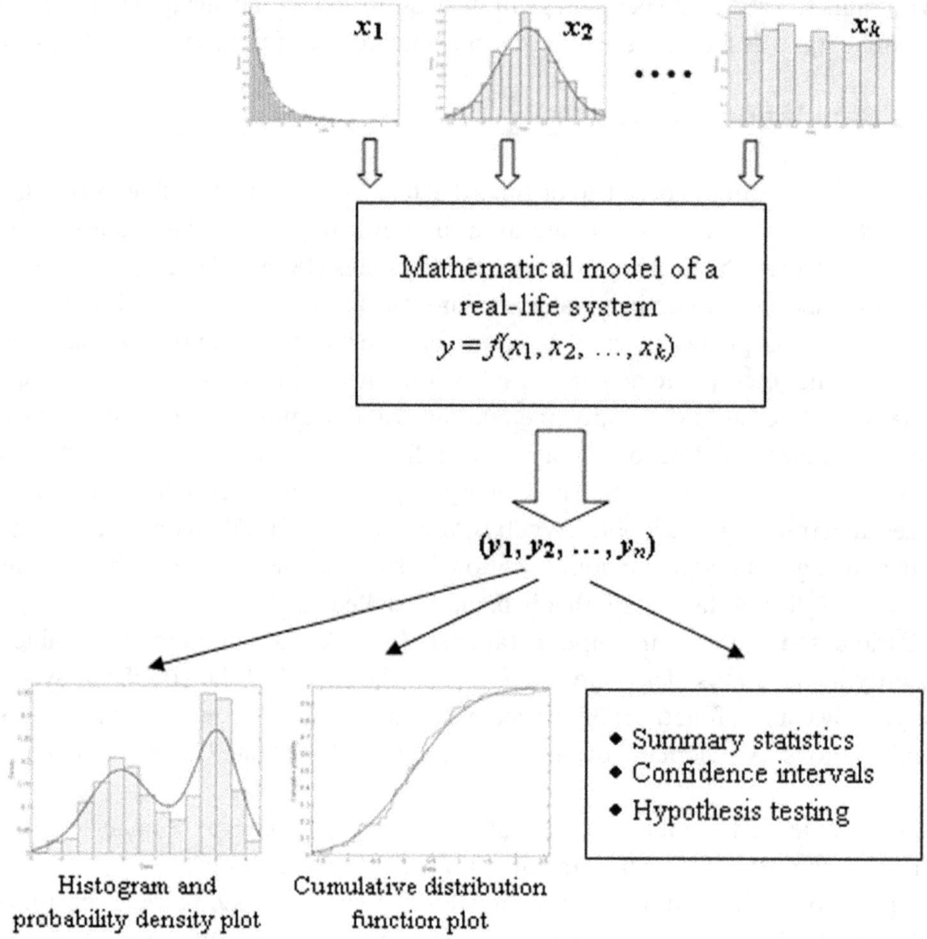

4. Determine the most appropriate probability distribution for modelling the variation of the geometric parameters, caused by the fabrication tolerances
5. Generate a set of input geometrical parameters randomly according to the defined variation laws
6. Perform a FEM simulation of the MEMS device using as input parameters the values generated at step 5
7. Collect the numerical simulation result in a vector of output values.
8. Repeat steps 6,7 and 8 until a termination condition is met
    a. A given number of iterations is performed
    b. The convergence of the Monte Carlo method is achieved
    c. A given simulation time is exhausted
9. Analyze by statistical methods the collection of output data
    a. Perform visualizations of the output data by histograms and cumulative distribution plots and decide on a possible probability distribution of the result
    b. Estimate the parameters of the chosen probability distribution by data summarization – calculate empirical mean, empirical variance

c. Fit the chosen probability distribution with the estimated parameters to the output data set
d. Perform hypothesis testing to confirm the chosen probability distribution
e. Estimate the range of variation by calculating tolerance intervals for the output data

The output of the Monte Carlo simulation is a set of data which can be viewed as outcome of random variables, random vectors, time series, or stochastic processes. In general, the output data set of the simulation, (y_1, y_2, …,y_n), forms an independent and identically distributed sample from some distribution. The objective of the simulation is to draw conclusions on various characteristics of the output data set of the simulation, especially on the probability distribution, but also on other characteristics such as data range or correlations.

Analysis of Monte Carlo Simulation Results

There is no specific rule on how to choose the number of iteration that is needed for a Monte Carlo simulation. As a statistical analysis is needed to be performed on the output data, the larger the number of iterations the more precise the obtained result is. On the other hand, a higher number of iterations implies a higher simulation time and a higher consume of computational resources. Stanciu I. (2017) has presented a method of determining a sufficient number of iterations is the convergence of averages. Suppose that the independent and identically distributed sample (y_1, y_2, …,y_n) comes from a simulation experiment. This sample is obtained by executing n iterations of the Monte Carlo simulation. The output y_i is produced by the i-th iteration. The statistical mean of the distribution of the output data set, E(y), has as an unbiased estimator the sample empirical mean:

$$\overline{y} = \frac{1}{n}\sum_{i=1}^{n} y_i \tag{1}$$

From the central limit theorem, it results that for a large number of iterations, $\overline{y}$ has a normal distribution. It results that a confidence interval for E(y) can be calculated from:

$$\left(\overline{y} - z_{1-\frac{a}{2}} \cdot \sqrt{\frac{s^2}{n}},\quad \overline{y} + z_{1-\frac{a}{2}} \cdot \sqrt{\frac{s^2}{n}}\right), \tag{2}$$

Where $z_{1-\frac{\alpha}{2}}$ is the $1-\frac{\alpha}{2}$- quantile of the $N(0, 1)$ distribution, s^2 is the empiric variance of the sample and n is the volume of the sample.

Stanciu I. (2017) presented the algorithm for determining whether the Monte Carlo simulation has converged with the following steps:

1. Generate a sample of independent and identically distributed data (y_1, y_2, …,y_n) from n independent simulation iterations.
2. Calculate the estimate $\overline{y}$ and a confidence interval (2) of the mean E(y).

3. If the confidence interval doesn't change for a certain number of iterations, then convergence has been achieved
4. If the confidence interval varies from one iteration to another, than generate more output data (y_{n+1}, y_{n+2}, ...,y_{n+k}) and repeat steps 2, 3 and 4

When a sufficient number of iterations have been performed the output data set is formed from the results of the simulation runs. Before performing a statistical analysis of the data it is useful to detect patterns in the data through visualization. The visualization of data can be done by various graphical techniques such as histograms, density plots and empirical cumulated distribution functions.

From the histogram and the empirical cumulative distribution plot it can be deducted the shape of the probability distribution and of the cumulative distribution function respectively. However, this is just an intuitive observation and it does not offer a rigorous mathematical proof. These observations can be used in a goodness of the fit procedure. A statistical test can be applied on the output data set to verify whether the assumed probability distribution is indeed a good fit of the data. Among the statistical tests that can be applied in this situation there are the Kolmogorov – Smirnov test, χ^2 goodness of the fit test, Lilliefors test, etc (Stanciu I., 2017).

TEST CASES

Stochastic Simulation of a Micro-Comb Resonator

Codreanu I. (2009), presents the implementation of a combined Monte Carlo – FEM simulation for studying the influence of the variability of device parameters, as result of fabrication tolerances, on the performance of a micro-comb resonator.

Microresonators are structures designed to vibrate at a particular frequency. They are employed as actuators in applications such as microengines, RF systems, microgyroscopes, or as sensing elements in a wide variety of sensors: mass sensors, pressure sensors, flow sensors, strain sensors, etc (Lin R.M.& Wang W.J., 2006, Codreanu I., Martowicz A., Gallina A., Pieczonka L. & Uhl T., 2009).

The geometry of the structure for which the stochastic modeling has been performed is presented in fig. 2. This microresonator is formed by two comb structures anchored to the substrate and a mobile shuttle mass suspended by folded flexure springs. The lateral sides of the mobile mass are combs interwoven with the fixed ones (Codreanu I., 2009).

The micro-comb drive resonator is usually actuated by an electrostatic field on one side to oscillate at its fundamental resonance frequency, and the capacitive sensing principle is used on the other side. For the actuation of this device the tangential electrostatic forces are used, while the effect of the normal force is eliminated by the symmetrical arrangement of the stationary electrodes on both sides of each movable finger so that the normal forces from both sides cancel out. The movable structure is electrically grounded and the left and right electrodes are biased to a DC voltage V_{bias} superposed with a differential voltage v. The differential voltage v is considerably smaller than the bias voltage ($|v| < |V_{bias}|$) and it generates a harmonic AC signal that has the role of exciting the mobile mass of the comb drive (Mawardi A., Pitchumani R., 2005).

In order to obtain the highest displacement of the shuttle mass, hence to have the best performance of the device in which the resonator serves as component, the alternative signal should be tuned to

Figure 2. Layout of a micro-comb resonator (Codreanu I., 2009)

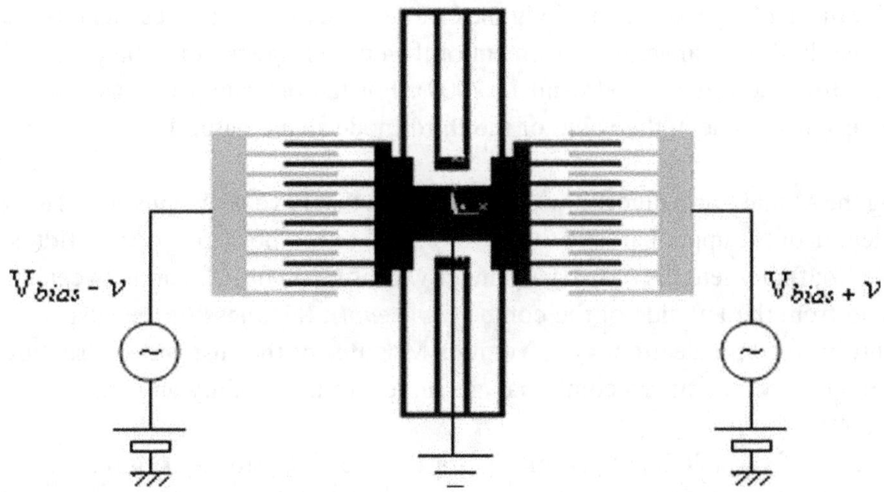

the natural frequency of the mobile mass to make it resonate, by moving on a parallel direction to the substrate. Therefore, it is necessary to determine the corresponding mode shape and calculate the value of the natural frequency of this mode shape from the design phase using theoretical considerations and numerical simulations.

The typical fabrication process for manufacturing a micro-comb device is surface micromachining. This technique is usually used to build up mechanical features or free moving structures on top of a wafer surface. The process flow includes successive deposition and patterning of sacrificial layers and structural layers, then the removal of the sacrificial layers to release the produced structures. The patterning of the suspended structure is realized with a photolithography process. The tolerance of a photolithography process can rise up to 10% (Codreanu I., 2012). The deposition of the suspended structure is usually done using the LPCVD (Low Pressure Chemical Vapor Deposition) technique. During this manufacturing step, due to the high temperatures involved, the material properties (such as Young's modulus) can be affected and present variations. Also the thickness of the deposited layer can have some variations (Codreanu I., 2009).

The variation of the geometry of the micro-comb resonator can lead to significant variations of the mass of the mobile structure, which, at its turn can lead to significant variations of the value of the natural frequency at which the micro-comb is operated to vibrate. This can lead to a significant drop in the device performance, which should be taken into consideration during the design phase. The combined Monte Carlo – FEM simulation is an appropriate tool for analyzing and quantifying these variations.

The FEM step of the stochastic simulation of the micro-comb device has been performed using the ANSYS simulation software (Codreanu I., 2009). The coupled electromechanical field analysis has been done with the help of the TRANS126 elements that model the fully coupled system. These elements have been used to perform a prestressed modal analysis to determine the system eigen-frequencies. Of interest for the micro-comb device is the frequency shift when a DC voltage is applied on the electrodes of the transducer. This effect has been simulated by, first, performing a static analysis of the device with the DC bias voltage applied to the transducer, and then performing a "prestressed" modal analysis of the structure. The results showed that the first mode shape is the torsional mode along the Oy axis, the second mode shape is the bending out of plane mode along the 0*z* axis and the third mode shape is the flexible mode for the in plane movement along the 0*x* axis. The mode that presents interest for the comb drive resonator the third one, since the movement of the comb fingers along the 0*x* axes is the one that generates the electrostatic forces (Codreanu I., 2009). For the deterministic case (with the geometry defined by the designed values), the value of the third mode shape natural frequency was obtained to be 135659 Hz.

For applying the Monte Carlo algorithm, the parameters that have been considered to be variable are: flexure beams length of the upper part and the lower part of the comb (*UBL, LBL*); widths of the flexure beams (*bw*1- *bw10*); finger lengths (*fl*1 -*fl*16); finger widths (*fw*1- *fw*16); gap between the fingers from the right side and from the left side of the comb (*gapr, gapl*); thickness of the suspended structures – mobile mass and fixed comb electrodes (*t*); Young's Modulus of the suspended structure (*E*). Table 2. presents the parameters of the micro-comb that are subject of uncertainty and their ranges of variation (Codreanu I., 2009).

The total number of variable input parameters for the Monte Carlo algorithm is 48. The considered law of variation is the truncated Gauss distribution in their defined ranges of variation and with the values of the standard deviation presented in Table 1. The eigen-frequency of the third mode of vibration is the output parameter of the stochastic simulation.

ANSYS Probabilistic Design System (PDS) has been used to implement the Monte Carlo algorithm. The stochastic simulation has been run for 1000 loop. One loop consisted in an execution of the finite element analysis using the TRANS126 elements for performing the pre-stressed modal analysis to calculate the value of the eigen-frequency of the third mode shape.

The scatter of the output parameter of the stochastic simulation is presented in the histogram plot (Fig. 3). The mean value of the result is 136268 Hz and the standard deviation is 906.82.

The sample histogram of the simulation results shows a variation of about 3.4% of the natural frequency from the deterministic value. This might lead to a decrease of the device's performance below desired limits.

Stochastic Simulation of a Passive Micromixer

Stanciu I. (2015) presents a stochastic simulation based methodology for estimating the influence of geometrical parametric deviations, caused by the fabrication dispersion, on the performance of passive micromixers with modified tesla elements (Hong C.C., Choi J.W. & Ahn C.H., 2003).

Table 1. Variation ranges for the uncertain parameters (Codreanu I., 2009)

Uncertain Parameters	Nominal value	Variation range	Standard deviation
UBL, LBL	100 µm	± 1 µm	0.3
*bw*1- *bw*10	4 µm	± 0.1 µm	0.03
*fl*1- *fl*16	50 µm	± 0.5 µm	0.16
*fw*1- *fw*16	4 µm	± 0.1 µm	0.03
gapr, gapl	3 µm	± 0.1 µm	0.06
t	3 µm	± 0.2 µm	0.06
E	165.3 GPa	± 5 GPa	1.6e+3

Figure 3. Sample histogram for the eigenfrequency of the third mode shape (Codreanu I., 2009)

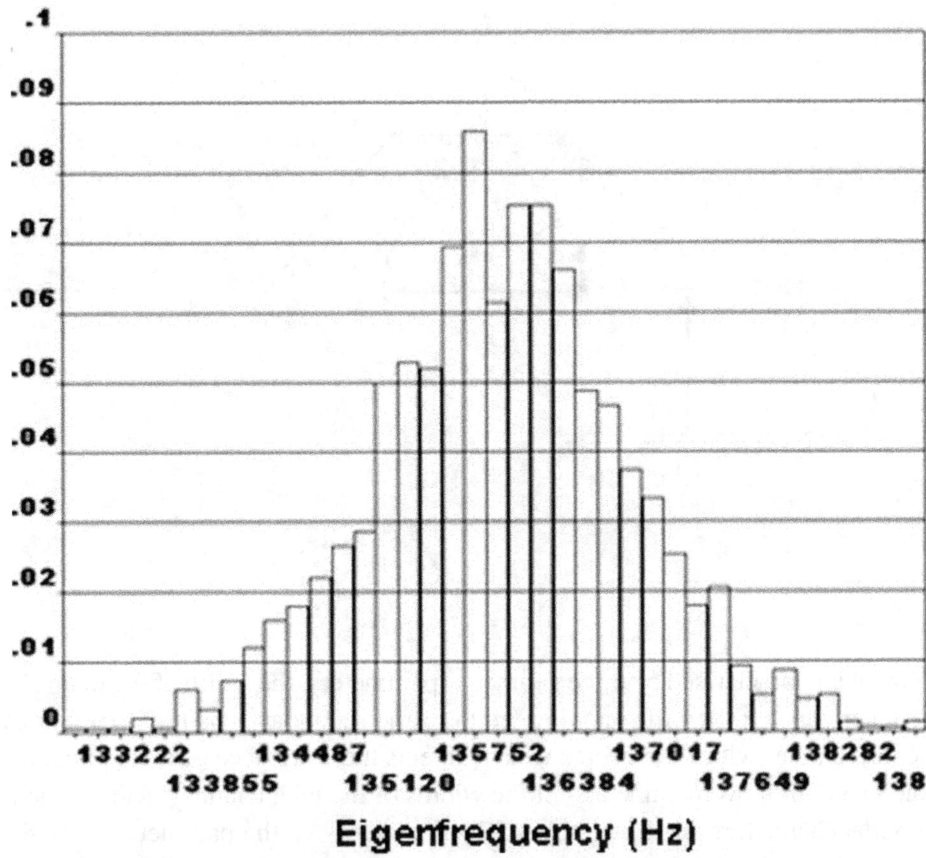

Passive micromixers are devices that use the geometry as the mixing enhancement factor. This type of devices is cost effective, from the fabrication point of view, and also, easy to integrate in complex microfluidic systems, therefore, they have a high potential of applicability.

The layout of a micromixer with a modified tesla element is presented in figure 4 (Hong C.C., Choi J.W. & Ahn C.H., 2003, Stanciu I., 2015). The fluids are introduced in the device through two separate inlet channels that meet in a main microchannel. A secondary channel leaves the main channel at an acute angle, while the main channel has an 180° turn that forms the Tesla element and meets again with the secondary channel. The fluids that are introduced in the micromixer are water, through the upper inlet and a NaCl solution with a concentration of 9mg/ml (154mol/m^3), through the lower inlet. The geometry of the passive micro-mixer is defined by the following parameters: width of the main channel (w1=500um), height of the main channel (h1=500um), radius of the inner curve (r=125um), opening of the secondary channel (sc=100um), width of the outlet channel (w2=500um) and height of the outlet channel (h2=500um). The widths of the two inlet channels are given by the parameters wi1=wi2=500um.

For implementing the stochastic simulation method, a parameterization of the geometry of the micromixer is necessary as well as definition of probabilistic models of each parameter. Stanciu I. (2015) proposed the following parameterization model:

Figure 4. Parameterization of the micromixer geometry (Stanciu I., 2015)

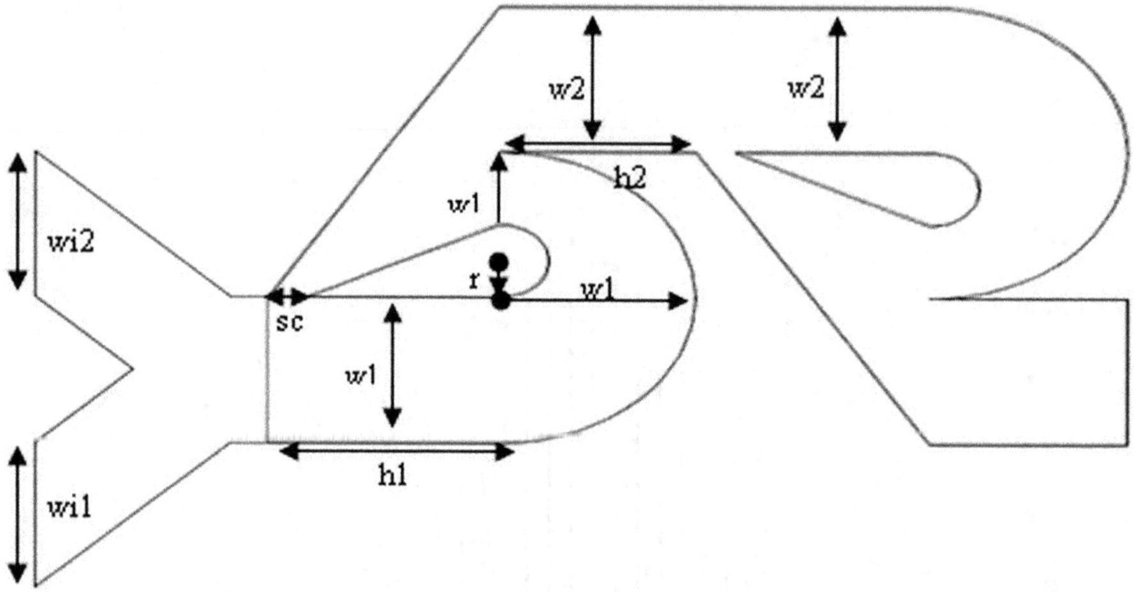

Each tesla element is characterised by three geometric parameters: the width of the main channel (w), the radius of the inner curve (r) and the opening width of the secondary channel (sc). For the parameterization of the micromixer with 8 connected tesla elements there are necessary 24 parameters for the tesla elements and in addition two parameters for the widths of the inlet channels and one parameter for the width of the outlet channel, resulting a total of 27 parameters. All the parameters variations follow a normal distribution, with the standard deviation, σ, calculated such that the variation range is ±5% (Stanciu I., 2015).

Table 2. Variation models of the geometrical parameters of the micromixer with 8 connected tesla elements (for each parameter, σ is calculated such that the variation of the parameter is up to 10% from its nominal value) (Stanciu I., 2015)

Parameter name	Model of variation	Description of the parameter
w1-w8	$500 + r,\ r \sim N(0, \sigma)$	Widths of the main channel of the 8 tesla elements
r1-r8	$125 + r,\ r \sim N(0, \sigma)$	Radius of the inner curves of the 8 tesla elements
sc1-sc8	$100 + r,\ r \sim N(0, \sigma)$	Openings of the secondary channels of the 8 tesla elements
wi1, wi2	$500 + r,\ r \sim N(0, \sigma)$	Widths of the two inlet channels of the micromixer
w9	$500 + r,\ r \sim N(0, \sigma)$	Width of the outlet channel

For an iteration of the Monte Carlo algorithm, a set of random parameters is generated using the probability laws presented in table 2 and a finite element simulation of the mixing process inside the passive micromixer is performed. In order to have a more precise result of the statistical analysis it is preferable to have a volume of data as large as possible. Therefore, a trial of 100 iterations has been performed (Stanciu I., 2015).

The FEM simulation of the passive micromixer with modified tesla elements has been performed with the COMSOL simulation software. In order to apply both models needed for the description of the mixing process a coupled simulation is needed. The coupled simulation is performed by using the Incompressible Navier-Stokes application mode of the MEMS module of COMSOL and the Convection and Diffusion application mode of the Multiphysics module. Because the geometry of the micromixer is prismatic (it has a constant section along the 0z axis) a 2D simulation is sufficient for the study of the mixing process. The simulation settings that have been used are presented in table 3.

The FEM simulation results showed that one tesla element is insufficient for and efficient mixing. In order to obtain a more uniform concentration in the outlet channel it is necessary to enhance the geometry of the micromixer by adding more tesla elements. The tesla elements are connected in series such that the outlet of one tesla element coincides with the inlet of the following tesla element. In order to determine the minimum number of tesla elements required for an efficient mixing, further analyses have been performed.

Table 3. Comsol simulation settings and input parameters used for performing the numerical simulation of the mixing process inside the micromixer with tesla elements (Stanciu I., 2015)

Comsol Application Mode	Subdomain settings	Boundary settings		
Incompressible Navier-Stokes	$\rho = 10^3\, kg/m^2$ $\eta = 10^{-3}\, Pa \cdot s$	**Boundary type**	**Boundary condition**	**Parameters**
		inlet	Laminar inflow	Average velocity: $u_0 = 10\, \mu m/s$ Entrance length: $L_{entr} = 10^{-6}\, m$
		wall	No slip	
		outlet	Pressure, no viscous stress	
Convection and Diffusion	$D = 10^{-10}\, m^2/s$	**Boundary type**	**Boundary condition**	**Parameters**
		inlet 1	Concentration	$c_0 = 154\, mol/m^2$.
		inlet 2	Concentration	$c_0 = 0\, mol/m^2$
		wall	Insulation/Symmetry	
		outlet	Convective flux	

The results of the COMSOL simulation of a micromixer with 8 connected tesla elements show that along the seventh and the eighth tesla element of the micromixer the concentration is visibly more uniform than in the previous tesla elements (figure 5).

For defining the performance of the micromixer, in (Stanciu I., 2015) it has been used the mixing efficiency coefficient, defined by:

$$e = \left(1 - \frac{\int_0^w c - c_c}{\int_0^w c_i - c_c}\right) \cdot 100 \tag{3}$$

where, *e* represents the mixing efficiency, *c* is the concentration distribution across a transversal section of the micromixer, c_c represents the concentration of complete mixing, c_i represents the initial concentration, calculated at a cross section of the micromixer, near the inlets and *l* represents the width of the considered cross section (Wang H., Iovenitti P., Harvey E., & Masood S., 2003, Jeon N. L., Dertinger S. K. W., Chiu D. T., Choi I. S., Stroock A. D., & Whitesides G. M., 2000). The numerical values of the concentration distribution obtained by the finite element simulation are exported in Matlab, and the mixing efficiency is computed using numerical integration.

The results of the computed mixing efficiencies for all the tesla elements of the passive micromixer are presented in table 4. It can be observed, that for the first tesla element, the mixing efficiency is very low: 36.65%, and that it is increasing with the number of tesla elements. Also, it can be observed that there is a very slight improvement of the mixing efficiency from the fifth up to the eighth tesla element: from 95.98% it raises to 98.83%. These results show that a design with minimum five tesla elements should ensure a satisfactory mixing efficiency (above 95%) (Stanciu I., 2015).

Figure 5. Comsol simulation of the concentration distribution along a micromixer with 8 connected tesla elements (Stanciu I., 2015)

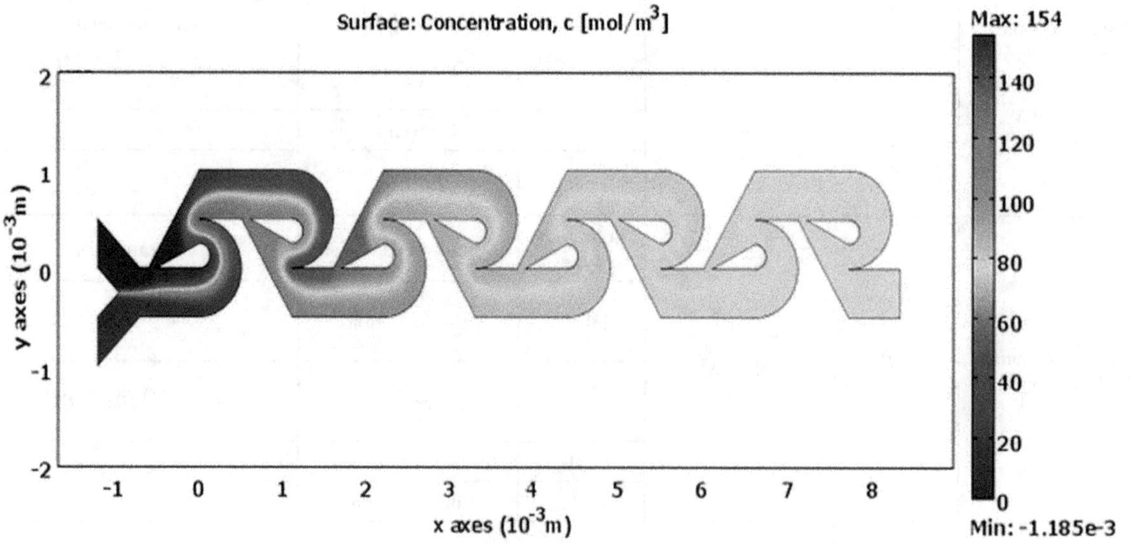

Table 4. Computed mixing efficiency coefficients at the outlet of every tesla element of the passive micromixer (Stanciu I., 2015)

	1st tesla element	2nd tesla element	3rd tesla element	4th tesla element	5th tesla element	6th tesla element	7th tesla element	8th tesla element
Mixing efficiency	36.65%	66.90%	85.12%	90.99%	95.98%	96.70%	98.93%	98.83%

Stanciu I. (2015) computed the mixing efficiency parameter at the outlet channels of the final three tesla elements of the passive micromixer, for each iteration of the Monte Carlo algorithm. There have been obtained three sample data sets. Figure 6 presents the histogram representation of the distributions of these data sample sets. It can be observed that the distribution of the mixing efficiency parameters are not normal, therefore, for the statistical analyses of these data non-parametric methods are required.

Since the objective of this study is to quantify the performance of the passive micromixer, regardless of the fabrication tolerances, there have been computed prediction intervals for the mixing efficiency coefficients from the data samples obtained by Mote Carlo simulation. A non-parametric method, based on order statistics (Randles R.H.& Douglas A. W., 1979) has been used for this purpose. The jth order statistic of a statistical sample is equal to its jth-smallest value. If $X_{(j)}$ and $X_{(k)}$ are order statistics of the sample with $j < k$ and $j + k = n + 1$, then $[X_{(j)}, X_{(k)}]$ is a prediction interval for X_{n+1} with significance level equal to $(n + 1 - 2j)/(n + 1)$. Hence, if it is sought a prediction interval with a significance level higher then α, then, the lower limit of the prediction interval can be found by choosing the jth order statistic with

$$j = \text{int}\left(\frac{1}{2(n+1)(1-\alpha)}\right). \quad (3)$$

Figure 6. Histograms of the results of the Monte Carlo simulation of the tesla micromixer at the outlets of the 6th up to the 8th tesla elements (Stanciu I., 2015)

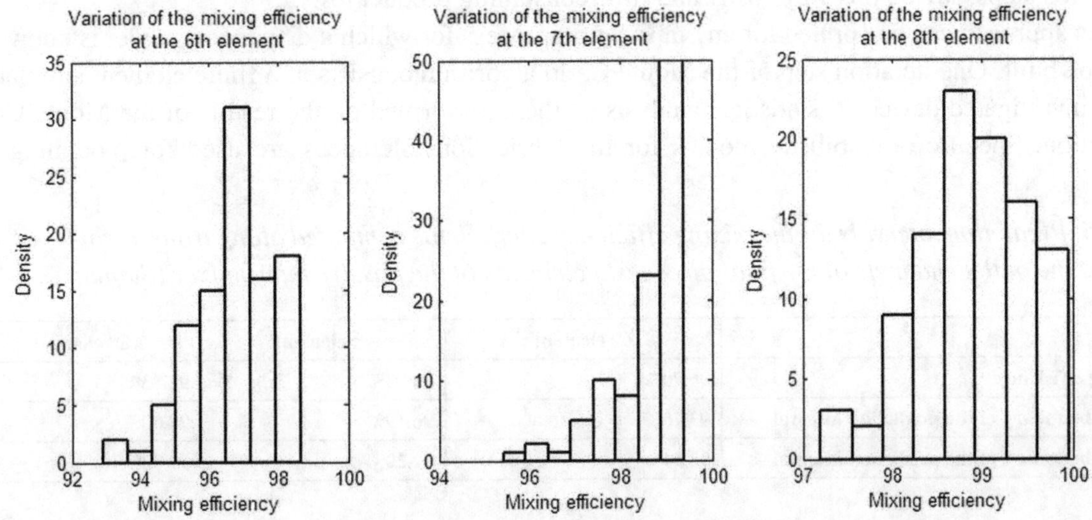

The results for computing the prediction interval for the three mixing efficiency coefficients are shown in table 5.

The first line of table 5 displays the value of the mixing efficiency found by deterministic simulation. The second line displays the lower boundary of the prediction interval and the third line displays the upper boundary. It can be observed that the value of the mixing efficiency can vary with up to two percent from the value computed by deterministic design. The prediction intervals offer a more realistic estimation of the expected mixing efficiency in a fabricated micromixer. The results of the stochastic simulation of the passive micromixer with tesla elements show that variations of up to 5% of the geometric parameters lead to a probability of 11.1% that the mixing efficiency will be lower than 95% for devices with a design with 6 tesla elements, while for a design with 7 tesla elements, this probability is 2.08%. All these considered, it can be concluded that a design with at least 7 tesla elements would be more reliable than a design with 6 tesla elements (Stanciu I., 2015).

CONCLUSION

Stochastic simulation methods are more efficient from point of view of time and cost than traditional experimental testing techniques. Depending on the application, the estimations of fabrication tolerances on functional parameters of the designed device can be obtained in only a few minutes up to a few hours without any fabrication and testing cost. It is useful both in prototype developing and in designing ready to use commercial micro-devices. In prototyping, it helps understand the cause of certain malfunctions or functional parameters deviations of the designed device. This is usually a very difficult task when approaching it with experimental testing methods. When a fabricated test structure is being tested, the exact value of all the geometrical parameters is usually not known. In a design with a large number of parameters it is very difficult if not impossible to determine the exact source of error. By simulation it is possible to investigate the complete design space and to obtain a detailed model of the influence of any parameter without physically measuring, disturbing or contaminating the device. In the development process of commercial micro-devices this method can be a key step, because it can predict the failure rate. Reliability risk analysis and process control can be performed even from the modelling and simulation steps, saving this way costly and time consuming product tests.

This approach can be applied for any fabrication process for which a dispersion model is known or it can be built. One iteration step of the Monte Carlo algorithm consists in a finite element simulation of the investigated device. A statistical analysis is, then, performed on the results of the Monte Carlo simulation. Specific probabilistic models for the fabrication tolerances are used for modelling the

Table 5. Prediction intervals for the mixing efficiency coefficients computed along transversal cross sections at the outlet channels of the final three tesla elements of the passive micromixer (Stanciu I., 2015)

	6th element	**7th element**	**8th element**
Mixing efficiency	96.70%	98.93%	98.83%
Lower boundary of the prediction interval	93.76%	96.19%	97.50%
Lower boundary of the prediction interval	98.54%	99.20%	99.93%

variations of the input geometrical parameters. This method is intended to help improve the design of micro-devices in order to reduce the appearance of parametric faults. By modelling the deviations of geometrical parameters caused by the fabrication tolerances it is possible to compute the propagated variations in the functional parameters of the simulated devices. This way it is possible to predict the percentage of devices that could fail due to parametric faults; hence it helps to determine if the design and the microfabrication process are suited for the desired application.

The two test cases presented in this chapter demonstrate the versatility of this type of simulation methods through two different implementation methods, using various specialised simulation software solutions.

REFERENCES

Ayyub, B. M., & Klir, G. J. (2006). *Uncertainty modelling and analysis in engineering and the sciences*. Chapman & Hall/CRC. doi:10.1201/9781420011456

Banks, D. (2006). *Microengineering, MEMS, and Interfacing A Practical Guide*. Birmingham, UK: Taylor & Francis Group. doi:10.1201/9781420015416

Chakrabarty, K., & Zeng, J. (2014). *Design automation methods and tools for microfluidics-based biochips*. Springer Netherlands.

Codreanu, I., Martowicz, A., Gallina, A., Pieczonka, L., & Uhl, T. (2009). Study of the effect of process induced uncertainties on the performance of a micro-comb resonator. *Solid State Phenomena, 147-149*, 716-725.

Fair, R. B. (2007). Digital microfluidics: Is a true lab-on-a-chip possible? *Microfluidics and Nanofluidics*, *3*(3), 245–281. doi:10.100710404-007-0161-8

Frangi, A. (2008). *Advances in multiphysics simulation and experimental testing of MEMS*. Imperial College Press. doi:10.1142/p522

Frenkel, D. (2004). *Introduction to Monte Carlo Methods. In Computational Soft Matter: From Synthetic Polymers to Proteins, Lecture Notes, John von Neumann Institute for Computing* (Vol. 23, pp. 29–60). Julich: NIC Series.

Gianchandani, Y. B., Tabata, O., & Zappe, H. P. (2008). *Comprehensive Microsystems*. Elsevier.

Heike, H. D., Tarcolea, C., Schüte, M., & Manescu, B. (1999). *Grundlagen der Statistik und Wahrscheinlichkeitsrechnung*. Oldenbourg.

Helton, J. C., & Davis, F. J. (2003). Latin hypercube sampling and the propagation of uncertainty in analyses of complex systems. *Reliability Engineering & System Safety*, *81*(1), 23–69. doi:10.1016/S0951-8320(03)00058-9

Hong, C. C., Choi, J. W., & Ahn, C. H. (2003). A novel in-plane passive microfluidic mixer with modified tesla structures. *Lab on a Chip*, *4*(2), 109–113. doi:10.1039/b305892a PMID:15052349

Jeon, N. L., Dertinger, S. K. W., Chiu, D. T., Choi, I. S., Stroock, A. D., & Whitesides, G. M. (2000). Generation of solution and surface gradients using microfluidic systems. *Langmuir*, *16*(22), 8311–8316. doi:10.1021/la000600b

Kroese, D. P., Taimre, T., & Botev, Z. I. (2011). *Handbook of Monte Carlo Methods*. Wiley. doi:10.1002/9781118014967

Lin, R. M., & Wang, W. J. (2006). Structural dynamics of microsystems – current state of research and future direction. *Mechanical Systems and Signal Processing*, *20*(5), 1015–1043. doi:10.1016/j.ymssp.2005.08.013

Mawardi, A., & Pitchumani, R. (2005). Design of Microresonators under Uncertainty. *Journal of Microelectromechanical Systems*, *14*(1), 63–69. doi:10.1109/JMEMS.2004.839021

Moens, D., & Vandepitte, D. (2005). A survey of non-probabilistic uncertainty treatment in finite element analysis. *Computer Methods in Applied Mechanics and Engineering*, *194*(12-16), 1527–1555. doi:10.1016/j.cma.2004.03.019

Qin, Y. (2010). *Micromanufacturing Engineering and Technology*. William Andrew.

Randles, R. H., & Douglas, A. W. (1979). *Introduction to the theory of nonparametric statistics* (Vol. 1). New York: Wiley.

Rubinstein, R. Y., & Kroese, D. P. (2008). *Simulation and the Monte Carlo Method*. Wiley-Interscience.

Stanciu, I. (2012). Dispersion of SU-8 Photolithography Process for Fabrication of Microfluidic Devices. *Scientific Bulletin, Series C: Electrical Engineering and Computer Science, University Politehnica Bucharest*, *74*(2), 127–142.

Stanciu, I. (2015). Uncertainty Analysis of Mixing Efficiency Variation in Passive Micromixers due to Geometric Tolerances. *Modelling and Simulation in Engineering*, *2015*, 343087. doi:10.1155/2015/343087

Stanciu, I. (2017). *Uncertainty Analysis for Improved Design of Microfluidic Devices*. LAP LAMBERT Academic Publishing.

Wang, H., Iovenitti, P., Harvey, E., & Masood, S. (2003). Passive mixing in microchannels by applying geometric variations. In *Micromachining and Microfabrication* (pp. 282–289). International Society for Optics and Photonics. doi:10.1117/12.472888

Chapter 8
Utilization of Social Networks for Human Resources Purposes

Maryam Haghshenas
Tehran University, Iran

Abouzar Sadeghzadeh
University of Bradford, UK

ABSTRACT

Social media is revolutionizing the way people connect and share information. LinkedIn, Facebook, YouTube, Twitter, and other social media are changing the way we interact, and many organizations are struggling to respond. Given the potential risks and benefits of social media in the workplace, it is critical for managers to develop policies and procedures governing its appropriate use. This chapter identifies key issues and poses strategic questions to help guide managers in making more informed decisions when navigating social media issues in their organizations. After a brief introduction, current most popular social websites and tools are described concisely. Relationships between social media and human resources are then discussed. Utilizing social media in organizations are not without risks, which are thoroughly talked about further along with the benefits of such websites for recruitment. Finally, recommendations are made for companies that are considering utilizing social media and for companies that have already benefitted from such networks to improve their strategies.

INTRODUCTION

There has always been a high demand on human resources decision makers and managers. The change in demographic and expert scarcity has dramatically increased competition for expert employees. Companies must constantly move forward in order to keep in line with advancements in the employee market whilst being up to date with latest market trends. The most prominent key trend that has emerged among young people is the internet which has transformed from being a one-way information platform to becoming Web 2.0 that features interactive user base information exchange. With the transformation of the mobile phone to smart phones, social networks are accessible at anytime from anywhere and have become a major part of everyday life. Social media applications are being launched every day. Social media platforms are being utilized by companies to showcase themselves and as a means of communication with their customers and employees.

DOI: 10.4018/978-1-5225-5045-7.ch008

Companies must fundamentally change their culture to reap the benefits of social networks. A lot of companies portray a young trend-oriented image of themselves on social network platforms. Also, social media enables companies' communication reach to expand considerably by taking advantage of users' networks. The only downside of the introduction of social media is the lack of strategy, internal knowledge and access.

Platform choice is vital to social media success. A company will decide on utilizing a social network based on how widely their platform is used. However, other factors are also decisive in the choice of social networks. The intended use is another distinguished factor for platform choice. A company's success depends on the implementation of the utilized social network platform.

The utilization of social networks for human resources purposes is still a new concept for both companies and employees. However, applicants are mostly familiar with the human resources applications presented to them by companies. Such applications are normally based on early employee cycle stages notably employer branding and recruiting as it is easier to assess and attribute success directly. Companies should get involved in social media in an active and timely manner and increase efforts in mobile recruitment systems and integrate new technology into current human resources processes.

Once social media are correctly utilized, they add considerable value in the field of human resources. However, a properly thought through decision making and implementation process is required to guarantee success. The way in which company efforts to attract expert employees that harnesses the potential of the digital world is crucial in today's competitive workforce market.

Social Media Websites and Tools

Facebook

A social networking service where users create personal profiles, add other users as friends and exchange messages, including automatic notifications when they update their own profile. Additionally, users may join common-interest user groups, organized by common characteristics (e.g. workplace). Users can instant message each other through the website.

Blog

A discussion or informational site published on the internet that consists of entries ('posts') typically displayed in reverse chronological order, i.e. the most recent post appears first. Until 2009 blogs were usually the work of a single individual, occasionally of a small group, and often covered a single subject. More recently multi-author blogs" have developed, with posts written by large numbers of authors and professionally edited. Blogs consist of regular entries of commentary and descriptions of events (blogging). The content of a micro blog is simply smaller in size.

Google

An American multinational corporation specializing in Internet-related services; most famously its core search engine, but also a large chain of products including Google+, the second largest social networking site in the world (as of January 2013).

LinkedIn

A business-related social networking site mainly used for professional networking. Users maintain a list of contact details of people with whom they have some level of relationship, called connections. This list of connections can then be used to build up a contact network, follow different companies and find jobs, people and business opportunities.

MySpace

An online community of personal profiles. These typically include photographs, information about personal interests and blogs. Users send one another messages and socialize within the MySpace community.

Podcasts

A type of digital media consisting of a series of audio, video, PDF, or electronic files subscribed to and downloaded or streamed online to a computer or mobile device.

Twitter

A popular micro blogging service enabling its users to send and read publicly visible messages called tweets. Tweets are text-based posts of up to 140 characters displayed on the user's profile page. Users may subscribe to other users" tweets.

YouTube

A (Google-owned) video-sharing website on which users can upload, share, and view videos. A wide variety of user-generated video content is displayed, including film and TV clips as well as amateur content such as video blogging. Media corporations including the BBC also offer some of their material via the site. Most videos enable users to leave and exchange comments. Organizations are increasingly using YouTube as a way of sharing information about their company and its vacancies with potential applicants.

Yahoo

An American multinational internet corporation widely had known for its social networking services and user-generated content including online discussion boards (Yahoo! Groups), community-driven Q&A site (Yahoo! Answers) and photo sharing (Flicker).

Why Use Social Media

Given the emergence of Web 2.0, the use of social networks by companies takes far more than a simple go-ahead from management. The changing requirements of the online community demand a deep transformation to a digital corporate culture.

Social media is well-established as a consumer and brand oriented set of tools. Increasingly, social media is being offered as an innovative solution for internal effectiveness. When strategy and governance are integrated with social media tools, they can positively impact any organization. Using social media places new demands on corporate culture. And given the nature of Web 2.0, these demands cannot be ignored. Transparency and trust are essential in order to provide employees with the information and skills they need to manage a platform, because whoever is entrusted with this task requires comprehensive powers and knowledge of business activities. The willingness to share information and to disclose this to a wider audience is an essential prerequisite to successfully embed social networks within the organization. This also involves relinquishing a certain amount of control, however. The sheer volume of data created by a user-based web alone makes it virtually impossible to control the flow of company-related data. Listed companies in particular are put off by this, and often shy away from using social networks as a result. Nevertheless, the trend seems to be moving in the other direction: away from control towards greater trust, combined with rules on how to correctly use social media. A digital corporate culture is characterized by flat hierarchies and fast decision-making. Employees can experiment and take on risk within predefined boundaries. Mixed teams work interdepartmentally and pursue innovation and continual improvement, seeking to overcome challenges. The market, as well as customer and employee needs, set the tempo and priorities.

The influence of a multifaceted digital corporate culture on a company's approach to social media is heavily underestimated today. Most Companies mainly focus on using the applications. Just as important, however, are changes in behavior, role models and networks. This helps employees to think, feel and act outside the box.

Social networks are currently en vogue among the younger generation. Companies also want to improve their recruitment processes and attract trend-setting and innovative employees through as many different communication channels as possible. Companies who count the use of social networks among their core competencies are generally considered forward-looking. At the same time they're accelerating the development of social media. Expanding the number of communication channels is another key driver for using social media. And with good reason: younger generations are used to having the choice between a wide range of communication channels. In addition to the phone and email, they now use chats and video conferencing. The shift towards multichannel communication seems to be advancing within the business world. This means that today's job seekers often have a choice between printed magazines, online job markets and social networks. It's interesting that, even though using social networks is much cheaper than other tools, cost-saving doesn't appear to be a driver behind the decision to use social media.

In addition to the external advantages, these key social media leaders become even more engaged and personally bonded to the company, its mission, and goals if done correctly. To date, no standard guidelines have been developed to assist business leaders in managing social media.

The web has a clear advantage for job seekers: users can react quickly and flexibly to enquiries and can search and filter job offers in line with their own requirements. Moreover, it is significantly cheaper to post online advertisements than publish printed ones. It is little wonder, then, that traditional newspaper advertisements have long been on the way out, and that more than 80% of students now search for jobs using online job portals.

Human Resources and Social Media Relationship

World class leaders enable prominent employees to become evangelists for the company on social media platforms, assisting customers, brand building, attracting talent and giving a personal face to the firm. Apart from external advantages, social media leaders are becoming more and more engaged and bonded to the company, its mission and goals if done precisely. Such employees can come from anywhere in the company, including interns, line personnel, managers, products leaders and executives. Social media is well known as a consumer and brand oriented set of tools. Social media is being offered as an innovative solution for internal effectiveness. Once strategy and governance are coordinated with social media tools, they will impact the company positively. Any organization that utilizes social media thoughtfully will find opportunities in four areas: speeding up talent acquisition and talent management, mobility and collaboration, attaining a multigenerational workforce and the ability to manage potential risks.

Here are some recent statistics regarding various companies' utilization of social media:

- 76 percent of U.S. companies used LinkedIn's 100 million registered users for recruiting.
- 84 percent of job seekers have a Facebook profile, and 48 percent of them have done at least one job-hunting activity on Facebook in the past year.
- 61 percent of millennials don't go to the traditional company support groups first; they prefer to turn to the Web and other external resources.
- Corporate e-mail is growing 20-25 percent per year, with data storage costs soaring.
- 41 percent of 2011 university graduates used social media in their job search.
- 40 percent of companies admit to having no training or governance of social media.
- Social networking is still the fastest growing active social media behavior online, increasing from 36 percent of global Internet users to 59 percent managing their profile on a monthly basis by the end of 2011.
- China is the most socially engaged market in the world, with 84 percent of Internet users contributing at least once a month to either social networking, blogging, video uploading, photo sharing, microblogging, or forums – they are followed by Russia, Brazil, and India.
- 36 percent of social media users post brand-related content.
- 60 percent of employees would like help from employers to share relevant content.

Talent Acquisition and Management

Acquiring talent is one the main areas within many companies to take advantage of social media. If utilized correctly, this can include speeding up the recruitment process from posting openings to sourcing candidates, reviewing resumes, making offers and hiring. Talent acquisition advances may be challenging for companies internal communication and human resources. Newly hired employees with great online experience may be surprised to find that the internal systems and culture are not as sophisticated or advanced and may not even have access to those same recruiting sites at work. Once employees are on board, social media can be leveraged for connecting employees to interest groups (e.g. volunteerism, mentoring, and diversity). More sophisticated companies are using technology and social media tools to create deep expert directories that are populated by employees with information about themselves and others. Progressive organizations are dispensing with the annual performance cycle and moving to more real-time feedback and coaching. Social media tools make true 360-degree feedback a reality.

Some of the main questions arising for companies human resources in talent acquisition using social media include:

- How can we take advantage of LinkedIn and Facebook to expand the candidate pool and connect with active and passive candidates?
- How can we use social media tools to get real-time feedback on employee performance
- Our company policy is to provide employment verification but not references, but there are all kinds of LinkedIn endorsements
- Our company is committed to ethical behavior of our employees. What should we do about some of those Facebook vacation pictures?
- Intellectual property and research breakthroughs are vital to our company's success. But there are prototypes posted for collaboration on Pinterest.
- How should we adapt our working life policies when employees are connected 24/7.
- How can we take advantage of YouTube to train employees?
- One of our employees (at a hotel or airline) just posted that a famous person boarded the flight. Will this be ok?
- When we consider a candidate, should we search social media for him or her as part of a background check?

Mobility and Collaboration

The combination of new technology and mobile devices with the changing needs of employees presents a significant challenge for organizations. The ability to access information and applications from a laptop, BlackBerry®, Android™-based phone, or iPhone® from around the world has created an expectation of immediacy for today's workforce. Increased availability of "on-demand" tools can be a double-edged sword: the information is easily accessible. Many innovative and fast-growing companies place a premium on collaboration. Finding the right expert to repair a piece of equipment, reusing an existing piece of market research, convening a geographically dispersed "team" are faster and less expensive with social media tools.

Many advanced companies use internal, secure versions of instant messaging, YouTube, Twitter, and virtual classrooms to collaborate. As if aligning intricate business strategy to social media objectives, evaluating leadership and employee change readiness, and guiding the entire workforce through a change aren't enough, a multigenerational workforce adds yet another layer of complexity. Four generations make up today's workplace demographics for the first time in recent history and at the extremes, their working styles and understanding of each other are different and incongruent. This is clearly evident in workplace adoption of social media. To effectively create a social media strategy that will be adopted by employees, the unique perspectives and needs of entire workforces must be considered.

Key questions in addressing mobility and collaboration:

- Are our employees skilled in facing with vast information?
- Are our policies up to date related to reimbursement for data plans and devices, and increasingly flexible work schedules?
- Is our company using crowd-sourcing for innovation?

- Are project teams using social media for collaboration and communication?
- Are senior leaders accessible and transparent in their communication?
- Can employees access the key tools from wherever they are working?

Views of different generations in regard to social media:

- 16-30 years
- I like to ping my social media network or just send a quick text. It's all about efficiency and speed
- I think Facebook, Twitter, blogs, etc are great ways of collaboration and innovation in companies
- I don't need to see my peers to build a great relationship with them
- 31-46 years
- I prefer to do business over email. I find it more effective.
- Face to face meetings are required on occasion but in practice things can be done by email.
- I don't have a problem adjusting to Facebook and other social media sites but I don't need to use them
- 47-65 years
- Business is most effectively conducted in face to face meetings or at least over the phone
- Face to face interactions build better rapport and stronger relationships
- I have started experimenting with Facebook and Twitter but I don't understand what all the hype is about
- 66-86 years
- I have only experienced business through face to face meetings. That is just how it works best
- I don't see the value of social media as a workplace tool. It lowers productivity.
- I don't need to share everything I know with a younger workforce since I've worked long and hard to gain the knowledge I have today.

Multigenerational Workforce

While the 16-30 generation employee may have no issues being asked to suddenly use a certain social media tool, a 47-65 generation employee may feel overwhelmed and frustrated by such a request. The 66-86 and 47-65 generations alike prefer face-to-face meetings or phone calls, while the 31-46 and 16-30 generations don't necessarily need, or want, face-to-face encounters and would prefer to pick up the phone to send a text versus make a call. Relationship building for this group often occurs through simply pinging their social media networks.

Implementation of a policy or procedure centered on social media can prove to be challenging with a diverse workforce. One way to better understand how a social media policy change may affect the broader organization is through assessing how each stakeholder group may be affected by the change. Such analysis helps to identify high priority impacts and highlights key stakeholder requirements. Age is often not the basis for the issue; it's more the acceptance, readiness, and adaption to change. In fact, research has shown that the fastest-growing demographic for Facebook and other social media tools is those aged 50 and above.

However, introducing a new piece of technology to this population should be different than introducing it to those who have grown up with technology from their early years. As adoption of social media tools readily grows, for both internal and external communications, it becomes important to clearly define and articulate social media policy. While the 47-65 and 66-86 generations may be more resistant to initially using social media and accepting social media policy, the 16-30 and 31-47 generation will almost positively grow frustrated if an organization does not use social media or fails to properly manage or recognize its use. Contrary to popular belief, social media has shown to increase productivity and employee engagement if controlled appropriately. It can also foster innovation through collaboration. As a result, more than half of the organizations using social media technology have achieved best-in class performance.

To effectively implement and cultivate the appropriate usage of social media in a multigenerational workforce, it is essential for organizations to develop effective communication plans that allow for ongoing dialogue and feedback. Communication serves as a strategic change lever by facilitating the transition from awareness to the actual ownership and adoption of social media guidelines and integration of social media tools. Effective communication helps put into practice the internal and external guidelines required to govern its use. A carefully planned communication strategy often starts with a communication assessment, focusing on three components: a communication channel/vehicle analysis, an audience analysis, and a communication culture analysis. In practice, creating such a communication plan may involve analyzing current communications around social media policy and adjusting those based on the new requirements and identified stakeholder groups, while developing an implementation strategy that includes a variety of the preferred communication channels/vehicles.

Key questions in regard to attaining a multigenerational workforce:

- What type of training is available to the 47-65 and 66-86 generation?
- How does the organization use cross-generational mentoring to barter technology tutoring with business experience?
- What type of multichannel communication is used to reach different types of employees?
- Does the organization provide kiosks and big screen monitors to broadcast updates for employees without laptops or smart devices?
- Are communications and collaboration message and style consistent with employee's external presence?
- Can employees access the key tools from wherever they are working?

Benefits and Risks of Using Social Media for HR

Despite the potential risks and benefits of social media in the workplace, a recent study conducted by the Society for Human Resource Management (SHRM) found that 72 percent of organizations do not have a clear strategy or goals for their social media activities. Social media continues to evolve rapidly. This makes it difficult to produce rigorous academic studies evaluating its effectiveness for organizational use. According to a recent *Emerging Workforce Study* by Spherion, companies in 2012 are nearly twice as likely to have a social networking strategy in place as they were in 2009. However, less than half (just 45 percent) of today's organizations actually have a formal social media policy for social and digital media initiatives. Despite a greater adoption of social/digital networking initiatives and strategy, the relative success of such programs was only slightly improved in 2012. Approximately one quarter of companies in 2009 rated their efforts at achieving social networking/media goals as "successful" or "very successful." Three years later, that number increased only slightly to 30 percent of companies, indicating that organizations are still struggling to move from simply utilizing these strategies to actually reaping the benefits of social media.

Benefits

- Improved sourcing of high quality candidates
- Enhanced candidate quality
- Greater organizational identification and commitment
- Reduced turnover
- Improved employment compatibility and job satisfaction
- Reducing training costs
- Greater training effectiveness
- Continuous skill enhancement and self directed learning
- Fast and inexpensive dissemination of information
- Easy knowledge sharing from few up to many
- Ability to reach new customers
- Enhancement of customer loyalty
- Increased cost efficiency
- Increased accuracy for problem solving
- Innovation spread throughout organization with minimum effort
- Reinforcement of organizational culture
- Culture change through communication

Risks

- Employment discrimination (e.g. obtaining personal information that is not job related from web-sites such as Facebook and using it in hiring decisions).
- Variability in applicant use of social media disadvantages those who do not use it (some age groups or racial/ethnic groups maybe be underrepresented on social media sites).
- Spread of inappropriate or damaging information (e.g. negative information posted by disgruntled employees could discourage new hires).
- Creation of division rather than inclusion (e.g. establishment of cliques or subcultures).
- No positive effect on learning or development (e.g. skills to be learned are not transferred to the job).
- Unequal access or usage (e.g. those who do not use social media have fewer opportunities for training or development).
- Too much information (e.g. work time consumed with reading social network posts).
- Quick spread of negative or private information (e.g. co-workers learn of an employee's serious medical condition).
- No relevant effect on customer metrics (e.g. investment in social media does not generate sales or customer growth).

- Hidden costs with maintenance of media (e.g. staff time to monitor social network sites and constantly provide fresh content).
- Consensus process slows decision making (e.g. the higher number of people involved in decision making increases time needed to resolve an issue).
- Consensus leads to pursuit of poor choices; lack of critical thinking (e.g. groupthink or an influential employee generating support for a bad idea).
- Broad resistance to organizational change (e.g. employers bond together to resist change).
- Facilitation of undesirable company culture (e.g. employees use social media to form coalitions counter to the organization culture).

Issues That Hinder the Use of Social Media

Various surveys carried out in large organizations that were hesitant in utilizing social media indicated that a lack of strategy for using social media across the company as a whole was the main reason for their hesitation. Time consumption (e.g. content update), legal considerations, little added value compared with traditional channels, lack of internal know how, and lack of employee interest were also some of the other reasons organizations are reluctant in using social media.

A clear strategy prevents a detached presence on social networks being perceived as the enclave of an inward-looking HR department. Users should see a consistent company image across all communication channels. This requires a cross-departmental HR and marketing strategy that also incorporates the use of social media. In order to take advantage of the value added by social networks, a company needs a multichannel strategy that allows a degree of freedom as well as opportunities to adjust layout and design. This also means accepting that there are uncontrollable factors involved. The amount of time required to effectively manage social platforms was almost as frequently cited as a reason for not using social media. It's worth weighing up the time required against the value added: if HR benefits in terms of quantity or quality from using social networks, it justifies the associated time investment.

More than 70% of companies using social media perceive the danger that they will not be able to keep content sufficiently up to date due to a lack of staff or skills. Continually updating content and providing speedy responses to user input is a time-consuming process. Nevertheless, it is an absolutely essential one, because users will only visit a profile on a regular basis if it is actively managed.

Social media are often not used at all because of a lack of internal know-how. 50% of companies that use social media cite an incorrect approach due to a lack of know-how as one of the major hurdles to successful implementation. This conflict is due first and foremost to a generational issue: the necessary know-how regarding how to use social networks is exhibited by younger employees in their private spheres, whereas within companies it is primarily the older generation that makes the relevant decisions. This expertise simply needs to be made accessible to the business as a whole, without this leading to misunderstandings and fears about control on the part of managerial staff.

Depending on how social media are used, the return on investment can be measured in various ways. These include using the number of job applications within a specific period of time or through a specific (social media) channel, measuring the cost per channel and per candidate, or examining reduced response times or the qualifications of job applicants. Nevertheless, there are also a number of variables that can't be measured directly but can be positively influenced by social networks. For example, it's difficult to put concrete figures on the link between employee loyalty or satisfaction and the integration of social media. The added value of social media is similar to that of other HR tools such as training measures, company events, internal communication, etc.: it's impossible to put an exact figure on the return on investment for these tools, but they're still an indispensable element of successful HR.

Selecting the Right Platform

It is vital that each area within HR choose the social media platform that is right for them. Several factors come into play here: on the one hand, the number of users varies widely from platform to platform, while on the other, the differences in the way these platforms operate mean that some are more suitable for HR purposes than others.

In addition to size, there are several other criteria that determine which platform is the most suitable. The number of users determines the reach each social media website could have and what its potential audience may be. However, the nature and purpose of a platform are generally far more significant factors when selecting a platform. In addition to how well known a platform is, there are clearly a number of other factors that are central for recruitment. Companies consider Facebook to be only partially suitable, while LinkedIn, for example, is much less well known but regarded as more suitable. There are two possible reasons for this: private sphere and structure.

Facebook is mainly used privately. Users post photos from their leisure time, talk about their hobbies, and in most cases do not use the site with any business intentions in mind. This environment makes it difficult to build business connections. Nevertheless, private communication platforms such as Facebook and Twitter can also be extremely helpful for HR purposes since information can go viral much more easily. The difficulty lies predominantly in achieving a degree of personal involvement. Not all platforms are equally suitable for uses that require specific tools. Job searches require a search function, while teambuilding and on boarding benefit from a group chatting function. Microblogs such as Twitter are only suitable for short, ad hoc messages with a maximum of 140 characters. A HR team would therefore be well advised to examine in advance what's required of a platform and what its strengths are.

The most vital factor when evaluating a platform will always be how well known it is. Some interesting networks such as Equipia and Kununu have large potential but aren't used as much by companies because of their limited target audience. A platform's popularity isn't the only factor that decides its suitability for HR purposes, however; the purpose for which it is used plays a critical role in how business content is accepted. The structure of a platform should also be taken into account in terms of its intended use by HR. A reliable procedure for selecting social media platforms which is applicable to all situations has not yet been developed. Therefore, success ultimately depends on how the choice is used in practice. Also, to choose the right platform, companies should be guided by a certain amount of instinct and experience.

Managing Potential Risks Related to Social Media

Many companies that use social media struggle to successfully integrate it into their daily business routines. The challenge is frequently found within the creation, execution, and enforcement of social media policies. Many companies do not establish policies to govern social media usage or can find their policy difficult to enforce due to lack of employee engagement and training on the topic. Given these challenges, it is not surprising that many executives prefer to avoid the topic by saying that they have no need for social media or policies to manage it. However, as companies begin to realize the true value to be gained by social media adoption, consideration of workforce risk is essential for avoidance of loss of employee goodwill, information breaches, and reputational damage. Many companies' social media policies are negatively oriented, directing what not to do. Consider the impact to organizations of a positive social media policy with a positive orientation.

Successful organizations clients have successfully delivered real value through formally engaging their workforce with enhanced social media policies and engagement models. When these are defined and communicated to employees, a new set of evangelists are created. They can aid the organization in carrying positive messages to the marketplace regarding the quality of their products and services as well as the nature of the workplace itself being a collaborative environment where creativity and expression of new ideas is both welcomed and encouraged.

Organizations are progressing to the next level of adoption but are still struggling with the question, "How do we best manage our social media programs?" These same organizations are challenged with fully adopting social media and capitalizing on all of its benefits. The rapid evolution of social media and diversity of related applications creates a highly challenging legal environment that will need to be continually monitored. Organizations must consider workers' rights as they customize social media policies. The ultimate goal should be to provide guidance that does not contravene legal frameworks. Successful firms' professionals have found that the leaders who are charged with creating social media policies are often not a typical social media user. Policy-making executives should start with learning more about social media tools and how both the organization and its employees are using them. This knowledge gap can be a considerable challenge to overcome, but in many cases, is addressable through third-party courses or the use of management consulting services.

Once leaders are better informed on the nature of social media utilization in their organization, they can prepare a strategy to leverage social media as a tool that aligns with the overall vision, mission, and strategy of the organization. Managers responsible for the performance of others should also now consider how tools such as LinkedIn and Facebook are being employed by their employees to build professional networks and how performance in these areas can be quantified and used for performance evaluations.

While social media policy creation has challenges, it is not so different from previous challenges such as the adoption of corporate-wide e-mail in the 1990s. Many organizations already have the framework of a policy that governs communications and this framework can provide an excellent foundation to consider where social media might affect the business. Therefore, a good place to start is to create a governance body of key stakeholders. Stakeholders can be defined as any person/group of individuals, internal or external to the organization who will be impacted by the proposed changes, or who could have an impact on the success of the adoption. Traditionally, this would include a cross-functional representation of IT, Legal, HR, Compliance, Marketing, and Risk Management that reviews current and planned use of social media against current communications-related policies. It is important to have HR and Legal involved in this discussion to make sure the emerging policies strike an appropriate balance between workplace and personal use while maintaining some level of corporate oversight. Above all, the forums created to consider social media use in the workplace must strive to provide policies that are enforceable while being absolutely clear on what is mandatory and what the consequences are for non compliance. These policies should result in guidelines that easily create awareness and understanding of your organization's position on social media adoption. Your unique framework can be considered successful when it can demonstrate that it has been effective in building organizational knowledge of the benefits and risk of social media, and is supported through ongoing tailored communication and training.

Overall, there are two sets of risks associated with organizations in terms of social media:

Internal

- Employees accidently or deliberately leak privileged information that belongs to the organization.
- The creation of public and therefore discoverable public record related to internal employment issues.
- Ownership issues regarding public forums/networks where employees might engage with customers to discuss business issues.
- Introduction of sensitive personal information into the workplace (politics, religion, sexual orientation, lifestyle issues).

External

- The potential for viral growth of negative sentiment either in response to organization communications or simply a market discussion on products or services.
- Inappropriate or inaccurate commentary on organization financial performance.
- Misrepresentation of organization's position on public issues.
- Damage to company's reputation or desired brand image.
- Data privacy breach related to lost personal identifiable information.

Key questions regarding managing potential social media risks:

- Where guidelines for managing social media should be defined?
- What areas of the organization are affected?
- How does the organization manage a rapidly evolving environment?
- How can policies be developed that are encompassing and applicable to everyone yet specific enough to be effective?

Social Recruiting

Social websites are generally the go-to social media site for social recruiting efforts. However, the company needs to consider specific target audience. If the type of candidate trying to reach is more likely to have an active presence or Facebook or Twitter than on LinkedIn, then it's important to refocus or at least expand.

One of the benefits of social media is reaching a potentially very large and varied audience. Taking a blanket approach can be extremely helpful in recruiting efforts, particularly if the goals are to use social media efforts to promote company or put out a broad and general hiring announcement. However, the drawback of this unfocused method is that while may get a large quantity of applicants, the majority won't necessarily be from high-quality candidates.

There are a number of ways that can fine-tune social recruitment efforts to target the candidates that the company wants, starting with using the right keywords. In any social media job posting or announcement, keep Search Engine Optimization best practices in mind use key phrases and words that describe both the position and potential candidates, and that would likely be used in a search. LinkedIn allows creating job postings with specific industry, experience, and educational requirements, and also gives employers the ability to search for and reach out to individual, handpicked candidates. Using Facebook ads for job postings are another option for reaching a very specific audience, as they'll only shows up on the profiles of candidates that match company's specified requirements.

The recruitment board may have created an exciting job posting that target the ideal demographic and catches the eye of highly qualified candidates, but if application process is convoluted and complicated, it may all be for nothing. LinkedIn provides the opportunity to allow candidates to apply directly from the job posting. Link directly to an online application in Facebook or Twitter posts (or clearly provide contact info, such as an email address to submit resumes).

Interested candidates are going to take the time to research about company before applying, and candidates reach through social media are likely going to turn company social media accounts to learn more. In addition to website; social media profiles need to suggest the same. Bare-bones profiles with few details and little activity aren't likely to create much excitement therefore using social media profiles should highlight company qualities, achievements and benefits.

Examples of Recruiting Through Popular Social Media Websites

Recruitment Through Facebook

Facebook is now the most visited place on the web. Facebook is a relatively closed network it can make direct sourcing a challenge. Depending on the user's security settings, visitors are allowed certain access making sourcing on this platform somewhat difficult. Even still, there are ways to grow candidate pool, source, and fill positions recruiting on Facebook. Facebook's ad network offers recruiters a unique opportunity to target job advertisements to Facebook users by keyword, education, location, and even age. Facebook users are providing detailed information and news about their own interests, hobbies, education, and activities throughout the day and at staggering rates. The average Facebook user now spends 15 hours and 33 minutes per month on Facebook. Facebook's ads capitalize on their members and the information that is voluntarily submitted by offering advertisers very small and specific areas in which to target for not a lot of money. To keep up with the competition for top talent, companies must recruit with Facebook ads. Facebook Fan Page can serve as a recruiting and sourcing tool just like a company blog or opportunities to directly engage and educate target candidate market. Approximately 81% of Facebook's monthly active users are outside the U.S. and Canada. Facebook has seen a 41% growth in active users from Russia, South Korea, Japan, India and Brazil during 2012. 70% of Facebook-using job seekers are male, 63% are under the age of 40, 40% earn more than $75,000 and 36% are college graduates.

Recruitment Through LinkedIn

LinkedIn had more than 187 million members in over 200 countries and territories. 63% of LinkedIn members are located outside of the United States new members sign up to LinkedIn every second. There are more than 1 million LinkedIn groups. 38% of job seekers use LinkedIn to help find work in 2012,.19% of job seekers had a contact share a job on LinkedIn,11% of job seekers searched for jobs on LinkedIn and 79% of LinkedIn users are 35 or older; the average LinkedIn user is 40.5 years old. 60% of LinkedIn-using job seekers are male, 62% are under the age of 40, 51% earn more than $75,000 and 50% are college graduates.

Recruitment Through Twitter

Recruiters and hiring managers can also leverage recruiting on Twitter as a Social Media candidate source in a number of different ways. Searching for candidates via hash tags can net a solid list of job search and recruiting leads long before candidate has responded to a resume-mining database like CareerBuilder or Monster. Hash tag feeds on Twitter allow companies to go beyond their audience (followers) when posting, and when reading hash tagged Tweets, companies can target thought leaders in the industry. Like any job board or social network, finding top talent is the top priority of e-recruiting. 67% of Twitter-using job seekers are male, 69% are under the age of 40, 46% earn more than $75,000 and 44% are college graduates.

Recruitment Through Google+

Google+, Google's social media platform is the fastest growing social network. It provides recruiters and hiring managers another wide channel to recruit, source, and search. Google+ offers a unique opportunity for recruiters to blaze new trails and connect with candidates in different ways. Users add individuals they wish to engage by categorizing them into circles, which they create. Through circles, one can contact individuals or groups privately or publicly to recruit on Google+.

Strategic Planning

Choices about how to engage or respond to social media must first be addressed from a strategic perspective. When considering strategic issues, it is helpful to create a cross functional task force represented by key stakeholders. These stakeholders may include leaders from relevant areas such as finance, human resources, legal, IT, marketing, accounting and operations. Because social media has the potential to affect all parts of the organization, it is also important to get feedback from a diverse cross-section of the employee population and perhaps even customers and shareholders or investors as well.

Key questions that should be addressed in strategic planning include:

- Should the organization consider using social media?
- What are the organizational benefits and risks of utilizing or not utilizing social media?
- What strategic goals are involved?
- In what ways can social media support the organization's business strategy?
- Who are the stakeholders affected by social media?
- Who is responsible for social media strategy?
- Are there enough resources to implement social media effectively?
- What mechanisms are there to oppose risks?
- How will the organization as a whole engage in social media?
- How should the organization implement social media?
- For organizations planning to use social media, the process must begin with identifying goals and metrics
- Scope of engagement should be defined. For example, with the talk about the talk about social media benefits, there is little evidence that show it results in quality hires or is faster and less expensive that traditional recruiting methods

- Which types of social media should be used? For example, professional networking sites such as LinkedIn may be of more use compared to sites such as Facebook
- New social media policy should be aligned with existing corporate privacy policies
- The amount of time needed to implement strategic plans should be determined
- Ownership must be determined. Who has budget responsibility? Who is responsible for reviewing content? Who is responsible for posting information? Who has final authority for making content changes?
- Progress monitoring. The effectiveness of social media activities should be measured against defined metrics regularly and corrections should be make when necessary
- How can the organization maximize success while minimizing risks?
- A formal social media policy should be developed. Organizations should create at least a basic formal policy or a set of guidelines even if social media is not yet utilized. Such policies should be vetted with stakeholders along with legal and human resources. It will likely be necessary to implement an organizational policy and then be more specific with functional policies such as human resources or marketing. The organization's position on privacy issues must also be addressed.
- A monitoring process should be developed. Who is the entity or person responsible for ensuring compliance with social media policy?
- Directives for policy violations. Consequences for policy violations should be identified.
- Senior management should enforce defined policies and consistently support them
- How can the organization maximize the value of specific social medial applications such as recruitment and selection?
- Equivalent alternatives exist for those who do not have social media access. Recruitment activities should be limited exclusively to social media websites
- Social media should be used for targeted recruiting, sourcing passive and active applications, knowledge sharing, training, reinforcing identification with the organization and promoting the company brand
- Social media should not be used for making human resources decisions such as candidate selection, promotion, disciplinary or termination decisions

Recommendations

If social media are implemented correctly, they can add value in the field of HR. Nevertheless, a well thought-through decision-making and implementation process is needed to ensure success. The results also bring to light which hurdles and stumbling blocks an HR team may encounter while trying to successfully implement social media, and what they should pay particular attention to. Here are some prominent recommendations for action, which are of use for every HR department, irrespective of sector or focus.

- Build top management consensus to drive digitization across human resources processes. The success of a large-scale digitization effort depends on top-management support. Top-management resistance towards digitization in HR is often linked with a lack of understanding of the business benefits of digitization. To build top-management consensus, HR should develop a clear vision for digitization that takes a long-term, holistic view of talent acquisition and engagement. The vision should be clearly aligned with the overall business goals of the organization so that HR is positioned for a more strategic role that of a profit driver rather than an administrative cost center. The vision should be backed by a strong business case for technology investments. As part of the vision, HR must also lay down clear guidelines for the use of social media by employees, to ensure that the employer brand is not negatively impacted by the information shared by employees on external media.

- Upgrade to a flexible, interoperable IT structure. In order to roll out initiatives that cut across departments and geographies, HR will need IT systems that are interoperable. Since the current HR IT landscape in most organizations consists of fragmented, localized systems that do not support end-to-end process integration, this means HR will need to graduate to a more flexible IT architecture. Additionally, HR IT systems will need to support mobile and social media integration and at the same time, be flexible enough to accommodate future advancements in technology. The shift to a cloud-based model can address many of these requirements as cloud based platforms are easily scalable and upgradable, can be managed centrally, and implemented at lower costs than a full-scale legacy upgrade.
- Deliver a seamless multichannel experience. HR should use social and mobile platforms to build stronger relationships with employees as well as prospects. For the user, the experience of interacting with HR should be seamless across channels. Mobile websites and applications should be designed such that they are easy to use and simplify the applicant's access to information. HR should also aim to personalize the experience based on the user's profile. Building a digital HR strategy will need to be an ongoing process for organizations. HR will need to continuously monitor digital channels to ensure that it is listening and responding to the needs of its target audience. In a competitive marketplace, it is steps like these that will differentiate an organization and enable it to attract and retain the best talent.
- Use data to guide decision making. A shift to a data-driven approach to decision making is a crucial element of digitization that HR needs to address. Data should be used across all HR processes — including recruitment, learning and development, and performance management. As a first step, HR should develop a unified view of all employee data. To do so, HR will need to consolidate data that currently resides in multiple systems spread across business units and geographies into a centralized repository. The next step should be to invest in analytics tools and skill sets so that the data can be utilized effectively. Finally, HR should establish metrics and KPIs that will determine the effectiveness of its initiatives. These steps will enable HR to prioritize its investments and develop targeted programs. By not digitizing fast enough, organizations risk being left out in the race to attract and retain the best talent. Digital technologies offer a host of innovative ways to enhance employee experience and organizations that have realized this, have reaped rich rewards. For the others, it is time to take swift action towards envisioning, defining and implementing a comprehensive digital HR strategy.
- Get an overview of the platforms and their possible uses so you can find the most suitable platform for your purposes. Your company may already be represented or being discussed on certain platforms – perhaps among customers, perhaps among (potential) employees.
- Create space so that your company can gather experiences with social networks. Don't try to ban social networks and online communities from the workplace – thanks to smart phones, you'll never succeed anyway.

- Develop a strategy for using social media. Set rules so that social media are used consistently throughout the company. Don't only consider HR, but work closely with your PR, marketing, production and compliance departments as well. Don't set your strategy in stone; otherwise you won't be able to respond to short-term developments.
- Set clear boundaries for the use of social media. It's time-consuming and therefore costly to update social media. Concentrate on the platforms that are relevant for you – that way you can make more out of less.
- Make use of the know-how within your company, hold training and brainstorming sessions and allow your employees to feed in their personal experiences with social networks. You'll be surprised how much knowledge you already have within the company.
- Provide sufficient human resources and equip staff with the necessary skills. It takes time to keep content up to date and communicate with users. Employees entrusted with these tasks need to be able to process input from outside the company quickly and flexibly.
- Integrate social media into your existing HR processes. The new media shouldn't be managed in isolation, but should be actively linked to on boarding, development and alumni processes, for example. This helps create a consistent and uniform experience.

CONCLUSION

It is clear that social media tools are transforming the way we work and will continue to do so in ways we can only imagine. It is critical for Human Resource professionals to understand what these tools are used for and assess the risks and opportunities they may present to an organization. Social media is redefining how companies innovate by connecting people and ideas in ways that have previously not been explored. Organizations now have the ability to utilize knowledge from external audiences around the globe that they could never reach before. It is changing the way we do business today. Additional benefits include the ability to enhance your public image, improve interactions with customers, and promote employee engagement. But it is not without internal and external risk.

Key External Forces

- Social media technologies are constantly evolving.
- They enable access to vast amounts of information from anywhere.
- Flexibility, mobility, transparency and diversity are expected by employees.
- Multigenerational employees have a wide range of skills, expectations and loyalty.
- Privacy concerns are increased

Social Media Considerations

- Complete guidance strategy to manage all risks consistently.
- Collaboration with appropriate skills to succeed in the working environment.
- Overall internal strategy to aid the wider organizational approach.

Opportunity for Positive Impact

- Social media enables the acceleration of employee engagement and communication.
- The whole recruiting process is transforming along with candidate-company relations.
- Social media can be leveraged by learning opportunities.
- Social media can potentially transform talent and performance management.
- Human resources delivery can be improved via newer technologies.

New technologies emerge quickly, and there is little legal precedent or research to steer appropriate action. Truly world-class leaders empower key employees to become evangelists for the firm on social media platforms, helping customers, building the brand, attracting talent, and giving a personal face to the company. Social networks offer a powerful tool for recruiters to reach a pool of qualified candidates that they might not otherwise be able to reach. Social media is here to stay and employers can gain a significant advantage by adopting hiring methods that have a social media element. There is no doubt that social media has improved the recruitment process by making it more open and democratic; increasing the visible talent pool from which to engage and recruit. Having an intimate knowledge of someone's capabilities or knowing who the best person for a role is, however, can only be gained through personal knowledge of an individual and of a particular industry sector. The social media sites also recognize this, and new tools are becoming available all of the time that allow recruiters and employers to work together to take advantage of social media hiring but it is not without internal and external risk.

REFERENCES

Barney, J. (1991). Firm resources and sustained competitive advantage. *Journal of Management*, *17*(1), 99–120. doi:10.1177/014920639101700108

Baron, R. M., & Kenny, D. A. (1986). The moderator-mediator variable distinction in social psychological research: Conceptual, strategic, and statistical considerations. *Journal of Personality and Social Psychology*, *51*(6), 1173–1182. doi:10.1037/0022-3514.51.6.1173 PMID:3806354

Bartram, D. (2000, December). Internet recruitment and selection: Kissing frogs to find princes. *International Journal of Selection and Assessment*, *8*(4), 261–274. doi:10.1111/1468-2389.00155

Becker, B., & Huselid, M. A. (1998). High performance work systems and firm performance: A synthesis of research and managerial implications. In G. R. Ferris (Ed.), *Research in personnel and human resources management* (Vol. 16, pp. 53–101). Greenwich, CT: JAI Press.

Bernard, M. (2013). *People Management: Data and Analytics Will Change It Forever, November 2013*. Retrieved from https://www.linkedin.com/pulse/20131111071432-64875646-2-keys-to-a-very-different-people-management

Bliese, P. D. (1998). Team size, ICC values, team-level correlations: A simulation. *Organizational Research Methods*, *1*(4), 355–373. doi:10.1177/109442819814001

Burt, R. S. (1982). *Toward a Structural Theory of Action: Network Models of Social Structure, Perception and Action (Quantitative studies in social relations)*. New York: Academic Press. doi:10.1016/B978-0-12-147150-7.50014-4

Cohen, D., & Prusak, L. (2001). *In good company: How social capital makes organizations work*. Boston: Harvard Business School Press.

Doherty, R. D. (2010). Getting social with recruitment, Strategic HR Review. 9(6), 11-15. *International Journal of Selection and Assessment*, *8*(4), 261–274.

Feldman, D. C., & Klaas, B. S. (2002). Internet job hunting: A field study of applicant experiences with on-line recruiting. *Human Resource Management*, *41*(2), 175–192. doi:10.1002/hrm.10030

Flap, H. D., & Volker, B. (2004). Social networks and performance at work. In H. D. Flap & B. Volker (Eds.), *Creation and returns of social capital* (pp. 172–196). London, UK: Routledge.

Florian, W. (2009). E-Recruitment Vs. Traditional Recruitment: A Descriptive Analysis. GRIN Verlag.

Friedken, N. E., & Johnson, E. C. (1998). *A structural theory of social influence*. Cambridge, MA: Cambridge University Press. doi:10.1017/CBO9780511527524

Giniger, S., Dispenzieri, A., & Eisenberg, J. (1983). Age, experience and performance on speed and skill jobs in an applied setting. *The Journal of Applied Psychology*, *68*(3), 469–475. doi:10.1037/0021-9010.68.3.469

Globoforce. (2012). SHRM/Globforce Employee Recognition survey FALL 2012 Report. In *The business impact of employee recognition, 2012*. Author.

Granovetter, M. S. (1985). Economic action and social structure: The problem of embeddedness. *American Journal of Sociology*, *91*(3), 481–510. doi:10.1086/228311

Haimeecode. (2012). *The State of Employer Branding, October 2012*. Retrieved from https://www.slideshare.net/haimeecode/state-of-employer-brand

Hair, J. F., Black, W. C., Babin, B. J., Anderson, R. E., & Tatham, R. L. (2006). *Multivariate data analysis* (6th ed.). Upper Saddle River, NJ: Prentice-Hall.

Jason, B. (2013a). *20 Mobile Recruiting Insights #mrec13, September 2013*. Retrieved from http://www.talenthq.com/2013/09/20-mobile-recruiting-insights

Jason, B. (2013b). *45 Mobile Recruiting Tips and Trends From #mrec, September 2013*. Retrieved from http://www.talenthq.com/2013/09/45-mobile-recruiting-tips-and-trends-from-mrec

Jeanne, M. (2013). *Gamification In Leadership Development: How Companies Use Gaming To Build Their Leader Pipeline, September 2013*. Retrieved from https://www.forbes.com/sites/jeannemeister/2013/09/30/gamification-in-leadership-development-how-companies-use-gaming-to-build-their-leader-pipeline/#7bfa74325a5

Jen, D. (2012). *How to use Social Media for Employer Branding, July 2012*. Retrieved from https://www.lever.co/blog/how-to-use-social-media-for-employer-branding

Jobvite. (2012). *2012 Social Job Seeker Survey*. London: Forum Jobs, Inc.

Juilo, R. (2013). *HMV tweets their own sacking, January 2013*. Retrieved from https://www.marketing-technews.net/news/2013/feb/01/hmv-tweets-their-own-sacking

Lars, S. (2012). *Three Days, Three Key Themes: My Review of Talent Connect 2012, October 2012*. Retrieved from https://business.linkedin.com/talent-solutions/blog/2012/10/talent-connect-2012-3-themes

Luis, S. (2013). *2013: The Year of Social HR, January 2013*. Retrieved from http://www.elsua.net/2013/01/09/2013-the-year-of-social-hr-2

Mike, W. (2012). *Making the Quantum Leap: UPS Social Media Recruitment ROI 2012, January 2012*. Retrieved from https://www.slideshare.net/MikeVangel/real-social-media-recruitment-roi-from-ups-11278020

Oxford Economics. (2012). *The Digital Transformation of People Management", March 2012*. Retrieved from https://www.oxfordeconomics.com/my-oxford/projects/129033

Ployhard. (2006). Social media in the workplace: issues and strategic questions. SHRM Foundation: shaping the future of HR. In *Human resources and social media*. KMPG International Cooperative.

Ryan, H. (2013). *Inside HMV's Twitter Disaster: What went wrong (and how to keep your company safe), February 2013*. Retrieved from https://www.linkedin.com/pulse/20130214190038-2967511-inside-hmv-s-twitter-disaster-what-went-wrong-and-how-to-keep-your-company-safe

Sarah, P. (2012). *With 40% of Traffic from Outside the U.S., Glassdoor Raises $20M Series D to Expand Internationally, October 2012*. Retrieved from https://techcrunch.com/2012/10/31/with-40-of-traffic-from-outside-u-s-glassdoor-raises-20-million-series-d-to-expand-internationally

Snell, S. A., Youndt, M. A., & Wright, P. M. (1996). Establishing a framework for research in strategic human resource management: Merging resource theory and organizational learning. In G. R. Ferris (Ed.), *Research in personnel and human resource management* (Vol. 14, pp. 61–90). Greenwich, CT: JAI Press.

Sykes, T. A., Venkatesh, V., & Gosain, S. (2009). Model of acceptance with peer support: A social network perspective to understand employees' system use. *Management Information Systems Quarterly*, *33*(2), 371–393.

Thompson, J. D. (1967). Organizations in Action: Social Science Bases of Administrative Theory Classics in Organization and Management Series. New York: McGraw-Hill.

Tsai, W. (2002). Social structure of "coopetition" within a multiunit organization: Coordination, Competition, and intra organizational knowledge sharing. *Organization Science*, *13*(2), 179–190. doi:10.1287/orsc.13.2.179.536

Tsai, W., & Ghosal, S. (1998). Social capital and value creation: The role of intrafirm networks. *Academy of Management Journal*, *41*(4), 464–476. doi:10.2307/257085

Vivian, G. (2013). *Gamification' Techniques Increase Your Employees' Ability To Learn By 40%, September 2013*. Retrieved from http://www.businessinsider.com/gamification-techniques-increase-your-employees-ability-to-learn-by-40-2013-9

Waldman, D. A., & Avolio, B. J. (1986). A meta-analysis of age differences and job performance. *The Journal of Applied Psychology*, *71*(1), 33–38. doi:10.1037/0021-9010.71.1.33

Walker, G., Kogut, B., & Shan, W. (1997). Social capital, structural holes and the formation of an industry network. *Organization Science*, *8*(2), 109–125. doi:10.1287/orsc.8.2.109

Wasko, M. M., & Faraj, S. (2005). Why should I share? Examining social capital and knowledge contribution in electronic networks of practice. *Management Information Systems Quarterly*, *29*(1), 35–57.

Welbourne, T. M., & Andrews, A. O. (1996). Predicting the performance of initial public offerings: Should human resource management be in the equation? *Academy of Management Journal*, *39*(4), 891–919. doi:10.2307/256716

Wright, P. M., Dunford, B. B., & Snell, S. A. (2001). Human resources and the resource-based view of the firm. *Journal of Management*, *27*(6), 701–721. doi:10.1177/014920630102700607

Yang, S., Lee, H., & Kurnia, S. (2009). Social capital in information and communications technology research: Past, present, and future. *Communications of the AIS*, *25*, 184–220.

Chapter 9
Stochastic Methods Applied to Structural Mechanics:
Reliability and Optimization Methods

Siham Ouhimmou
University Hassan II, Morocco

ABSTRACT

Uncertainty modelling with random variables motivates the adoption of advanced PTM for reliability analysis to solve problems of mechanical systems. Probabilistic transformation method (PTM) is readily applicable when the function between the input and the output of the system is explicit. When these functions are implicit, a technique is proposed that combines finite element analysis (FEA) and probabilistic transformation method (PTM) that is based on the numerical simulations of the finite element analysis (FEA) and the probabilistic transformation method (PTM) using an interface between finite element software and Matlab. Structure problems are treated with the proposed technique, and the obtained results are compared to those obtained by the reference Monte Carlo method. A second aim of this work is to develop an algorithm of global optimization using the local method SQP. The proposed approach MSQP is tested on test functions comparing with other methods, and it is used to resolve a structural problem under reliability constraints.

INTRODUCTION

The lack of penetration of rational uncertainty analysis methods in engineering practice is hardly due to insufficient theoretical foundations or the scarcity of efficient algorithms. Indeed, uncertainty and reliability analysis in structural engineering have been vibrant topics of research for several decades (Mohamed &Lemaire, 1995; Der Kiureghian & Ke, 1988; Lemaire, Mohamed, & Flores-Macias, 1997). While a large portion of the associated efforts have focused on shedding light on the fundamental and theoretical aspects and on the application of the uncertainty analysis methods to strongly simplified, reduced-order models of structures, significant progress has also been made recently in the rational treatment of uncertainties in large FE models of complex structures. The structural designer must veri-

DOI: 10.4018/978-1-5225-5045-7.ch009

fy, within a prescribed safety level, the serviceability and ultimate conditions, commonly expressed by this inequality $S_d \prec R_d$

Where S_d represents the action effect, and R_d is the resistance. The intrinsic random nature of material properties and actions is actually considered by the Eurocodes (Mohamed & Lemaire, 1995; Kadry, 2006), which classify the methods available to deal with this randomness in three levels:

- Semi-probabilistic or level 1 method, the most used in common practice, where the probability of failure is indirectly considered through the definition of characteristic values and the application of partial safety indexes.
- Approximate probabilistic or level 2 methods such as the first order or second order reliability methods (FORM/SORM) where the probability of failure is based on the reliability index β (Muscolino, Ricciardi, & Impollonia, 2000).
- Exact probabilistic or level 3 methods, where the probability of failure is computed from the joint probability distribution of the random variables associated with the actions and resistances.

A fundamental problem in structural reliability analysis is the computation of the probability integral,

$$P_f = \Pr ob\left[G\left(X\right) \leq \left(0\right)\right] = \int_{G\left(X\right)\leq 0} f\left(X\right)dX \qquad (1)$$

Where $X = [X_1,..,X_n]^T$ in which T is the transpose, is a vector of random variables representing the uncertain parameters of considered structure, $f(X)$ is the probability density function of X, $G(X)$ is the Limit State Function defined such that: $G(X) \leq 0$ is the domain of integration denoted the failure set, and P_f is the probability of failure. The difficulty of computing this integration led to development of various methods of reliability analysis such as Monte Carlo, FORM and SORM (Hasofer & Lind, 1974, Muscolino, Ricciardi, & Impollonia, 2000), and Probabilistic Transformation Method PTM (Lemaire, Mohamed, & Flores-Macias, 1997).

Considering the properties of the structural model realistically it is necessary to take into account some uncertainty. This uncertainty can be conveniently described in terms of probability measures, such as distribution functions. It is a major goal of reliability methods to relate the uncertainties of the input variables to the uncertainty of the structural performance. Based on their meaning in reliability methods, the sources of uncertainty may be the actions (e.g. loads, stress), or system data (geometry, boundary conditions, mass density).

The Probabilistic Transformation Method (PTM) is an efficient reliability method to solve problems of mechanical systems with uncertain parameters. The advantage of this method is finding the "exact" expression of the Probability Density Function (PDF) of the solution when the PDF of the input variable is known. In many cases, the structural load effect cannot be expressed explicitly and some finite element calculations are necessary. Coupling the Finite Element Analysis (FEA) with The Probabilistic Transformation Method (PTM) is therefore necessary. In this paper, a proposed Method: Finite Element Analysis (FEA) coupled with the Probabilistic Transformation Method (PTM) is applied in order to evaluate numerically the probabilistic and statistical characteristics of the response of stochastic mechanical system. It involves four main steps: (1) sampling on input random variables, (2) using Finite Element Analysis (FEA) to have the response variable of system (3) estimating the Probabilistic Density Function (PDF) of the response variable using the Probabilistic Transformation Method (PTM), (4) concluding the probability of failure and reliability of systems. To show the advantage of the proposed method, we have carried out different applications to cover several structural problems.

Notation

- **PTM:** Probabilistic Transformation Method.
- **FE:** Finite Element.
- **PDF:** Probability Density Function.
- **FEA:** Finite Element Analysis.
- **FEACPTM:** Finite Element Analysis Combined to Probabilistic Transformation Method.
- **FORM:** First Order Reliability Method.
- **SORM:** Second Order Reliability Method.

$G(.)$: Limit State Function.

P_f: Probability of failure.

N_f: The number of simulation samples in which

$G(.) \prec 0$.

N: is the total number of simulation samples.

CDF: Cumulative Density Function.

1. METHOD OF ANALYSIS

1.1. General Remarks

The reliability analysis of the structure with uncertain parameters requires two modeling steps, namely the construction of a mathematical mechanical model and the modeling of the uncertainties. For the mathematical mechanical model of structures, the Finite Element Analysis (FEA) is the standard tool for structure analysis and for the second model; the concepts and notions of probability theory have been used early for capturing uncertainties in computational mechanics. In the present paper, we are interesting on evaluating the reliability of the structures with uncertain parameters based on the combination of the FEA software with the reliability method program.

1.2. Modeling of Uncertainties in Structural Properties and Loading

A most parameters which characterize any numerical model utilized in reliability analysis of structures, is affected by uncertainty (Kadry, 2006, Muscolino, Ricciardi, & Impollonia, 2000). Examples of such parameters are the mass density, the section dimensions of the structure or the magnitude of a load assumed to act on the considered structure. A convenient and rational way to represent this uncertainty consists in modeling these parameters as random variables. Formally, the latter are collected in a vector of random variables

$$X = X(\theta) = \left[X_1, X_2, \ldots, X_d\right] X_i = X_i\left(\theta\right) \theta \in \varsigma \tag{2}$$

Where θ denotes the random event, ς the so-called sample space and d the dimension of the random vector X. The latter is characterized by its joint Cumulative Distribution Function (CDF)

$$F_X(X) = P\left[X_1 \leq x_1, X_2 \leq x_2, \ldots, X_d \leq x_d\right] \tag{3}$$

Where $\left[x_1, x_2, \ldots, x_d\right]$ is an arbitrary value of the vector X and $P\left[.\right]$ denotes the probability of the event enclosed in the brackets. In many practical applications numerous uncertain parameters can be assumed to be statistically mutually independent, thus simplifying the definition of the joint CDF in (3).

1.2.1. Finite Element Analysis

In solving partial differential equations, the primary goal is to create an equation that approximates the equation to be studied, but is numerically stable, meaning that errors in the input data and intermediate calculations do not accumulate and cause the resulting output to be meaningless. There are many ways of doing this, all with advantages and disadvantages. The Finite Element Analysis is a good choice for solving partial differential equations over complex domains like mechanical system, when the domain changes (as during a solid state reaction with a moving boundary), when the desired precision varies over the entire domain, or when the solution lacks smoothness. The finite element method is the standard tool for certain classes of partial differential equations arising in various fields of engineering and in particular for those arising in solid mechanics. For linear systems enforcing global static or dynamic equilibrium the FE method leads to a system of linear equations, respectively

$$KU = F, M\ddot{U}(t) + C\dot{U}(t) + KU\left(t\right) = F\left(t\right) \tag{4}$$

Where the matrice K is the global stiffness matrice and M is the mass matrice, these matrices obtained by adding the contributions of all element matrices. There are symmetrical and positive vectors.

$$K = \sum_e K^e \tag{5}$$

$$M = \sum_{e} M^{e} \quad (6)$$

The latter matrices have the form,

$$K^{e} = \int_{\Omega^{e}} D^{e} B^{e} d\Omega^{e} \quad (7)$$

$$M^{e} = \int_{\Omega^{e}} \rho^{e} H^{eT} H^{e} d\Omega^{e} \quad (8)$$

Where B^{e} is the matrix relating element displacements and strains, D^{e} is the elasticity matrix relating stresses and strains, ρ^{e} is the mass density, H^{e} is the shape functions and Ω^{e} is the spatial domain of the element. The global damping matrix C is typically formulated in terms of M and K.

1.3. Structural Reliability Analysis

1.3.1. Monte Carlo Simulation

The Monte Carlo method is used to build the PDF of the response of system, or to evaluate his probability of failure. This method involves random sampling from the distribution of input and successive model runs until a statistically significant distribution of output is obtained (Sudreta & Der Kiureghian, 2002). This simulation technique can be used to estimate the probability of failure defined in (1). The Monte Carlo simulation consists of drawing samples of the basic variables according to their probabilistic characteristics and then feeding them into the performance function.

$$P_{f} = \frac{1}{N} \sum_{i=1}^{n} I(X_{1}, X_{2}, .., X_{n}) \quad (9)$$

Where $I(X_{1}, X_{2}, .., X_{n})$ is a function defined by The Monte Carlo method allows the determination of an estimate of the probability of failure, given by

$$I(X_{1}, X_{2}, .., X_{n}) = 1 \text{ If } G(X_{1}, X_{2}, .., X_{n}) \leq 0 \quad (10)$$

$$I(X_{1}, X_{2}, .., X_{n}) = 1 \text{ If } G(X_{1}, X_{2}, .., X_{n}) \succ 0 \quad (11)$$

An estimate of the probability of failure P_{f}

(Madsen, Kenk, & Lind, 1986; Transformation, 2006) can be computed by

$$P_f = \frac{N_f}{N} \qquad (12)$$

Where N_f: the number of simulation samples in which $G(.) \prec 0$, and N is the total number of simulation samples. As N approaches infinity, P_f approaches the true probability of failure. The main advantage of the Monte Carlo method is not only valid for static, but also for dynamic models and for probabilistic model with continuous or discrete variables. The main drawback of this method is that involves a long and onerous computer time.

1.3.2. Probabilistic Transformation Method (PTM)

The Probabilistic Transformation Method is based on the following theorem:

- **Theorem**: Suppose that X is a continuous random variable with PDF $f(x)$ and $A \subset \Re$ is the one-dimensional space where $f(x) \succ 0$, is differentiable and monotonic. Consider the random variable $Y = u(X)$, where $y = u(x)$ defines a one-to-one transformation that maps the set A onto a set $B \subset \Re$ so that the equation $y = u(x)$ can be uniquely solved for x in terms of y, say $x = u^{-1}(y)$. Then, the PDF of Y is

$$f_Y(Y) = f_X[u^{-1}(y)]\left|J\right| \qquad (13)$$

Where, $J = \frac{dx}{dy} = \frac{du^{-1}(y)}{dy}$ is the Jacobean transformation, which must be continuous for all points $y \in B$.

The PTM is based on one-to-one mapping between the random output(s) and input(s) where the transformation Jacobean J can be computed. The PDF of the output(s) is then computed through the known joint PDF of the inputs multiplied by the determinant of the Jacobean matrix.

The idea of PTM is based on the following formula (Rackwitz, 2001):

$$f_u(u) = \left|J\right| . f_z(z) \qquad (14)$$

$$\left|J\right| = \left|\frac{\partial z}{\partial u}\right| \qquad (15)$$

$f_u(u)$ is the probability density function of the variable u.

$f_z(z)$ is the probability density function of the variable z.

The general steps in the application of the Probabilistic Transformation Method (PTM): (1) the random variable input is generating and the stochastic equation of equilibrium is solved first using Finite Element Analysis software. (2) this solution is used to compute numerically the function between the input and the output, which is then inverted for the calculation of the determinant of the Jacobean transformation. Finally, the PDF of the response at any point in the domain can be deduced by using the formula (14). This is simply defined by multiplying the input PDF by the Jacobean of the inverse mechanical function. This approach has the advantage of giving a closed-form of the density function of the response, which is very helpful for reliability analysis of mechanical systems (Rackwitz, 2001; Eurocode 3, 1992).

The Probabilistic Transformation Method (PTM) is one of the most widely used methods in reliability analysis. However, this method has drawbacks in the solution of reliability problems. It requires the evaluation of the explicit response functions with respect to the random variables that is very difficult in analysis of complicated structures. To overcome these drawbacks, an interface between the Finite Element Analysis and The Probabilistic Transformation Method (PTM) is proposed in this paper.

1.4. Structural Reliability Analysis Methods: Explicit Problems

1.4.1. Proposed Technique: PTM-FEM

This technique is proposed by S.KADRY (2005), it's a combination of the deterministic finite element method and the random variable transformation technique (PTM) cited in previous section. In this technique, the differential equation is solved firstly by using the deterministic theory of finite element (a) which yields to accurate nodal solutions. These solutions are then used to obtain the "exact" PDF using the random variable transformation (b) between the input random variables and the output variable. The accuracy of the solution increased when increasing the number of elements in the FEM as usual. The organization chart of this technique is shown in Figure 1.

Figure 1. Organization chart of proposed method

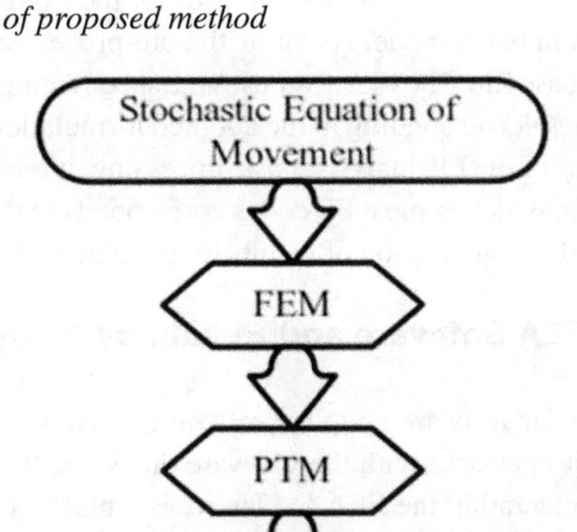

2. RELIABILITY ANALYSIS METHODS FOR STRUCTURAL PROBLEMS: PROPOSED METHOD

This paper deals with the case of coupling Finite Element Analysis (FEA) software, and Probabilistic Transformation Method (PTM) program to evaluate the Probability Density Function (PDF) of response where the expression between input and output of the considered structure is implicit. This technique is based on the numerical simulations of the Finite Element Analysis (FEA) and the Probabilistic Transformation Method (PTM) by making an interface between Finite Element Analysis software and Probabilistic Transformation Method program. Some problems of structures are treated in order to demonstrate the applicability of the proposed technique.

2.1. Finite Element Analysis Coupled With Probabilistic Transformation Method: FEACPTM

The Finite Element Analysis (FEA) software is used to perform the structural analysis to obtain the structure weight, maximal displacement, and maximal stress, corresponding to a set of given design variables. These analysis results are sent to the reliability program to conduct the Probabilistic Density Function (PDF), and the probability of failure and generate new random variables. The newly generated variables are then used to update the input file. The (FEA) software is then invoked again to perform the structural analysis with the new input parameters. This process is repeated until satisfactory results are obtained.

2.2. Finite Element Software

It exist large number FE software codes, but they are different implementations of the same numerical modeling and analysis methodology. In particular, most FE software codes involve three main phases, namely (i) the pre-processing, (ii) the assembly and solution, and (iii) the post-processing. The entire set of definitions is usually gathered into one or more input files for the finite element program, which can be used to execute the analysis in batch mode. As far as the pre-processor concerns, the finite element model is usually saved in a database file. The FE solver uses the specifications of the pre-processing phase to assemble the element matrices corresponding to the adopted formulation, e.g. the mass, damping and stiffness matrices. The final step of the FE analysis, post-processing, it recovers the derived quantities of interest from the solution vector, which in most FE codes corresponds to the vector of the displacements at the nodal DOFs. It involves the visualization of results to facilitate their interpretation by the analyst.

2.3. Interface Between FEA Software and Reliability Analysis Program

A fundamental characteristic of a software code for reliability analysis of structural engineering applications consists in the way it interfaces with the software that gives the finite element modeling and solution. In this type of implementation the FEA is viewed as a black box in the analysis process and the FE code is communicated with a generic interface, through the input _les of the latter.

The reliability analysis program controls the FE code by automatically modifying the input files, set identifiers, which govern the automatic generation of input file samples by the stochastic solver, using pattern matching and replacement

2.4. Coupled Method

By the direct coupling method, we mean any reliability procedure based on a Probability of failure search algorithm using directly the FEA software each time the output of system has to be evaluated. For the application of PTM algorithm we need to compute the function between the input and output variables of system and the determinant of Jacobean of variables, in the case where the function between input and output variables is implicit then we applied the Spline Interpolation to approximate this function, and calculating numerically the Jacobean of the input and output variables. These steps are discussing in previews sections.

2.4.1. Spline Interpolation

In the mathematical field of numerical analysis, spline interpolation is a form of interpolation where the interpolate is a special type of piecewise polynomial called a spline. Spline interpolation is preferred over polynomial interpolation because the interpolation error can be made small even when using low degree polynomials for the spline. Thus, spline interpolation avoids the problem of Runge's phenomenon which occurs when using high degree polynomials.

Given n+1 distinct knots x_i such that

$$x_0 < x_1 < \ldots < x_{n-1} < x_n \tag{16}$$

For a data set x_i of n + 1 points, we can construct a cubic spline with n piecewise cubic polynomials between the data points. If

$$f(x) = \begin{cases} f_0(x) & x \in [x_0, x_1] \\ f_1(x) & x \in [x_1, x_2] \\ \vdots & \\ f_n(x) & x \in [x_{n-1}, x_n] \end{cases} \tag{17}$$

Represents the spline function interpolating the function f, we require the interpolating property,

$$f(x_i) = S(x_i) \tag{18}$$

The splines to join up are

$$f_{i-1}(x_i) = f_i(x_i), i = 1, \ldots, n-1 \tag{19}$$

The function f is twice continuous differentiable,

$$f'_{i-1}(x_i) = f'_i(x_i) \ and \ f''_{i-1}(x_i), i = 1, \ldots, n-1 \tag{20}$$

For the n cubic polynomials comprising S, this means to determine these polynomials, we need to determine 4n conditions (since for one polynomial of degree three, there are four conditions on choosing the curve). However, the interpolating property gives us n+1conditions, and the conditions on the interior data points give us

n + 1 - 2 = n - 1 data points each, summing to

4n – 2 conditions.

We require two other conditions, and these can be imposed upon the problem for different reasons. One such choice results in the so-called clamped cubic spline, with

$$f'(x_0) = u \tag{21}$$

$$f'(x_k) = v \tag{22}$$

For the given values u and v.
Alternately, we can set

$$f''(x_0) = f''(x_n) = 0 \tag{23}$$

Interpolation using natural cubic spline can be defined as

$$f_i(x) = \frac{Z_{i+1}(x - x_i) + Z_i(x_{i+1} - x)}{6h_i} + \left(\frac{y_{i+1}}{h_i} - \frac{h_i z_{i+1}}{6}\right)(x - x_i) \tag{24}$$

And

$$h_i = x_{i+1} - x_i \tag{25}$$

The coefficients can be found by solving this system of equations

$$Z_0 = 0 \tag{26}$$

$$h_{i-1}Z_{i-1} + 2(h_{i-1} + h_i)Z_i + h_i Z_{i+1} = 6\left(\frac{y_{i+1} - y_i}{h_i} - \frac{y_i - y_{i-1}}{h_{i-1}}\right) \quad i = 1, \ldots, n-1 \tag{27}$$

In order to interpolate the function between the input variable and the output variable we are choosing a B-Spline curve, this schema converge faster towards the target curve and produces a better approximation curve than existing methods relying on data point parameterization. In order to applique the PTM we need to calculate the determinant of Jacobean of input and output variables.

2.4.2. Calculate of the Determinant of Jacobean of Variables

In our case we have the determinant of Jacobean of one input and one output variable

$$\left|J\right| = \left|\frac{\partial z}{\partial u}\right| \tag{15}$$

The input variable is generating, and for each input variable the output variable is calculating using FEA software and communicated to reliability program for approximating the function between these variables by spline interpolation. When we have the interpolate function, numerical schema used for computing the determinant of the Jacobean.

2.5. Algorithm of FEACPTM

The outline of a Proposed Method FEACPTM is as follows:

1. Generate the input random variables;
2. Calculate the value of output variables by FE software, for each value of input the correspondent value of output is estimating using FEA and stocking it in solution file;
3. Approximate the function between input and output variables using Spline Interpolation;
4. Calculate of the determinant of Jacobean of input and output variables ;(15)
5. Applique the basic relation of PTM ;(14)
6. Evaluate the graphic of PDF of output variable in function of this output variables (in our case the PDF of displacement in function of displacement);
7. Approximate the Probability of failure P_f.

The algorithm of proposed technique is illustrating by this Figure 2.

2.6 Applications

2.6.1 Truss

This application treats structural analysis of the arrow of a crane of construction that one assimilates to a spatial truss. It is constituted by bars that are identical. To the extremity of this truss is applied a load M = 5t. The bars forming the structure are in steel of which Young's modulus E = [100 GPa, 300 GPa] and the Poisson coefficient $\sigma = 0.29$. The bars have circular sections of which the big one is worth

Figure 2. Algorithm of FEACPTM

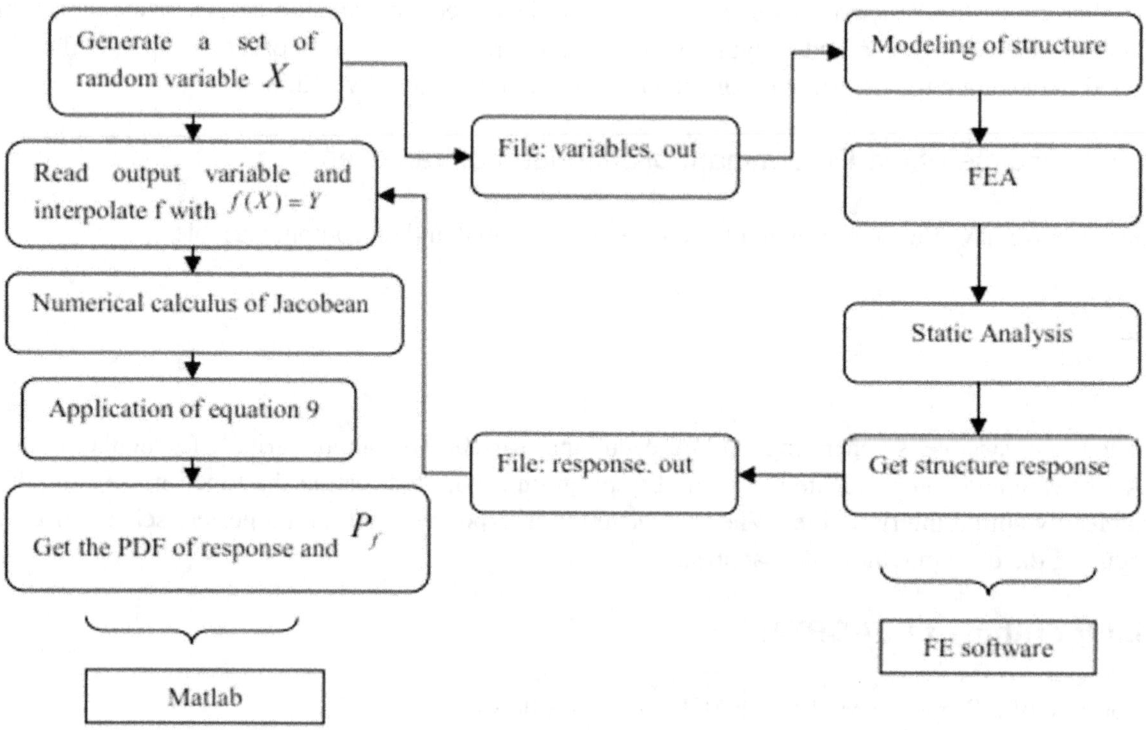

A1= 22mm2 and the small one A2 = 11mm2. The weight of each bar of the truss is not negligible in front of the load M. The goal of this analysis is to determine the efforts, the constraints in the different elements of the truss and the maximum displacement generated by the applied load to his extremity.

The deformation of the truss can be seen in the Figure 4, with FE software.

The Young's modulus E is uniformly distributed in the range E= [100 GPa, 300 GPa]. Using the proposed technique FEACPTM, we obtain the following graph

From the Figure 5, the PDF's of the normalized vertical displacement u_y of the truss calculated by our proposed technique FEACPTM are independent and uniformly distributed in the range [0.00704, 0.02027]. For Reliability Analysis, let us suppose that the limit displacement is

ulimit = 0.015 mm, It is required to find the failure probability $P_f = P\left(u \leq u_{\lim it}\right)$.

The numerical values of probabilistic characteristics of the displacement of this truss are listed in Table 1.

The evaluated failure probability and the values of displacement show that the results are the same with the FEACPTM and the Monte Carlo simulation. It is therefore proved by comparison that the results are validated. In other hand the values of MSE do not exceed 1.296e-07, and the values of APE do not exceed 4%. It can be clearly seen the performance of our technique proposed.

Figure 3. The space truss

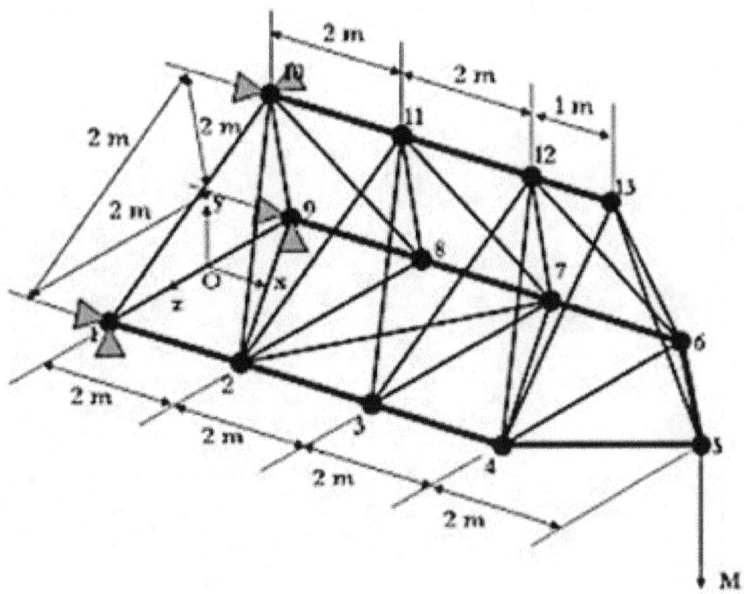

Figure 4. The truss treated by FE Software

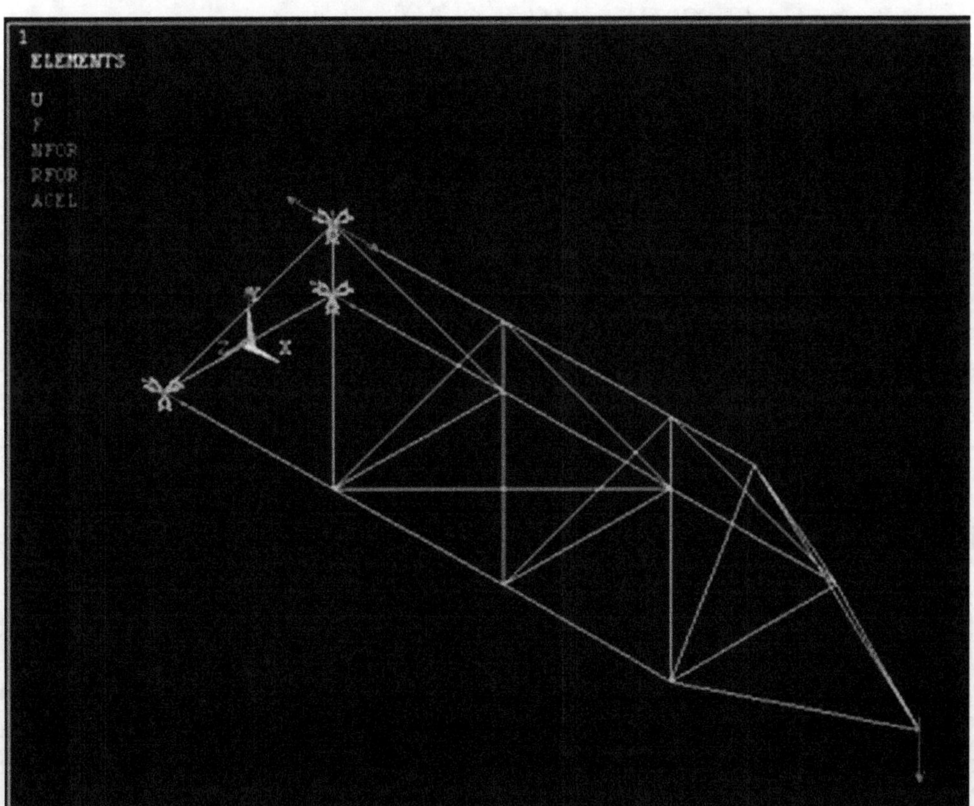

Figure 5. The spacial truss

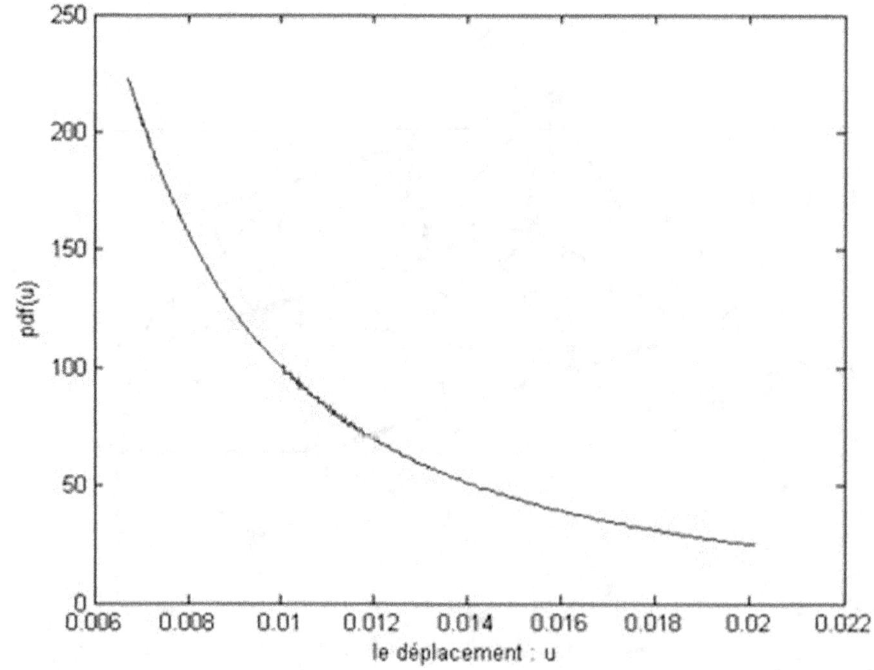

Table 1. Results obtained by FEACPTM and Monte Carlo simulation

	FEACPM	Monte Carlo	MSE	APE (%)
$u_{\min}$	7.04e-03	7.0394e-3	3.6e-13	0.0085
$u_{\max}$	2.027e-02	2.11e-02	6.889e-7	3.9300
u_{mean}	1.12e-02	1.156e-2	1.296e-07	3.110
var	3.6e-03	3.75e-03	2.25e-08	4.00
P_f	2.031e-01	2.03092e-01	6.40e-011	0.0039

2.6.2 Pylon

We are going to analyze the reliability of the pylon of a line of transportation of electricity that one assimilates to a truss plan. Two identical loads F of 1.8KN are applied to the two superior extremities of the following pylon an angle of $\theta = 15$. The bars forming the pylon are in steel of which Young's modulus E= [100GPa, 300GPa] and the Poisson coefficient $\sigma = 0.29$. The section of every bar is worth A= 27.90cm^2. The hypothesis for this problem is that the weight of each bar of the pylon is negligible in front of the applied efforts (see Figure 6).

Figure 6. The pylon

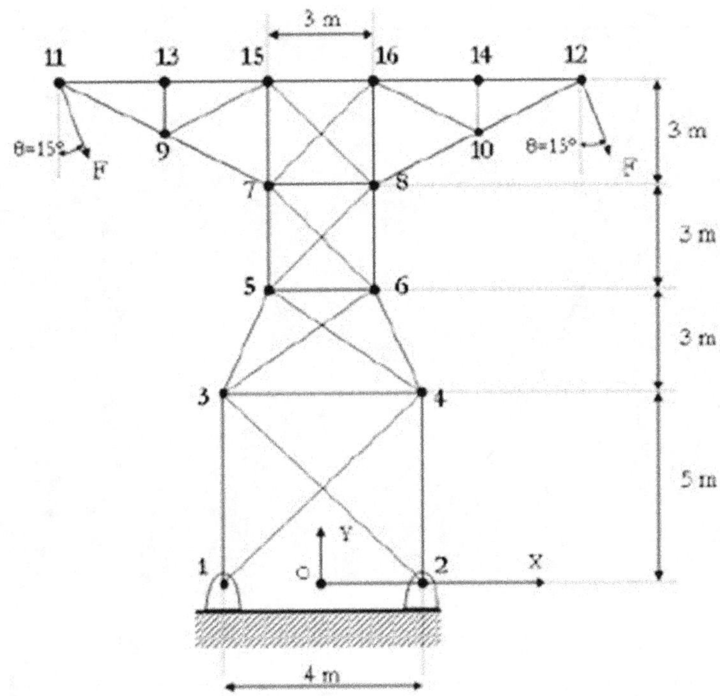

This structure is analyzed using the FEA software for structural modelling, and static analysis in which the FEA software is used to approximate the structural response, this response is used by the PTM program for computing the probability of failure. For that purpose, statistic models must be defined for each random variable involved. In this case, the Young's modulus E is uniformly distributed in the range [100GPa, 300GPa]. Using the proposed technique FEACPTM, we obtain the following graph (see Figure 7).

The PDFs of the normalized vertical displacement u_y are plotted in Figure 7 assuming that the variable u_y is independent and uniformly distributed in the range [2.9489e04, 8.8466e04]. Also in this case the results are accurate as shown by comparison with classical Monte Carlo simulation. Let us suppose the limit displacement is u= 6.635e04mm. It is required to find the failure probability $P_f = P\left(u \le u_{\lim it}\right)$

Mean Squared Error (MSE), and Absolute Percentage Error (APE) are calculated by using the following equation

$$MSE = \left(u_1 - u_{mc}\right)^2 \tag{28}$$

$$APE = \left| \frac{\left(u_1 - u_{mc}\right)}{u_{mc}} \times 100 \right| \tag{29}$$

Figure 7. PDF (u) when E is uniformly distributed

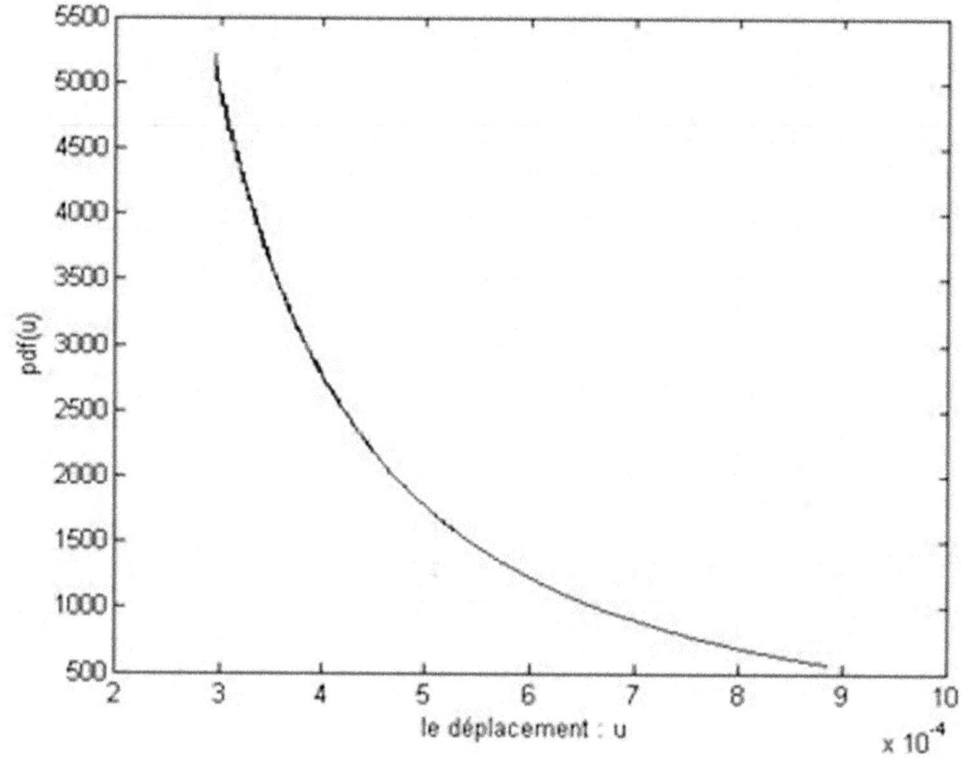

Where u_1 the value of displacement obtained with FEACPTM is, u_{mc} is the predicted value with Monte Carlo simulation.

The numerical values of probabilistic characteristics of the displacement of this pylon are listed in Table 2.

Table 2 reports the results obtained by our technique and the Monte Carlo simulation (10000 iterations). This Table also illustrates the efficiency of the FEACPTM, since a number of 800 (less than 1000)

Table 2. Results obtained by FEACPTM and Monte Carlo simulation

	FEAPTM	Monte Carlo	MSE	APE (%)
$u_{\min}$	2.9489e-04	2.9490284e-04	1.6487e-16	0.0044
$u_{\max}$	8.8466e-04	8.8460969e-04	2.5311e-15	0.0057
u_{mean}	4.8605e-04	4.8594847e-04	1.0308e-14	0.0209
Var	1.5756e-04	1.5725825e-04	9.1053e-14	0.1919
P_f	1.8856e-01	1.66693e-01	4.7817e-04	13.1181

iterations suffice to obtain results close to those obtained by Monte Carlo simulation. To compare the results, the MSE and APE are calculated. The values of MSE and APE are very small, which shows the accuracy and efficiency of FEACPTM.

Current and Future Developments

This paper has considered the FEA software combined to probabilistic approaches to determine the solution of a class of mechanical engineering systems whose have the function between the input and the output as implicit. Emphasis has been given the FEACPTM technique to determine the solution characteristics of structural response. This approach involves two stages. In the first stage, the finite element analysis (FEA) software is performed to get the mechanical response as implicit function of the random variables. In the second stage, the spline interpolation used to get the formula between the input and output of structural system, and the probabilistic transformation can be successfully applied to get the pdf of the mechanical response under certain mathematical conditions concerning the existence of a closed form joint pdf for the input stochastic variables.

Different applications in structural mechanics have shown the validity of the proposed technique. For simple structures, and for large-scale structures. The comparison with Monte Carlo simulations allowed us to verify the accuracy of the proposed method.

The good results obtained by using FEACPTM technique encourage us, in the future, to apply this technique on more complicated stochastic problems, especially dynamical large-scale systems.

It can also be interesting to combine the SFEM (Stochastic Finite Element Method) with the proposed technique in order to take account for stochastic fields, particularly when nonlinear behavior is considered.

3. OPTIMIZATION METHOD OF ANALYSIS

3.1. Sequential Quadratic Programming (SQP) Method

We present Sequential quadratic programming (SQP) method for optimization problems involving general linear and nonlinear constraints. This method has proved highly effective for many such problems. It typically finds a (local) optimum from an arbitrary starting point, and they require relatively few evaluations of the problem functions and gradients. The method SQP consist in to resolve an optimization problem for a limited number of variables under the following general form (Siarry, Dréo, Pétrowski, & Taillard, 2003; Fletcher, 2000):

$$\begin{cases} \mathit{Minimize} & f(X), \quad X \in \Re \\ \mathit{Subject\ to} & h(X) = 0 \\ & g(X) \leq 0 \end{cases} \tag{29}$$

Where $f(X)$ is the objective function, X is the vector of independents variables of optimization and $h(X)$ is the equality constraints, $g(X)$ are called "inequality constraints". An SQP method obtains search directions from a sequence of (Quadratic Problem: QP) sub problems. Each QP sub problem minimizes a quadratic model of a certain lagrangian function subject to linearized constraints. Some merit function is reduced along each search direction to ensure convergence from any starting point.

$$\begin{cases} Minimize & f(X), & X \in \Re \Rightarrow \min_q(d) = \frac{1}{2} d^T H_i d + \nabla f(X_i)^T d \\ Subject\ to & h(X) = 0 \Rightarrow & \nabla g(X_i)^T d + g(X_i) \leq 0 \\ & g(X) = 0 \Rightarrow & \nabla h(X_i)^T d + h(X_i) = 0 \end{cases} \tag{30}$$

At each iterate i a SQP sub problem is used to generate a search direction d towards the next iterate X_{i+1}

$$X_{i+1} = X_i + \alpha_i d \tag{31}$$

Where the value of α_i is determinate in each iteration i using one-dimensional minimization method.

3.2. Multi Start Algorithm

The multi start method for global optimization can overcome some of the limitations of the local Solving method. This method will automatically run the local method from a number of starting points and will display the best of several locally optimal solutions found, as the probable global optimal solution. The multi start algorithm attempts to find a global solution by starting a local solver from multiple starting points in the space of research S. This method generates uniformly distributed points in S, and starts local solver from each of these. This converges to a global solution.

3.3. Proposed Algorithm: MSQP

The MSQP Algorithm is a global optimization algorithm; this algorithm is the result of the combination of Multi start algorithm and the SQP method. It resumes the principal mechanisms of the SQP method to which are added other mechanisms destined to treat multimodales problems. The solutions found by the SQP method during the execution of each iteration are improved, so that always we keep the global solution and the local solution is ignored. Thus, at the end of the treatment, we obtain the global solution, that is the solution of the problem multimodal. In MSQP algorithm, the aspect of the global search of start Multi algorithm is used to maintain the diversity. In fact, we considering that MSQP algorithm will converge after a small number of iterations to a local solution if we find another local solution, we can consider that it is not useful to continue the search from this moment, and it is better to start a new search.

The Basic Steps of the MSQP Algorithm

```
Initialize i=1
Until (Stopping condition is not satisfied)
```

```
    Step 1: (generation) Construction of the solution X_i
     For i = 1 to n do
     Step 2:(research)
         Apply SQP algorithm (fmincon) to find X_opt
         Let X_opt be the solution obtained
         If X_opt improves the best solution
    Update the best
 i= i+ 1
    End For
End Until
```

3.4. Results

In order to evaluate the performance of the method proposed, we compare the solutions obtained by the MSQP algorithm with the solutions reported in (pPSA: perturbed Particle Swarm Algorithm and pPSA best (Xinchao, 2010) for certain functions tests. The table presents different solutions obtained by algorithms pPSA, pPSA best and proposed algorithm MSQP (see Table 3).

The computational results of MSQP algorithm and the algorithms PSO and TRIBES (that are cited in (Cooren, 2008), are summarized in tables for each example problem. The different results are obtained for a dimension D= 10 for every function test (see Table 4).

In this section, the different results are reported. The proposed MSQP technique was tested on a number of benchmark multimodal mathematical functions and the performance compared with other Global Optimization approaches. The results finding by MSQP algorithm are the best ones for the most functions test.

Table 3. The solutions obtained by pPSA, pPSA best and MSQP

Function	npart	pPSA	pPSA best	MSQP
Sphere	2 20	e5.96-6 e6.12-6	e4.39-6 e4.7-6	e4.3216-15 e7.1525-15
Quadratique	2 20	e1.47+4 e3.56-5	e8.06-6 e9.04-6	e7.3985-17 e3.5531-8
Rastrigin	2 20	e1.78+2 85.9	e8.76+1 54.7	0.9950 7.9597
Ackley	2 20	e1.83+1 e7.59-1	e1.64+1 e1.59-3	e5.6056-7 e1.6215-5

Table 4. The solutions obtained by MSQP, TRIBES and PSO.

Function	PSO	TRIBES	MSQP
Rastrigin	(100000)4.02	(100000)8.55	(100000)2.9849
Rosenbrock	(100000)1.88	(100000)0.06	(100000)0.7698
Ackley	(100000)20.11	(100000)20.32	(100000)17.0918

3.5. Structural Optimization Under Reliability Constraints

To illustrate the efficacy of the presented algorithm to solve the problems of structural design optimization under reliability constraint, we choose as example the truss constituted by six identical bars. For this problem, the objective function to minimize is the section A of each bars of the considered structure; the variables of conception are the dimensions of the section (the height and the width) (see Figure 3).

The geometric and material properties are:

The initial section of every bar A= 0.0015m2

The length L= 10m,

The load F= 10N,

The Young's modulus E= 2*108 N/ m2

The minimization of the section is done under constraints on the design variables and under a reliability constraint (the probability of failure of the structure must not surpass a value limit). The formulation of the optimization problem for the section of each bar of truss is the following:

$$\begin{aligned} &\min_{w,t}\ f = w * t \\ &\quad \textit{Subject to} \qquad w \geq 1 \text{ and } t \leq 5 \\ &\qquad\qquad\qquad 6.32 \leq w + t \leq 6.33 \\ &\qquad\qquad\qquad P_f = \frac{0.22}{w * t} \leq 0.001 \end{aligned} \tag{32}$$

Where w is the width and t is the height.

Figure 8. Truss constituted by 6 bars

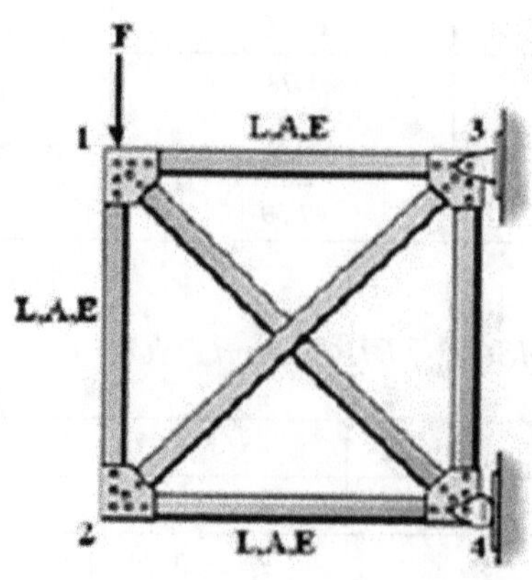

To resolve this problem of structural optimization under reliability constraint, we used the algorithm MSQP developed in the preceding sections. Therefore, we obtained as results of this problem the following values: w= 1.32cm, t= 5cm and A= 6.6e-4m2.

CONCLUSION

In this chapter, an efficient, accurate, robust algorithm is proposed to solve the reliability problem with implicit response functions. The proposed algorithm integrates the treatment by the Finite Element Analysis method with FEA Software and the Probabilistic Transformation Method PTM (reliability program). In the proposed method, the Finite Element Analysis is used to approximate the structural response function. Once the implicit response function is found numerically, the Probabilistic Transformation Method PTM can easily be applied to solve complicates structural reliability problems. The accuracy and efficiency of the proposed method is demonstrated through numerical examples of structures. A new global optimization algorithm MSQP is proposed. The new algorithm can be widely applied to a class of global optimization problems. The numerical results for some test functions show that the present algorithm has proved the robustness and high performance. The proposed algorithm MSQP is applied to solve structural problems under reliability constraints.

REFERENCES

Mohamed, A., & Lemaire, M. (1995). Linearzed mechanical model to evaluate reliability of offshore structures. *Journal of Structural Safety*, *17*(3), 167–193. doi:10.1016/0167-4730(95)00009-S

Sudreta, B., & Der Kiureghian, A. (2002). Comparison of finite element reliability methods. *Probabilistic Engineering Mechanics*, *17*(4), 337–348. doi:10.1016/S0266-8920(02)00031-0

Der Kiureghian, A., & Ke, J.-B. (1988). The stochastic finite element method in structural reliability. *Probabilistic Engineering Mechanics*, *3*(2), 83–91. doi:10.1016/0266-8920(88)90019-7

Hasofer, A. M., & Lind, N. C. (1974). Exact and invariant second-moment code format. *ASCE Journal of the Engineering Mechanics Division*, *100*, 111–121.

Madsen, H. O., Kenk, S., & Lind, N. C. (1986). *Methods of structural safety*. Prentice-Hall Inc.

Rackwitz, R. (2001). Reliability analysis: A review and some perspectives. *Structural Safety*, *23*(4), 365–395. doi:10.1016/S0167-4730(02)00009-7

Lemaire, Mohamed, & Flores-Macias. (1997). The use of finite element codes for the reliability of structural systems. *Frangopol.*

Mohamed, A., & Lemaire, M. (1995). Linearzed mechanical model to evaluate reliability of o shore structures. *Structural Safety*, *17*(3), 167–193. doi:10.1016/0167-4730(95)00009-S

Kadry. (2006). A Proposed Technique to evaluate the Stochastic Mechanical Response based on Transformation with Finite Element Method. *International Journal of Applied Mathematics and Mechanics*.

Muscolino, G., Ricciardi, G., & Impollonia, N. (2000). Improved dynamic analysis of structures with mechanical uncertainties under deterministic input. *Structural Safety*, *15*, 199–212.

Eurocode 3: Design of steel structures. (1992). Brussels: European Committee for Standardization.

Siarry, P., Dréo, J., Pétrowski, A., & Taillard, E. (2003). *Métaheuristiques pour l'optimisation difficile*. Eyrolles.

Fletcher, R. (2000). *Practical Methods of Optimization* (2nd ed.). John Wiley and Sons. doi:10.1002/9781118723203

Xinchao, Z. (2010). A perturbed particle swarm algorithm for numerical optimization. *Applied Soft Computing*, *10*(1), 119–124. doi:10.1016/j.asoc.2009.06.010

Cooren, Y. (2008). *Perfectionnement d'un algorithme adaptatif d'Optimisation par Essaim Particulaire. Applications en génie médical et en électronique* (Ph.D. Thesis). 12 Val de Marne University, Paris, France.

Kadry, S. (2006). A Proposed Technique to evaluate the Stochastic Mechanical Response based on Transformation with Finite Element Method. *International. Journal of Applied Mathematics and Mechanics*, *2*, 94–108.

Ouhimmou,, S., El Hami, A., Ellaia, R., & Tkiouat, M. (2012). Contribution to development of Reliability and Optimization methods applied to Mechanical Structures. *Journal of Applied Mathematics*.

Ouhimmou, El Hami, Ellaia, & Tkiouat. (2010). Approximate solution of the structural problems using probabilistic transformation. *International Journal Key Engineering Materials, 446*, 91-99.

Ouhimmou, S., El Hami, A., Ellaia, R., & Tkiouat, M. (2010). Reliability method coupled with Finite Element Analysis for structural problems with implicit response functions. *International Journal of Advances and Applications in Mechanical Engineering and Technology*, *1*, 109–131.

Ouhimmou, S., El Hami, A., Ellaia, R., & Tkiouat, M. (2011). Reliability Problems with Implicit Response Functions. *Recent Patents on Mechanical Engineering*, *4*, 37–46.

Related References

To continue our tradition of advancing research on topics in the field of engineering, we have compiled a list of recommended IGI Global readings. These references will provide additional information and guidance to further enrich your knowledge and assist you with your own research and future publications.

Abawajy, J. H., Pathan, M., Rahman, M., Pathan, A., & Deris, M. M. (2013). *Network and traffic engineering in emerging distributed computing applications*. Hershey, PA: IGI Global. doi:10.4018/978-1-4666-1888-6

Abu-Faraj, Z. O. (2012). Bioengineering/biomedical engineering education. In Z. Abu-Faraj (Ed.), *Handbook of research on biomedical engineering education and advanced bioengineering learning: Interdisciplinary concepts* (pp. 1–59). Hershey, PA: IGI Global. doi:10.4018/978-1-4666-0122-2.ch001

Abu-Nimeh, S., & Mead, N. R. (2012). Combining security and privacy in requirements engineering. In T. Chou (Ed.), *Information assurance and security technologies for risk assessment and threat management: Advances* (pp. 273–290). Hershey, PA: IGI Global. doi:10.4018/978-1-61350-507-6.ch011

Abu-Taieh, E., El Sheikh, A., & Jafari, M. (2012). *Technology engineering and management in aviation: Advancements and discoveries*. Hershey, PA: IGI Global. doi:10.4018/978-1-60960-887-3

Achumba, I. E., Azzi, D., & Stocker, J. (2010). Low-cost virtual laboratory workbench for electronic engineering. *International Journal of Virtual and Personal Learning Environments*, *1*(4), 1–17. doi:10.4018/jvple.2010100101

Achumba, I. E., Azzi, D., & Stocker, J. (2012). Low-cost virtual laboratory workbench for electronic engineering. In M. Thomas (Ed.), *Design, implementation, and evaluation of virtual learning environments* (pp. 201–217). Hershey, PA: IGI Global. doi:10.4018/978-1-4666-1770-4.ch014

Addo-Tenkorang, R., & Eyob, E. (2013). Engineer-to-order: A maturity concurrent engineering best practice in improving supply chains. In Industrial engineering: Concepts, methodologies, tools, and applications (pp. 1780-1796). Hershey, PA: IGI Global. doi:10.4018/978-1-4666-1945-6.ch095

Aguilera, A., & Davim, J. (2014). *Research developments in wood engineering and technology*. Hershey, PA: IGI Global. doi:10.4018/978-1-4666-4554-7

Aharoni, A., & Reinhartz-Berger, I. (2013). Semi-automatic composition of situational methods. In K. Siau (Ed.), *Innovations in database design, web applications, and information systems management* (pp. 335–364). Hershey, PA: IGI Global. doi:10.4018/978-1-4666-2044-5.ch013

Ahmad, M., Jung, L. T., & Zaman, N. (2014). A comparative analysis of software engineering approaches for sequence analysis. In *Software design and development: Concepts, methodologies, tools, and applications* (pp. 1093–1102). Hershey, PA: IGI Global. doi:10.4018/978-1-4666-4301-7.ch053

Ahrens, A., Bassus, O., & Zaščerinska, J. (2014). Enterprise 2.0 in engineering curriculum. In M. Cruz-Cunha, F. Moreira, & J. Varajão (Eds.), *Handbook of research on enterprise 2.0: Technological, social, and organizational dimensions* (pp. 599–617). Hershey, PA: IGI Global. doi:10.4018/978-1-4666-4373-4.ch031

Akbar, D. (2012). Community engagement in engineering education: Needs and learning outcomes. In M. Rasul (Ed.), *Developments in engineering education standards: Advanced curriculum innovations* (pp. 301–317). Hershey, PA: IGI Global. doi:10.4018/978-1-4666-0951-8.ch017

Alam, F., Subic, A., Plumb, G., Shortis, M., & Chandra, R. P. (2012). An innovative offshore delivery of an undergraduate mechanical engineering program. In M. Rasul (Ed.), *Developments in engineering education standards: Advanced curriculum innovations* (pp. 233–245). Hershey, PA: IGI Global. doi:10.4018/978-1-4666-0951-8.ch013

Ali, D. F., Patil, A., & Nordin, M. S. (2012). Visualization skills in engineering education: Issues, developments, and enhancement. In A. Patil, H. Eijkman, & E. Bhattacharyya (Eds.), *New media communication skills for engineers and IT professionals: Trans-national and trans-cultural demands* (pp. 175–203). Hershey, PA: IGI Global. doi:10.4018/978-1-4666-0243-4.ch011

Aljawarneh, S. (2013). Cloud security engineering: Avoiding security threats the right way. In S. Aljawarneh (Ed.), *Cloud computing advancements in design, implementation, and technologies* (pp. 147–153). Hershey, PA: IGI Global. doi:10.4018/978-1-4666-1879-4.ch010

Alkhatib, G. (2012). *Models for capitalizing on web engineering advancements: Trends and discoveries*. Hershey, PA: IGI Global. doi:10.4018/978-1-4666-0023-2

Allee, T., Handorf, A., & Li, W. (2010). Electrospinning: Development and biomedical applications. In A. Shukla & R. Tiwari (Eds.), *Intelligent medical technologies and biomedical engineering: Tools and applications* (pp. 48–78). Hershey, PA: IGI Global. doi:10.4018/978-1-61520-977-4.ch003

Alsmadi, I. (2014). Website performance measurement: Process and product metrics. In *Software design and development: Concepts, methodologies, tools, and applications* (pp. 1801–1827). Hershey, PA: IGI Global. doi:10.4018/978-1-4666-4301-7.ch086

Altarawneh, H., Alamaro, S., & El Sheikh, A. (2012). Web engineering and business intelligence: Agile web engineering development and practice. In A. Rahman El Sheikh & M. Alnoukari (Eds.), *Business intelligence and agile methodologies for knowledge-based organizations: Cross-disciplinary applications* (pp. 313–344). Hershey, PA: IGI Global. doi:10.4018/978-1-61350-050-7.ch015

Altarawneh, H., & El-Shiekh, A. (2010). Web engineering in small Jordanian web development firms: An XP based process model. In A. Tatnall (Ed.), *Web technologies: Concepts, methodologies, tools, and applications* (pp. 1696–1707). Hershey, PA: IGI Global. doi:10.4018/978-1-60566-982-3.ch091

Alzoabi, Z. (2014). Agile software: Body of knowledge. In *Software design and development: Concepts, methodologies, tools, and applications* (pp. 96–116). Hershey, PA: IGI Global. doi:10.4018/978-1-4666-4301-7.ch006

Andrade-Campos, A. (2013). Development of an optimization framework for parameter identification and shape optimization problems in engineering. In J. Davim (Ed.), *Dynamic methods and process advancements in mechanical, manufacturing, and materials engineering* (pp. 1–24). Hershey, PA: IGI Global. doi:10.4018/978-1-4666-1867-1.ch001

Andreatos, A. (2012). Educating the 21st century's engineers and IT professionals. In A. Patil, H. Eijkman, & E. Bhattacharyya (Eds.), *New media communication skills for engineers and IT professionals: Trans-national and trans-cultural demands* (pp. 132–159). Hershey, PA: IGI Global. doi:10.4018/978-1-4666-0243-4.ch009

Annamalai, C., & Ramayah, T. (2013). Reengineering for enterprise resource planning (ERP) systems implementation: An empirical analysis of assessing critical success factors (CSFs) of manufacturing organizations. In Industrial engineering: Concepts, methodologies, tools, and applications (pp. 791-806). Hershey, PA: IGI Global. doi:10.4018/978-1-4666-1945-6.ch044

Antchev, M. (2010). Other applications of converters and systems of converters. In *Technologies for electrical power conversion, efficiency, and distribution: Methods and processes* (pp. 270–299). Hershey, PA: IGI Global. doi:10.4018/978-1-61520-647-6.ch011

Anzelotti, G., & Valizadeh, M. (2010). Purpose-oriented small software: A case study for some engineering subjects. In R. Luppicini & A. Haghi (Eds.), *Cases on digital technologies in higher education: Issues and challenges* (pp. 164–178). Hershey, PA: IGI Global. doi:10.4018/978-1-61520-869-2.ch012

Anzelotti, G., & Valizadeh, M. (2010). Sights inside the virtual engineering education. In D. Russell & A. Haghi (Eds.), *Web-based engineering education: Critical design and effective tools* (pp. 160–174). Hershey, PA: IGI Global. doi:10.4018/978-1-61520-659-9.ch012

Asadi, M., Mohabbati, B., Gaševic, D., Bagheri, E., & Hatala, M. (2012). Developing semantically-enabled families of method-oriented architectures. *International Journal of Information System Modeling and Design, 3*(4), 1–26. doi:10.4018/jismd.2012100101

Augusti, G., & Feyo de Azevedo, S. (2011). Qualification frameworks and field-specific approaches to quality assurance: Initiatives in engineering and technical education. *International Journal of Quality Assurance in Engineering and Technology Education*, *1*(1), 44–57. doi:10.4018/ijqaete.2011010104

Aung, Z., & Nyunt, K. K. (2014). Constructive knowledge management model and information retrieval methods for software engineering. In *Software design and development: Concepts, methodologies, tools, and applications* (pp. 253–269). Hershey, PA: IGI Global. doi:10.4018/978-1-4666-4301-7.ch014

Azad, A. K., Auer, M. E., & Harward, V. (2012). *Internet accessible remote laboratories: Scalable e-learning tools for engineering and science disciplines*. Hershey, PA: IGI Global. doi:10.4018/978-1-61350-186-3

Azar, A. T. (2013). Overview of biomedical engineering. In Bioinformatics: Concepts, methodologies, tools, and applications (pp. 1-28). Hershey, PA: IGI Global. doi:10.4018/978-1-4666-3604-0.ch001

Badr, K. B., Badr, A. B., & Ahmad, M. N. (2013). Phases in ontology building methodologies: A recent review. In M. Nazir Ahmad, R. Colomb, & M. Abdullah (Eds.), *Ontology-based applications for enterprise systems and knowledge management* (pp. 100–123). Hershey, PA: IGI Global. doi:10.4018/978-1-4666-1993-7.ch006

Baer, W., & Renfro, C. (2013). Information sources and collection planning for engineering. In S. Holder (Ed.), *Library collection development for professional programs: Trends and best practices* (pp. 128–144). Hershey, PA: IGI Global. doi:10.4018/978-1-4666-1897-8.ch008

Baghdadi, Y., & Kraiem, N. (2014). Business process modeling with services: reverse engineering databases. In R. Perez-Castillo & M. Piattini (Eds.), *Uncovering essential software artifacts through business process archeology* (pp. 177–200). Hershey, PA: IGI Global. doi:10.4018/978-1-4666-4667-4.ch007

Bakopoulou, A., Leyhausen, G., Geurtsen, W., & Koidis, P. (2013). Dental tissue engineering research and translational approaches towards clinical application. In A. Daskalaki (Ed.), *Medical advancements in aging and regenerative technologies: Clinical tools and applications* (pp. 279–312). Hershey, PA: IGI Global. doi:10.4018/978-1-4666-2506-8.ch013

Baporikar, N. (2012). Developing right graduate attributes through project-based teaching. In M. Rasul (Ed.), *Developments in engineering education standards: Advanced curriculum innovations* (pp. 64–79). Hershey, PA: IGI Global. doi:10.4018/978-1-4666-0951-8.ch004

Baraldi, E., & Nadin, G. (2012). "Network process re-engineering" in a home textile network: The importance of business relationships and actor bonds. In T. Choi (Ed.), *Fashion supply chain management: Industry and business analysis* (pp. 212–234). Hershey, PA: IGI Global. doi:10.4018/978-1-60960-756-2.ch012

Barbu, M. C., Hasener, J., & Bernardy, G. (2014). Modern testing of wood-based panels, process control, and modeling. In A. Aguilera & J. Davim (Eds.), *Research developments in wood engineering and technology* (pp. 90–130). Hershey, PA: IGI Global. doi:10.4018/978-1-4666-4554-7.ch003

Bas, T. G. (2013). Dual market(ing) in "bio-engineering high technology" new products: The risk of uncertainty and failure. *International Journal of Measurement Technologies and Instrumentation Engineering*, *3*(2), 63–74. doi:10.4018/ijmtie.2013040104

Bedi, P., Gandotra, V., & Singhal, A. (2014). Innovative strategies for secure software development. In *Software design and development: Concepts, methodologies, tools, and applications* (pp. 2099–2119). Hershey, PA: IGI Global. doi:10.4018/978-1-4666-4301-7.ch097

Bellatreche, L. (2010). *Data warehousing design and advanced engineering applications: Methods for complex construction*. Hershey, PA: IGI Global. doi:10.4018/978-1-60566-756-0

Bhattacharyya, S., & Dutta, P. (2013). *Handbook of research on computational intelligence for engineering, science, and business* (Vols. 1–2). Hershey, PA: IGI Global. doi:10.4018/978-1-4666-2518-1

Blicblau, A. S., & Richards, D. (2012). Development of "real world" project skills for engineering students. *International Journal of Quality Assurance in Engineering and Technology Education*, *2*(1), 1–13. doi:10.4018/ijqaete.2012010101

Boci, E. S., Sarkani, S., & Mazzuchi, T. A. (2013). Development of a complex geospatial/RF design model in support of service volume engineering design. In M. Bartolacci & S. Powell (Eds.), *Advancements and innovations in wireless communications and network technologies* (pp. 56–67). Hershey, PA: IGI Global. doi:10.4018/978-1-4666-2154-1.ch005

Boudreaux, A., & Primeaux, B. (2014). Modular game engine design. In *Software design and development: Concepts, methodologies, tools, and applications* (pp. 1179–1199). Hershey, PA: IGI Global. doi:10.4018/978-1-4666-4301-7.ch058

Boudriga, N., & Hamdi, M. (2014). *Security engineering techniques and solutions for information systems: Management and implementation*. Hershey, PA: IGI Global. doi:10.4018/978-1-61520-803-6

Brad, S. (2010). Competitive design of web-based courses in engineering education. In D. Russell & A. Haghi (Eds.), *Web-based engineering education: Critical design and effective tools* (pp. 119–148). Hershey, PA: IGI Global. doi:10.4018/978-1-61520-659-9.ch010

Brad, S. (2010). Designing effective web-based courses in engineering. In R. Luppicini & A. Haghi (Eds.), *Cases on digital technologies in higher education: Issues and challenges* (pp. 217–240). Hershey, PA: IGI Global. doi:10.4018/978-1-61520-869-2.ch016

Bradford, M., Gingras, R., & Hornby, J. (2010). Business process reengineering and ERP: Weapons for the global organization. In K. St.Amant (Ed.), *IT outsourcing: Concepts, methodologies, tools, and applications* (pp. 211–228). Hershey, PA: IGI Global. doi:10.4018/978-1-60566-770-6.ch012

Brennan, R. W., Hugo, R., & Rosehart, W. D. (2012). CDIO as an enabler for graduate attributes assessment: A Canadian case study. *International Journal of Quality Assurance in Engineering and Technology Education*, *2*(2), 45–54. doi:10.4018/ijqaete.2012040105

Burnett, M. (2012). End-user software engineering and why it matters. In A. Dwivedi & S. Clarke (Eds.), *End-user computing, development, and software engineering: New challenges* (pp. 185–201). Hershey, PA: IGI Global. doi:10.4018/978-1-4666-0140-6.ch009

Burns, G. U., & Chisohlm, C. (2011). Engineering professional development related to sustainability of quality. *International Journal of Quality Assurance in Engineering and Technology Education, 1*(1), 15–29. doi:10.4018/ijqaete.2011010102

Byrne, D., Kelly, L., & Jones, G. J. (2014). Multiple multimodal mobile devices: Lessons learned from engineering lifelong solutions. In *Software design and development: Concepts, methodologies, tools, and applications* (pp. 2014–2032). Hershey, PA: IGI Global. doi:10.4018/978-1-4666-4301-7.ch093

Cervera, M., Albert, M., Torres, V., & Pelechano, V. (2012). A model-driven approach for the design and implementation of software development methods. *International Journal of Information System Modeling and Design, 3*(4), 86–103. doi:10.4018/jismd.2012100105

Chang, L., Levy, M., & Powell, P. (2011). Small firm process re-engineering success. In M. Tavana (Ed.), *Managing adaptability, intervention, and people in enterprise information systems* (pp. 138–155). Hershey, PA: IGI Global. doi:10.4018/978-1-60960-529-2.ch007

Chinyemba, F. (2011). Mobility of engineering and technology professionals and its impact on the quality of engineering and technology education: The case of Chinhoyi University of Technology, Zimbabwe. *International Journal of Quality Assurance in Engineering and Technology Education, 1*(2), 35–49. doi:10.4018/ijqaete.2011070104

Chiong, R. (2010). *Nature-inspired informatics for intelligent applications and knowledge discovery: Implications in business, science, and engineering*. Hershey, PA: IGI Global. doi:10.4018/978-1-60566-705-8

Chiprianov, V., Kermarrec, Y., & Rouvrais, S. (2014). Integrating DSLs into a software engineering process: Application to collaborative construction of telecom services. In *Software design and development: Concepts, methodologies, tools, and applications* (pp. 570–595). Hershey, PA: IGI Global. doi:10.4018/978-1-4666-4301-7.ch028

Chis, M. (2010). Introduction: A survey of the evolutionary computation techniques for software engineering. In M. Chis (Ed.), *Evolutionary computation and optimization algorithms in software engineering: Applications and techniques* (pp. 1–12). Hershey, PA: IGI Global. doi:10.4018/978-1-61520-809-8.ch001

Chiu, D. K. (2013). *Mobile and web innovations in systems and service-oriented engineering*. Hershey, PA: IGI Global. doi:10.4018/978-1-4666-2470-2

Chu, P. K., & Wu, S. (2012). Biomaterials. In Z. Abu-Faraj (Ed.), *Handbook of research on biomedical engineering education and advanced bioengineering learning: Interdisciplinary concepts* (pp. 238–283). Hershey, PA: IGI Global. doi:10.4018/978-1-4666-0122-2.ch006

Cimellaro, G. P. (2013). Optimal placement of controller for seismic structures. In N. Lagaros, V. Plevris, & C. Mitropoulou (Eds.), *Design optimization of active and passive structural control systems* (pp. 1–33). Hershey, PA: IGI Global. doi:10.4018/978-1-4666-2029-2.ch001

Clark, R., & Andrews, J. (2012). Engineering the future. In M. Rasul (Ed.), *Developments in engineering education standards: Advanced curriculum innovations* (pp. 143–155). Hershey, PA: IGI Global. doi:10.4018/978-1-4666-0951-8.ch008

Clark, T., & Willans, J. (2013). Software language engineering with XMF and XModeler. In M. Mernik (Ed.), *Formal and practical aspects of domain-specific languages: Recent developments* (pp. 311–340). Hershey, PA: IGI Global. doi:10.4018/978-1-4666-2092-6.ch011

Coll, R. K., & Zegwaard, K. E. (2012). Enculturation into engineering professional practice: Using legitimate peripheral participation to develop communication skills in engineering students. In A. Patil, H. Eijkman, & E. Bhattacharyya (Eds.), *New media communication skills for engineers and IT professionals: Trans-national and trans-cultural demands* (pp. 22–33). Hershey, PA: IGI Global. doi:10.4018/978-1-4666-0243-4.ch003

Coll, R. K., & Zegwaard, K. E. (2012). Enculturation into engineering professional practice: Using legitimate peripheral participation to develop communication skills in engineering students. In A. Patil, H. Eijkman, & E. Bhattacharyya (Eds.), *New media communication skills for engineers and IT professionals: Trans-national and trans-cultural demands* (pp. 22–33). Hershey, PA: IGI Global. doi:10.4018/978-1-4666-0243-4.ch003

Colomb, R. M. (2013). Representation of action is a primary requirement in ontologies for interoperating information systems. In M. Nazir Ahmad, R. Colomb, & M. Abdullah (Eds.), *Ontology-based applications for enterprise systems and knowledge management* (pp. 68–76). Hershey, PA: IGI Global. doi:10.4018/978-1-4666-1993-7.ch004

Cooklev, T. (2013). The role of standards in engineering education. In K. Jakobs (Ed.), *Innovations in organizational IT specification and standards development* (pp. 129–137). Hershey, PA: IGI Global. doi:10.4018/978-1-4666-2160-2.ch007

Costa, L., Loughran, N., & Grønmo, R. (2014). Model-driven engineering, services and interactive real-time applications. In *Software design and development: Concepts, methodologies, tools, and applications* (pp. 178–202). Hershey, PA: IGI Global. doi:10.4018/978-1-4666-4301-7.ch010

Dai, Y., Chakraborty, B., & Shi, M. (2011). *Kansei engineering and soft computing: Theory and practice*. Hershey, PA: IGI Global. doi:10.4018/978-1-61692-797-4

Daud, M. F., Taib, J. M., & Shariffudin, R. S. (2012). Assessing mechanical engineering undergraduates' conceptual knowledge in three dimensional computer aided design (3D CAD). In K. Yusof, N. Azli, A. Kosnin, S. Yusof, & Y. Yusof (Eds.), *Outcome-based science, technology, engineering, and mathematics education: Innovative practices* (pp. 350–363). Hershey, PA: IGI Global. doi:10.4018/978-1-4666-1809-1.ch017

Daugherty, A., Hires, W. E., & Braunstein, S. G. (2013). Collection development for the college of engineering at Louisiana State University libraries: Liaison responsibilities and duties. In S. Holder (Ed.), *Library collection development for professional programs: Trends and best practices* (pp. 291–305). Hershey, PA: IGI Global. doi:10.4018/978-1-4666-1897-8.ch017

Davim, J. (2013). *Dynamic methods and process advancements in mechanical, manufacturing, and materials engineering*. Hershey, PA: IGI Global. doi:10.4018/978-1-4666-1867-1

de Vere, I., & Melles, G. (2013). Integrating 'designerly' ways with engineering science: A catalyst for change within product design and development. In Industrial engineering: Concepts, methodologies, tools, and applications (pp. 56-78). Hershey, PA: IGI Global. doi:10.4018/978-1-4666-1945-6.ch005

del Sagrado Martinez, J., & del Aguila Cano, I. M. (2010). A Bayesian network for predicting the need for a requirements review. In F. Meziane & S. Vadera (Eds.), *Artificial intelligence applications for improved software engineering development: New prospects* (pp. 106–128). Hershey, PA: IGI Global. doi:10.4018/978-1-60566-758-4.ch006

Dhar-Bhattacharjee, S., & Takruri-Rizk, H. (2012). An Indo-British comparison. In C. Romm Livermore (Ed.), *Gender and social computing: Interactions, differences and relationships* (pp. 50–71). Hershey, PA: IGI Global. doi:10.4018/978-1-60960-759-3.ch004

Díaz, V. G., Lovelle, J. M., García-Bustelo, B. C., & Martinez, O. S. (2014). *Advances and applications in model-driven engineering*. Hershey, PA: IGI Global. doi:10.4018/978-1-4666-4494-6

Dogru, A., Senkul, P., & Kaya, O. (2012). Modern approaches to software engineering in the compositional era. In *Machine learning: Concepts, methodologies, tools and applications* (pp. 1903–1923). Hershey, PA: IGI Global. doi:10.4018/978-1-60960-818-7.ch803

Dogru, A. H., & Biçer, V. (2011). *Modern software engineering concepts and practices: Advanced approaches*. Hershey, PA: IGI Global. doi:10.4018/978-1-60960-215-4

Doll, W. J., & Deng, X. (2013). Antecedents of improvisation in IT-enabled engineering work. In A. Dwivedi & S. Clarke (Eds.), *Innovative strategies and approaches for end-user computing advancements* (pp. 242–264). Hershey, PA: IGI Global. doi:10.4018/978-1-4666-2059-9.ch013

Dominguez, U., & Magdaleno, J. (2011). Industrial training in engineering education in Spain. In P. Keleher, A. Patil, & R. Harreveld (Eds.), *Work-integrated learning in engineering, built environment and technology: Diversity of practice in practice* (pp. 72–84). Hershey, PA: IGI Global. doi:10.4018/978-1-60960-547-6.ch004

Dormido, S., Vargas, H., & Sánchez, J. (2013). AutomatL@bs consortium: A Spanish network of web-based labs for control engineering education. In *Industrial engineering: Concepts, methodologies, tools, and applications* (pp. 679–699). Hershey, PA: IGI Global. doi:10.4018/978-1-4666-1945-6.ch039

Dyro, J. F. (2012). Clinical engineering. In Z. Abu-Faraj (Ed.), *Handbook of research on biomedical engineering education and advanced bioengineering learning: Interdisciplinary concepts* (pp. 521–576). Hershey, PA: IGI Global. doi:10.4018/978-1-4666-0122-2.ch012

Easton, J. M., Davies, J. R., & Roberts, C. (2011). Ontology engineering the "what's", "why's", and "how's" of data exchange. *International Journal of Decision Support System Technology*, *3*(1), 40–53. doi:10.4018/jdsst.2011010103

Eijkman, H., & Kayali, O. (2011). Addressing the politics of accreditation in engineering education: The benefits of soft systems thinking. *International Journal of Quality Assurance in Engineering and Technology Education*, *1*(2), 1–10. doi:10.4018/ijqaete.2011070101

El-Khalili, N. H. (2013). Teaching agile software engineering using problem-based learning. *International Journal of Information and Communication Technology Education*, *9*(3), 1–12. doi:10.4018/jicte.2013070101

Favre, L. M. (2010). MDA-based object-oriented reverse engineering. In L. Favre (Ed.), *Model driven architecture for reverse engineering technologies: Strategic directions and system evolution* (pp. 199–229). Hershey, PA: IGI Global. doi:10.4018/978-1-61520-649-0.ch010

Favre, L. M. (2010). Reverse engineering and MDA: An introduction. In L. Favre (Ed.), *Model driven architecture for reverse engineering technologies: Strategic directions and system evolution* (pp. 1–14). Hershey, PA: IGI Global. doi:10.4018/978-1-61520-649-0.ch001

Fenton, N., Hearty, P., Neil, M., & Radlinski, L. (2010). Software project and quality modelling using Bayesian networks. In F. Meziane & S. Vadera (Eds.), *Artificial intelligence applications for improved software engineering development: New prospects* (pp. 1–25). Hershey, PA: IGI Global. doi:10.4018/978-1-60566-758-4.ch001

Ferreira, R., Brisolara, L., Mattos, J. C., Spech, E., & Cota, E. (2010). Engineering embedded software: From application modeling to software synthesis. In L. Gomes & J. Fernandes (Eds.), *Behavioral modeling for embedded systems and technologies: Applications for design and implementation* (pp. 245–270). Hershey, PA: IGI Global. doi:10.4018/978-1-60566-750-8.ch010

Ferreira da Silva Oliveira, M., Wehrmeister, M. A., Assis do Nascimento, F., & Pereira, C. E. (2010). High-level design space exploration of embedded systems using the model-driven engineering and aspect-oriented design approaches. In L. Gomes & J. Fernandes (Eds.), *Behavioral modeling for embedded systems and technologies: Applications for design and implementation* (pp. 114–146). Hershey, PA: IGI Global. doi:10.4018/978-1-60566-750-8.ch005

Ferris, T. L. (2013). Engineering design as research. In *Industrial engineering: Concepts, methodologies, tools, and applications* (pp. 1766–1779). Hershey, PA: IGI Global. doi:10.4018/978-1-4666-1945-6.ch094

Frank, M. (2010). Capacity for engineering systems thinking (CEST): Literature review, principles for assessing and the reliability and validity of an assessing tool. In M. Hunter (Ed.), *Strategic information systems: Concepts, methodologies, tools, and applications* (pp. 1171–1183). Hershey, PA: IGI Global. doi:10.4018/978-1-60566-677-8.ch076

Fraser, D. (2012). A case study of curriculum development in engineering: Insights gained over two decades. In K. Yusof, N. Azli, A. Kosnin, S. Yusof, & Y. Yusof (Eds.), *Outcome-based science, technology, engineering, and mathematics education: Innovative practices* (pp. 27–49). Hershey, PA: IGI Global. doi:10.4018/978-1-4666-1809-1.ch002

Fraser, D. (2012). Curriculum initiatives to help engineering students learn and develop. In K. Yusof, N. Azli, A. Kosnin, S. Yusof, & Y. Yusof (Eds.), *Outcome-based science, technology, engineering, and mathematics education: Innovative practices* (pp. 85–106). Hershey, PA: IGI Global. doi:10.4018/978-1-4666-1809-1.ch005

Fries, T. P. (2014). Reengineering structured legacy system documentation to UML object-oriented artifacts. In *Software design and development: Concepts, methodologies, tools, and applications* (pp. 749–771). Hershey, PA: IGI Global. doi:10.4018/978-1-4666-4301-7.ch036

Funabashi, M. (2014). Transdisciplinary science and technology and service systems. In M. Kosaka & K. Shirahada (Eds.), *Progressive trends in knowledge and system-based science for service innovation* (pp. 101–127). Hershey, PA: IGI Global. doi:10.4018/978-1-4666-4663-6.ch006

Gaetano, L., Puppato, D., & Balestra, G. (2012). Modeling clinical engineering activities to support healthcare technology management. In A. Kolker & P. Story (Eds.), *Management engineering for effective healthcare delivery: Principles and applications* (pp. 113–131). Hershey, PA: IGI Global. doi:10.4018/978-1-60960-872-9.ch005

Galvão, T. A., Neto, F., & Bonates, M. F. (2013). eRiskGame: A persistent browser-based game for supporting project-based learning in the risk management context. In Industrial engineering: Concepts, methodologies, tools, and applications (pp. 1243-1259). Hershey, PA: IGI Global. doi:10.4018/978-1-4666-1945-6.ch067

Garousi, V., Shahnewaz, S., & Krishnamurthy, D. (2013). UML-driven software performance engineering: A systematic mapping and trend analysis. In V. Díaz, J. Lovelle, B. García-Bustelo, & O. Martínez (Eds.), *Progressions and innovations in model-driven software engineering* (pp. 18–64). Hershey, PA: IGI Global. doi:10.4018/978-1-4666-4217-1.ch002

Ge, W., Yang, N., Wang, W., & Li, J. (2011). Interfacial interactions: Drag. In S. Pannala, M. Syamlal, & T. O'Brien (Eds.), *Computational gas-solids flows and reacting systems: Theory, methods and practice* (pp. 128–177). Hershey, PA: IGI Global. doi:10.4018/978-1-61520-651-3.ch004

Génova, G., Llorens, J., & Morato, J. (2014). Software engineering research: The need to strengthen and broaden the classical scientific method. In *Software design and development: Concepts, methodologies, tools, and applications* (pp. 1639–1658). Hershey, PA: IGI Global. doi:10.4018/978-1-4666-4301-7.ch079

Genvigir, E. C., & Vijaykumar, N. L. (2010). Requirements traceability. In M. Ramachandran & R. de Carvalho (Eds.), *Handbook of research on software engineering and productivity technologies: Implications of globalization* (pp. 102–120). Hershey, PA: IGI Global. doi:10.4018/978-1-60566-731-7.ch008

Ghosh, S. (2010). Online automated essay grading system as a web based learning (WBL) tool in engineering education. In D. Russell & A. Haghi (Eds.), *Web-based engineering education: Critical design and effective tools* (pp. 53–62). Hershey, PA: IGI Global. doi:10.4018/978-1-61520-659-9.ch005

Gill, A. Q., & Bunker, D. (2013). SaaS requirements engineering for agile development. In X. Wang, N. Ali, I. Ramos, & R. Vidgen (Eds.), *Agile and lean service-oriented development: Foundations, theory, and practice* (pp. 64–93). Hershey, PA: IGI Global. doi:10.4018/978-1-4666-2503-7.ch004

Giraldo, F. D., Villegas, M. L., & Collazos, C. A. (2014). The use of HCI approaches into distributed CSCL activities applied to software engineering courses. In *Software design and development: Concepts, methodologies, tools, and applications* (pp. 2033–2050). Hershey, PA: IGI Global. doi:10.4018/978-1-4666-4301-7.ch094

Girardi, R., & Leite, A. (2011). Knowledge engineering support for agent-oriented software reuse. In M. Ramachandran (Ed.), *Knowledge engineering for software development life cycles: Support technologies and applications* (pp. 177–195). Hershey, PA: IGI Global. doi:10.4018/978-1-60960-509-4.ch010

Goodhew, P. (2013). Why get your engineering programme accredited? In *Industrial engineering: Concepts, methodologies, tools, and applications* (pp. 18–20). Hershey, PA: IGI Global. doi:10.4018/978-1-4666-1945-6.ch002

Goossenaerts, J., Possel-Dölken, F., & Popplewell, K. (2011). Vision, trends, gaps and a broad roadmap for future engineering. In I. Management Association (Ed.), Global business: Concepts, methodologies, tools and applications (pp. 2229-2243). Hershey, PA: IGI Global. doi:10.4018/978-1-60960-587-2.ch802

Goyal, S. B., & Prakash, N. (2013). Functional method engineering. *International Journal of Information System Modeling and Design, 4*(1), 79–103. doi:10.4018/jismd.2013010104

Gunasekaran, A., & Shea, T. (2010). Requirements management for ERP projects. In *Organizational advancements through enterprise information systems: Emerging applications and developments* (pp. 29–45). Hershey, PA: IGI Global. doi:10.4018/978-1-60566-968-7.ch003

Gustavsson, I., Claesson, L., Nilsson, K., Zackrisson, J., Zubia, J. G., & Jayo, U. H. … Claesson, I. (2012). The VISIR open lab platform. In A. Azad, M. Auer, & V. Harward (Eds.) Internet accessible remote laboratories: Scalable e-learning tools for engineering and science disciplines (pp. 294-317). Hershey, PA: IGI Global. doi:10.4018/978-1-61350-186-3.ch015

Habib, M. K., & Davim, J. (2013). *Engineering creative design in robotics and mechatronics*. Hershey, PA: IGI Global. doi:10.4018/978-1-4666-4225-6

Haghi, A. K. (2013). *Methodologies and applications for chemoinformatics and chemical engineering*. Hershey, PA: IGI Global. doi:10.4018/978-1-4666-4010-8

Haghi, A. K., & Noroozi, B. (2010). Adapting engineering education to the new century. In D. Russell & A. Haghi (Eds.), *Web-based engineering education: Critical design and effective tools* (pp. 30–41). Hershey, PA: IGI Global. doi:10.4018/978-1-61520-659-9.ch003

Haghpanahi, M., Nikkhoo, M., & Peirovi, H. A. (2010). Computer aided tissue engineering from modeling to manufacturing. In A. Lazakidou (Ed.), *Biocomputation and biomedical informatics: Case studies and applications* (pp. 75–88). Hershey, PA: IGI Global. doi:10.4018/978-1-60566-768-3.ch004

Hepsø, I. L., Rindal, A., & Waldal, K. (2013). The introduction of a hand-held platform in an engineering and fabrication company. In T. Rosendahl & V. Hepsø (Eds.), *Integrated operations in the oil and gas industry: Sustainability and capability development* (pp. 246–260). Hershey, PA: IGI Global. doi:10.4018/978-1-4666-2002-5.ch015

Hernández-López, A., Colomo-Palacios, R., García-Crespo, Á., & Cabezas-Isla, F. (2013). Software engineering productivity: Concepts, issues and challenges. In J. Wang (Ed.), *Perspectives and techniques for improving information technology project management* (pp. 69–79). Hershey, PA: IGI Global. doi:10.4018/978-1-4666-2800-7.ch006

Herrmann, A., & Morali, A. (2012). Interplay of security requirements engineering and reverse engineering in the maintenance of undocumented software. In J. Rech & C. Bunse (Eds.), *Emerging technologies for the evolution and maintenance of software models* (pp. 57–91). Hershey, PA: IGI Global. doi:10.4018/978-1-61350-438-3.ch003

Hochrainer, M., & Ziegler, F. (2013). Tuned liquid column gas damper in structural control: The salient features of a general purpose damping device and its application in buildings, bridges, and dams. In N. Lagaros, V. Plevris, & C. Mitropoulou (Eds.), *Design optimization of active and passive structural control systems* (pp. 150–179). Hershey, PA: IGI Global. doi:10.4018/978-1-4666-2029-2.ch007

Hochstein, L., Schott, B., & Graybill, R. B. (2013). Computational engineering in the cloud: Benefits and challenges. In A. Dwivedi & S. Clarke (Eds.), *Innovative strategies and approaches for end-user computing advancements* (pp. 314–332). Hershey, PA: IGI Global. doi:10.4018/978-1-4666-2059-9.ch017

Höhn, S., Lowis, L., Jürjens, J., & Accorsi, R. (2010). Identification of vulnerabilities in web services using model-based security. In C. Gutiérrez, E. Fernández-Medina, & M. Piattini (Eds.), *Web services security development and architecture: Theoretical and practical issues* (pp. 1–32). Hershey, PA: IGI Global. doi:10.4018/978-1-60566-950-2.ch001

Holton, D. L., & Verma, A. (2010). Designing animated simulations and web-based assessments to improve electrical engineering education. In D. Russell & A. Haghi (Eds.), *Web-based engineering education: Critical design and effective tools* (pp. 77–95). Hershey, PA: IGI Global. doi:10.4018/978-1-61520-659-9.ch007

Honnutagi, A. R., Sonar, R., & Babu, S. (2012). Quality accreditation system for Indian engineering education using knowledge management and system dynamics. *International Journal of Quality Assurance in Engineering and Technology Education, 2*(3), 47–61. doi:10.4018/ijqaete.2012070105

Hua, G. B. (2013). Business process re-engineering. In *Implementing IT business strategy in the construction industry* (pp. 118–140). Hershey, PA: IGI Global. doi:10.4018/978-1-4666-4185-3.ch006

Hussein, B., Hage-Diab, A., Hammoud, M., Kawtharani, A., El-Hage, H., & Haj-Ali, A. (2014). Management response to improve the educational performance of engineering students: The case of the lebanese international university. In G. Khoury & M. Khoury (Eds.), *Cases on management and organizational behavior in an Arab context* (pp. 91–110). Hershey, PA: IGI Global. doi:10.4018/978-1-4666-5067-1.ch006

Hussey, M., Wu, B., & Xu, X. (2011). Co-operation models for industries and software education institutions. In *Software industry-oriented education practices and curriculum development: Experiences and lessons* (pp. 39–56). Hershey, PA: IGI Global. doi:10.4018/978-1-60960-797-5.ch003

Hussey, M., Wu, B., & Xu, X. (2011). Curriculum issues in industry oriented software engineering education. In *Software industry-oriented education practices and curriculum development: Experiences and lessons* (pp. 153–165). Hershey, PA: IGI Global. doi:10.4018/978-1-60960-797-5.ch010

Hussey, M., Wu, B., & Xu, X. (2011). Industry oriented curriculum and syllabus creation for software engineering series courses in the school of software. In *Software industry-oriented education practices and curriculum development: Experiences and lessons* (pp. 98–109). Hershey, PA: IGI Global. doi:10.4018/978-1-60960-797-5.ch006

Ilyas, Q. M. (2013). Ontology augmented software engineering. In K. Buragga & N. Zaman (Eds.), *Software development techniques for constructive information systems design* (pp. 406–413). Hershey, PA: IGI Global. doi:10.4018/978-1-4666-3679-8.ch023

Islam, S., Mouratidis, H., Kalloniatis, C., Hudic, A., & Zechner, L. (2013). Model based process to support security and privacy requirements engineering. *International Journal of Secure Software Engineering*, *3*(3), 1–22. doi:10.4018/jsse.2012070101

Jahan, K., Everett, J. W., Tang, G., Farrell, S., Zhang, H., Wenger, A., & Noori, M. (2010). Use of living systems to teach basic engineering concepts. In D. Russell & A. Haghi (Eds.), *Web-based engineering education: Critical design and effective tools* (pp. 96–107). Hershey, PA: IGI Global. doi:10.4018/978-1-61520-659-9.ch008

Jakovljevic, M. (2013). A conceptual model of creativity, invention, and innovation (MCII) for entrepreneurial engineers. In S. Buckley & M. Jakovljevic (Eds.), *Knowledge management innovations for interdisciplinary education: Organizational applications* (pp. 66–87). Hershey, PA: IGI Global. doi:10.4018/978-1-4666-1969-2.ch004

Jaroucheh, Z., Liu, X., & Smith, S. (2014). A software engineering framework for context-aware service-based processes in pervasive environments. In *Software design and development: Concepts, methodologies, tools, and applications* (pp. 71–95). Hershey, PA: IGI Global. doi:10.4018/978-1-4666-4301-7.ch005

Joiner, R., Iacovides, I., Darling, J., Diament, A., Drew, B., Duddley, J., ... Gavin, C. (2013). Racing academy: A case study of a digital game for supporting students learning of physics and engineering. In Y. Baek & N. Whitton (Eds.), *Cases on digital game-based learning: Methods, models, and strategies* (pp. 509–523). Hershey, PA: IGI Global. doi:10.4018/978-1-4666-2848-9.ch026

Jordan, M. E. (2014). Interweaving the digital and physical worlds in collaborative project-based learning experiences. In D. Loveless, B. Griffith, M. Bérci, E. Ortlieb, & P. Sullivan (Eds.), *Academic knowledge construction and multimodal curriculum development* (pp. 265–282). Hershey, PA: IGI Global. doi:10.4018/978-1-4666-4797-8.ch017

Juang, Y. (2010). WIRE: A highly interactive blended learning for engineering education. In D. Russell & A. Haghi (Eds.), *Web-based engineering education: Critical design and effective tools* (pp. 149–159). Hershey, PA: IGI Global. doi:10.4018/978-1-61520-659-9.ch011

Kadry, S. (2013). *Diagnostics and prognostics of engineering systems: Methods and techniques*. Hershey, PA: IGI Global. doi:10.4018/978-1-4666-2095-7

Kamboj, A., Kumar, S., & Singh, H. (2012). Design and development of hybrid stir casting process. *International Journal of Applied Industrial Engineering*, *1*(2), 1–6. doi:10.4018/ijaie.2012070101

Kamthan, P. (2010). A methodology for integrating the social web environment in software engineering education. In S. Dasgupta (Ed.), *Social computing: Concepts, methodologies, tools, and applications* (pp. 457–471). Hershey, PA: IGI Global. doi:10.4018/978-1-60566-984-7.ch031

Kamthan, P. (2010). A social web perspective of software engineering education. In S. Murugesan (Ed.), *Handbook of research on web 2.0, 3.0, and X.0: Technologies, business, and social applications* (pp. 472–495). Hershey, PA: IGI Global. doi:10.4018/978-1-60566-384-5.ch026

Kamthan, P. (2010). On the prospects and concerns of pattern-oriented web engineering. In G. Alkhatib & D. Rine (Eds.), *Web engineering advancements and trends: Building new dimensions of information technology* (pp. 97–128). Hershey, PA: IGI Global. doi:10.4018/978-1-60566-719-5.ch006

Kamthan, P. (2011). Using the social web environment for software engineering education. In L. Tomei (Ed.), *Online courses and ICT in education: Emerging practices and applications* (pp. 23–45). Hershey, PA: IGI Global. doi:10.4018/978-1-60960-150-8.ch003

Kamthan, P. (2013). An exploration of the social web environment for collaborative software engineering education. In N. Karacapilidis, M. Raisinghani, & E. Ng (Eds.), *Web-based and blended educational tools and innovations* (pp. 1–23). Hershey, PA: IGI Global. doi:10.4018/978-1-4666-2023-0.ch001

Kapos, G., Dalakas, V., Nikolaidou, M., & Anagnostopoulos, D. (2014). An integrated framework to simulate SysML models using DEVS simulators. In P. Fonseca i Casas (Ed.), *Formal languages for computer simulation: Transdisciplinary models and applications* (pp. 305–332). Hershey, PA: IGI Global. doi:10.4018/978-1-4666-4369-7.ch010

Karlsson, F., & Ågerfalk, P. J. (2011). Towards structured flexibility in information systems development: Devising a method for method configuration. In K. Siau (Ed.), *Theoretical and practical advances in information systems development: Emerging trends and approaches* (pp. 214–238). Hershey, PA: IGI Global. doi:10.4018/978-1-60960-521-6.ch010

Karpati, P., Sindre, G., & Matulevicius, R. (2012). Comparing misuse case and mal-activity diagrams for modelling social engineering attacks. *International Journal of Secure Software Engineering*, *3*(2), 54–73. doi:10.4018/jsse.2012040103

Keleher, P., & Patil, A. (2012). Conducting an effective residential school for an undergraduate materials science and engineering course. *International Journal of Quality Assurance in Engineering and Technology Education*, *2*(3), 41–46. doi:10.4018/ijqaete.2012070104

Ker, H. W. (2012). Engineering education and attitudes toward mathematics: A comparative study. *International Journal of Quality Assurance in Engineering and Technology Education*, *2*(1), 63–76. doi:10.4018/ijqaete.2012010105

Kinsner, W. (2012). Challenges in the design of adoptive, intelligent and cognitive systems. In Y. Wang (Ed.), *Software and intelligent sciences: New transdisciplinary findings* (pp. 47–67). Hershey, PA: IGI Global. doi:10.4018/978-1-4666-0261-8.ch004

Kirikova, M. (2011). Domain modeling approaches in IS engineering. In J. Osis & E. Asnina (Eds.), *Model-driven domain analysis and software development: Architectures and functions* (pp. 388–406). Hershey, PA: IGI Global. doi:10.4018/978-1-61692-874-2.ch018

Kljajic, M., & Farr, J. V. (2010). Importance of systems engineering in the development of information systems. In D. Paradice (Ed.), *Emerging systems approaches in information technologies: Concepts, theories, and applications* (pp. 51–66). Hershey, PA: IGI Global. doi:10.4018/978-1-60566-976-2.ch004

Kljajic, M., & Farr, J. V. (2010). The role of systems engineering in the development of information systems. In M. Hunter (Ed.), *Strategic information systems: Concepts, methodologies, tools, and applications* (pp. 369–381). Hershey, PA: IGI Global. doi:10.4018/978-1-60566-677-8.ch026

Knoell, H. D. (2010). User participation in the quality assurance of requirements specifications: An evaluation of traditional models and animated systems engineering techniques. In M. Hunter (Ed.), *Strategic information systems: Concepts, methodologies, tools, and applications* (pp. 1623–1638). Hershey, PA: IGI Global. doi:10.4018/978-1-60566-677-8.ch106

Kof, L. (2010). From textual scenarios to message sequence charts. In F. Meziane & S. Vadera (Eds.), *Artificial intelligence applications for improved software engineering development: New prospects* (pp. 83–105). Hershey, PA: IGI Global. doi:10.4018/978-1-60566-758-4.ch005

Köhler, B., Gluchow, M., & Brügge, B. (2014). Teaching basic software engineering to senior high school students. In *K-12 education: Concepts, methodologies, tools, and applications* (pp. 1634–1649). Hershey, PA: IGI Global. doi:10.4018/978-1-4666-4502-8.ch094

Kolker, A. (2012). Efficient managerial decision-making in healthcare settings: Examples and fundamental principles. In A. Kolker & P. Story (Eds.), *Management engineering for effective healthcare delivery: Principles and applications* (pp. 1–45). Hershey, PA: IGI Global. doi:10.4018/978-1-60960-872-9.ch001

Kolker, A., & Story, P. (2012). *Management engineering for effective healthcare delivery: Principles and applications*. Hershey, PA: IGI Global. doi:10.4018/978-1-60960-872-9

Kornyshova, E., Deneckère, R., & Claudepierre, B. (2013). Towards method component contextualization. In J. Krogstie (Ed.), *Frameworks for developing efficient information systems: Models, theory, and practice* (pp. 337–368). Hershey, PA: IGI Global. doi:10.4018/978-1-4666-4161-7.ch015

Krishnan, S. (2012). Problem-based learning curricula in engineering. In M. Rasul (Ed.), *Developments in engineering education standards: Advanced curriculum innovations* (pp. 23–40). Hershey, PA: IGI Global. doi:10.4018/978-1-4666-0951-8.ch002

Ku, H., & Thonglek, S. (2011). Running a successful practice school: Challenges and lessons learned. In P. Keleher, A. Patil, & R. Harreveld (Eds.), *Work-integrated learning in engineering, built environment and technology: Diversity of practice in practice* (pp. 131–163). Hershey, PA: IGI Global. doi:10.4018/978-1-60960-547-6.ch007

Lacuesta, R., Fernández-Sanz, L., & Romay, M. D. (2014). Requirements specification as basis for mobile software quality assurance. In *Software design and development: Concepts, methodologies, tools, and applications* (pp. 719–732). Hershey, PA: IGI Global. doi:10.4018/978-1-4666-4301-7.ch034

Lai, A., Zhang, C., & Busovaca, S. (2013). 2-SQUARE: A web-based enhancement of SQUARE privacy and security requirements engineering. *International Journal of Software Innovation*, *1*(1), 41–53. doi:10.4018/ijsi.2013010104

Lane, J. A., Petkov, D., & Mora, M. (2010). Software engineering and the systems approach: A conversation with Barry Boehm. In M. Hunter (Ed.), *Strategic information systems: Concepts, methodologies, tools, and applications* (pp. 333–337). Hershey, PA: IGI Global. doi:10.4018/978-1-60566-677-8.ch024

Lansiquot, R. D. (2013). *Cases on interdisciplinary research trends in science, technology, engineering, and mathematics: Studies on urban classrooms*. Hershey, PA: IGI Global. doi:10.4018/978-1-4666-2214-2

Laribi, S., Le Bris, A., Huang, L. M., Olsson, P., & Guillemoles, J. F. (2013). Phononic engineering for hot carrier solar cells. In L. Fara & M. Yamaguchi (Eds.), *Advanced solar cell materials, technology, modeling, and simulation* (pp. 214–242). Hershey, PA: IGI Global. doi:10.4018/978-1-4666-1927-2.ch012

Lee, C. K., & Sidhu, M. S. (2013). Computer-aided engineering education: New learning approaches and technologies. In V. Wang (Ed.), *Handbook of research on teaching and learning in K-20 education* (pp. 317–340). Hershey, PA: IGI Global. doi:10.4018/978-1-4666-4249-2.ch019

Lee, J., Ma, S., Lee, S., Wu, C., & Lee, C. L. (2014). Towards a high-availability-driven service composition framework. In *Software design and development: Concepts, methodologies, tools, and applications* (pp. 1498–1520). Hershey, PA: IGI Global. doi:10.4018/978-1-4666-4301-7.ch073

Leng, J., Rhyne, T., & Sharrock, W. (2012). Visualization: Future technology and practices for computational science and engineering. In J. Leng & W. Sharrock (Eds.), *Handbook of research on computational science and engineering: Theory and practice* (pp. 381–413). Hershey, PA: IGI Global. doi:10.4018/978-1-61350-116-0.ch016

Leng, J., & Sharrock, W. (2012). *Handbook of research on computational science and engineering: Theory and practice* (Vols. 1–2). Hershey, PA: IGI Global. doi:10.4018/978-1-61350-116-0

Li, Z. J. (2013). Prototyping of robotic systems in surgical procedures and automated manufacturing processes. In *Industrial engineering: Concepts, methodologies, tools, and applications* (pp. 1969–1987). Hershey, PA: IGI Global. doi:10.4018/978-1-4666-1945-6.ch106

Lladó, C. M., Bonet, P., & Smith, C. U. (2014). Towards a multi-formalism multi-solution framework for model-driven performance engineering. In M. Gribaudo & M. Iacono (Eds.), *Theory and application of multi-formalism modeling* (pp. 34–55). Hershey, PA: IGI Global. doi:10.4018/978-1-4666-4659-9.ch003

Loo, A. (2013). *Distributed computing innovations for business, engineering, and science*. Hershey, PA: IGI Global. doi:10.4018/978-1-4666-2533-4

Ma, Z. M. (2011). Engineering design knowledge management. In D. Schwartz & D. Te'eni (Eds.), *Encyclopedia of knowledge management* (2nd ed.; pp. 263–269). Hershey, PA: IGI Global. doi:10.4018/978-1-59904-931-1.ch025

Maiti, C. K., & Maiti, A. (2013). Teaching technology computer aided design (TCAD) online. In Industrial engineering: Concepts, methodologies, tools, and applications (pp. 1043-1063). Hershey, PA: IGI Global. doi:10.4018/978-1-4666-1945-6.ch057

Male, S. A. (2012). Generic engineering competencies required by engineers graduating in Australia: The competencies of engineering graduates (CEG) project. In M. Rasul (Ed.), *Developments in engineering education standards: Advanced curriculum innovations* (pp. 41–63). Hershey, PA: IGI Global. doi:10.4018/978-1-4666-0951-8.ch003

Malmqvist, J. (2012). A comparison of the CDIO and EUR-ACE quality assurance systems. *International Journal of Quality Assurance in Engineering and Technology Education*, *2*(2), 9–22. doi:10.4018/ijqaete.2012040102

Maree, D. J., & Maree, M. (2010). Factors contributing to the success of women working in science, engineering and technology (SET) careers. In A. Cater-Steel & E. Cater (Eds.), *Women in engineering, science and technology: Education and career challenges* (pp. 183–210). Hershey, PA: IGI Global. doi:10.4018/978-1-61520-657-5.ch009

Marichal, G. N., & González, E. J. (2012). Intelligent MAS in system engineering and robotics. In *Machine learning: Concepts, methodologies, tools and applications* (pp. 175–182). Hershey, PA: IGI Global. doi:10.4018/978-1-60960-818-7.ch204

Martinez, L., Favre, L., & Pereira, C. (2013). Architecture-driven modernization for software reverse engineering technologies. In V. Díaz, J. Lovelle, B. García-Bustelo, & O. Martínez (Eds.), *Progressions and innovations in model-driven software engineering* (pp. 288–307). Hershey, PA: IGI Global. doi:10.4018/978-1-4666-4217-1.ch012

Matanovic, D. (2014). The macondo 252 disaster: Causes and consequences. In D. Matanovic, N. Gaurina-Medjimurec, & K. Simon (Eds.), *Risk analysis for prevention of hazardous situations in petroleum and natural gas engineering* (pp. 115–131). Hershey, PA: IGI Global. doi:10.4018/978-1-4666-4777-0.ch006

Matanovic, D., Gaurina-Medjimurec, N., & Simon, K. (2014). *Risk analysis for prevention of hazardous situations in petroleum and natural gas engineering*. Hershey, PA: IGI Global. doi:10.4018/978-1-4666-4777-0

Matsuo, T., & Fujimoto, T. (2012). Analogical thinking based instruction method in IT professional education. In R. Colomo-Palacios (Ed.), *Professional advancements and management trends in the IT sector* (pp. 95–108). Hershey, PA: IGI Global. doi:10.4018/978-1-4666-0924-2.ch007

Mazo, R., Salinesi, C., Diaz, D., Djebbi, O., & Lora-Michiels, A. (2012). Constraints: The heart of domain and application engineering in the product lines engineering strategy. *International Journal of Information System Modeling and Design*, *3*(2), 33–68. doi:10.4018/jismd.2012040102

Mead, N. R. (2010). Benefits and challenges in the use of case studies for security requirements engineering methods. *International Journal of Secure Software Engineering, 1*(1), 74–91. doi:10.4018/jsse.2010102005

Méausoone, P., & Aguilera, A. (2014). Inventory of experimental works on cutting tools' life for the wood industry. In A. Aguilera & J. Davim (Eds.), *Research developments in wood engineering and technology* (pp. 320–342). Hershey, PA: IGI Global. doi:10.4018/978-1-4666-4554-7.ch009

Medlin, B. D., & Cazier, J. (2011). Obtaining patient's information from hospital employees through social engineering techniques: An investigative study. In H. Nemati (Ed.), *Pervasive information security and privacy developments: Trends and advancements* (pp. 77–89). Hershey, PA: IGI Global. doi:10.4018/978-1-61692-000-5.ch006

Meziane, F., & Vadera, S. (2010). *Artificial intelligence applications for improved software engineering development: new prospects*. Hershey, PA: IGI Global. doi:10.4018/978-1-60566-758-4

Meziane, F., & Vadera, S. (2012). Artificial intelligence in software engineering: Current developments and future prospects. In *Machine learning: Concepts, methodologies, tools and applications* (pp. 1215–1236). Hershey, PA: IGI Global. doi:10.4018/978-1-60960-818-7.ch504

Milanovic, N. (2011). *Engineering reliable service oriented architecture: Managing complexity and service level agreements*. Hershey, PA: IGI Global. doi:10.4018/978-1-60960-493-6

Miller, W. S., & Summers, J. D. (2013). Tool and information centric design process modeling: Three case studies. In *Industrial engineering: Concepts, methodologies, tools, and applications* (pp. 1613–1637). Hershey, PA: IGI Global. doi:10.4018/978-1-4666-1945-6.ch086

Mishra, B., & Shukla, K. K. (2014). Data mining techniques for software quality prediction. In *Software design and development: Concepts, methodologies, tools, and applications* (pp. 401–428). Hershey, PA: IGI Global. doi:10.4018/978-1-4666-4301-7.ch021

Mishra, S. (2011). Social and ethical concerns of biomedical engineering research and practice. In A. Shukla & R. Tiwari (Eds.), *Biomedical engineering and information systems: Technologies, tools and applications* (pp. 54–80). Hershey, PA: IGI Global. doi:10.4018/978-1-61692-004-3.ch003

Moeller, D. P., & Sitzmann, D. (2012). Online computer engineering: combining blended e-learning in engineering with lifelong learning. In M. Rasul (Ed.), *Developments in engineering education standards: Advanced curriculum innovations* (pp. 194–215). Hershey, PA: IGI Global. doi:10.4018/978-1-4666-0951-8.ch011

Moeller, D. P., & Vakilzadian, H. (2012). Technology-enhanced learning standard through integration of modeling and simulation into engineering study programs. In M. Rasul (Ed.), *Developments in engineering education standards: Advanced curriculum innovations* (pp. 157–177). Hershey, PA: IGI Global. doi:10.4018/978-1-4666-0951-8.ch009

Mohan Baral, L., Kifor, C. V., Bondrea, I., & Oprean, C. (2012). Introducing problem based learning (PBL) in textile engineering education and assessing its influence on six sigma project implementation. *International Journal of Quality Assurance in Engineering and Technology Education*, *2*(4), 38–48. doi:10.4018/ijqaete.2012100104

Moore, S., May, D., & Wold, K. (2012). Developing cultural competency in engineering through transnational distance learning. In R. Hogan (Ed.), *Transnational distance learning and building new markets for universities* (pp. 210–228). Hershey, PA: IGI Global. doi:10.4018/978-1-4666-0206-9.ch013

Mora, M., Gelman, O., Frank, M., Paradice, D. B., & Cervantes, F. (2010). Toward an interdisciplinary engineering and management of complex IT-intensive organizational systems: A systems view. In D. Paradice (Ed.), *Emerging systems approaches in information technologies: Concepts, theories, and applications* (pp. 1–24). Hershey, PA: IGI Global. doi:10.4018/978-1-60566-976-2.ch001

Mora, M., O'Connor, R., Raisinghani, M. S., Macías-Luévano, J., & Gelman, O. (2013). An IT service engineering and management framework (ITS-EMF). In P. Ordóñez de Pablos & R. Tennyson (Eds.), *Best practices and new perspectives in service science and management* (pp. 76–91). Hershey, PA: IGI Global. doi:10.4018/978-1-4666-3894-5.ch005

Moser, T., Biffl, S., Sunindyo, W. D., & Winkler, D. (2013). Integrating production automation expert knowledge across engineering domains. In N. Bessis (Ed.), *Development of distributed systems from design to application and maintenance* (pp. 152–167). Hershey, PA: IGI Global. doi:10.4018/978-1-4666-2647-8.ch009

Mouratidis, H. (2011). *Software engineering for secure systems: Industrial and research perspectives*. Hershey, PA: IGI Global. doi:10.4018/978-1-61520-837-1

Mourtos, N. J. (2012). Defining, teaching, and assessing engineering design skills. *International Journal of Quality Assurance in Engineering and Technology Education*, *2*(1), 14–30. doi:10.4018/ijqaete.2012010102

Moustafa, A. (2012). Damage assessment of inelastic structures under simulated critical earthquakes. In V. Plevris, C. Mitropoulou, & N. Lagaros (Eds.), *Structural seismic design optimization and earthquake engineering: Formulations and applications* (pp. 128–151). Hershey, PA: IGI Global. doi:10.4018/978-1-4666-1640-0.ch006

Nasir, M. H., Alias, N. A., Fauzi, S. S., & Massatu, M. H. (2012). Implementation of the personal software process in academic settings and current support tools. In S. Fauzi, M. Nasir, N. Ramli, & S. Sahibuddin (Eds.), *Software process improvement and management: Approaches and tools for practical development* (pp. 117–148). Hershey, PA: IGI Global. doi:10.4018/978-1-61350-141-2.ch007

Nhlabatsi, A., Nuseibeh, B., & Yu, Y. (2012). Security requirements engineering for evolving software systems: A survey. In K. Khan (Ed.), *Security-aware systems applications and software development methods* (pp. 108–128). Hershey, PA: IGI Global. doi:10.4018/978-1-4666-1580-9.ch007

Noroozi, B., Valizadeh, M., & Sorial, G. A. (2010). Designing of e-learning for engineering education in developing countries: Key issues and success factors. In D. Russell & A. Haghi (Eds.), *Web-based engineering education: Critical design and effective tools* (pp. 1–19). Hershey, PA: IGI Global. doi:10.4018/978-1-61520-659-9.ch001

Oh Navarro, E. (2011). On the role of learning theories in furthering software engineering education. In *Instructional design: Concepts, methodologies, tools and applications* (pp. 1645–1666). Hershey, PA: IGI Global. doi:10.4018/978-1-60960-503-2.ch709

Onwubiko, C. (2014). Modelling situation awareness information and system requirements for the mission using goal-oriented task analysis approach. In *Software design and development: Concepts, methodologies, tools, and applications* (pp. 460–478). Hershey, PA: IGI Global. doi:10.4018/978-1-4666-4301-7.ch023

Osis, J., & Asnina, E. (2011). Is modeling a treatment for the weakness of software engineering? In J. Osis & E. Asnina (Eds.), *Model-driven domain analysis and software development: Architectures and functions* (pp. 1–14). Hershey, PA: IGI Global. doi:10.4018/978-1-61692-874-2.ch001

Othman, R., & Awang, Z. (2012). Using multiple methods in assessing oral communication skills in the final year project design course of an undergraduate engineering program. In K. Yusof, N. Azli, A. Kosnin, S. Yusof, & Y. Yusof (Eds.), *Outcome-based science, technology, engineering, and mathematics education: Innovative practices* (pp. 263–287). Hershey, PA: IGI Global. doi:10.4018/978-1-4666-1809-1.ch013

Ozcelik, Y. (2010). IT-enabled reengineering: Productivity impacts. In K. St.Amant (Ed.), *IT outsourcing: Concepts, methodologies, tools, and applications* (pp. 371–376). Hershey, PA: IGI Global. doi:10.4018/978-1-60566-770-6.ch022

Ožvoldová, M., & Schauer, F. (2012). Remote experiments in freshman engineering education by integrated e-learning. In A. Azad, M. Auer, & V. Harward (Eds.), *Internet accessible remote laboratories: Scalable e-learning tools for engineering and science disciplines* (pp. 60–83). Hershey, PA: IGI Global. doi:10.4018/978-1-61350-186-3.ch004

Paay, J., Pedell, S., Sterling, L., Vetere, F., & Howard, S. (2011). The benefit of ambiguity in understanding goals in requirements modelling. *International Journal of People-Oriented Programming*, *1*(2), 24–49. doi:10.4018/ijpop.2011070102

Palmer, S., & Holt, D. (2010). Online discussion in engineering education: Student responses and learning outcomes. In L. Shedletsky & J. Aitken (Eds.), *Cases on online discussion and interaction: Experiences and outcomes* (pp. 105–122). Hershey, PA: IGI Global. doi:10.4018/978-1-61520-863-0.ch005

Pandian, A., Ismail, S. A., & Abdullah, A. S. (2012). Communication framework to empower 21st century engineers and IT professionals. In A. Patil, H. Eijkman, & E. Bhattacharyya (Eds.), *New media communication skills for engineers and IT professionals: Trans-national and trans-cultural demands* (pp. 34–54). Hershey, PA: IGI Global. doi:10.4018/978-1-4666-0243-4.ch004

Pasupathy, K. S. (2010). Systems engineering and health informatics: Context, content, and implementation. In S. Kabene (Ed.), *Healthcare and the effect of technology: Developments, challenges and advancements* (pp. 123–144). Hershey, PA: IGI Global. doi:10.4018/978-1-61520-733-6.ch009

Pasupathy, K. S. (2010). Transforming healthcare: Leveraging the complementarities of health informatics and systems engineering. *International Journal of Healthcare Delivery Reform Initiatives*, *2*(2), 35–55. doi:10.4018/jhdri.2010040103

Pasupathy, K. S. (2011). Systems engineering and health informatics. In *Clinical technologies: Concepts, methodologies, tools and applications* (pp. 1684–1705). Hershey, PA: IGI Global. doi:10.4018/978-1-60960-561-2.ch606

Patel, C., & Ramachandran, M. (2010). Story card process improvement framework for agile requirements. In M. Ramachandran & R. de Carvalho (Eds.), *Handbook of research on software engineering and productivity technologies: Implications of globalization* (pp. 61–54). Hershey, PA: IGI Global. doi:10.4018/978-1-60566-731-7.ch006

Patro, C. S. (2012). A study on adaptability of total quality management in engineering education sector. *International Journal of Quality Assurance in Engineering and Technology Education*, *2*(4), 25–37. doi:10.4018/ijqaete.2012100103

Peña de Carrillo, C. I., Choquet, C., Després, C., Iksal, S., Jacoboni, P., Lekira, A., … Thi-Ngoc, D. P. (2014). Engineering and reengineering of technology enhanced learning scenarios using context awareness processes. In Software design and development: Concepts, methodologies, tools, and applications (pp. 1289-1313). Hershey, PA: IGI Global. doi:10.4018/978-1-4666-4301-7.ch063

Peng, F., & Li, H. (2011). A steganalysis method for 2D engineering graphics based on the statistic of geometric features. *International Journal of Digital Crime and Forensics*, *3*(2), 35–40. doi:10.4018/jdcf.2011040103

Pérez, J. L., Rabuñal, J., & Abella, F. M. (2012). Soft computing techniques in civil engineering: Time series prediction. In *Computer engineering: Concepts, methodologies, tools and applications* (pp. 1982–1997). Hershey, PA: IGI Global. doi:10.4018/978-1-61350-456-7.ch811

Pérez-Castillo, R., Rodríguez de Guzmán, I. G., & Piattini, M. (2011). Architecture-driven modernization. In A. Dogru & V. Biçer (Eds.), *Modern software engineering concepts and practices: Advanced approaches* (pp. 75–103). Hershey, PA: IGI Global. doi:10.4018/978-1-60960-215-4.ch004

Pérez-Castillo, R., Rodríguez de Guzmán, I. G., & Piattini, M. (2012). Model-driven reengineering. In J. Rech & C. Bunse (Eds.), *Emerging technologies for the evolution and maintenance of software models* (pp. 200–229). Hershey, PA: IGI Global. doi:10.4018/978-1-61350-438-3.ch008

Petkov, D., Edgar-Nevill, D., Madachy, R., & O'Connor, R. (2010). Information systems, software engineering, and systems thinking: Challenges and opportunities. In M. Hunter (Ed.), *Strategic information systems: Concepts, methodologies, tools, and applications* (pp. 315–332). Hershey, PA: IGI Global. doi:10.4018/978-1-60566-677-8.ch023

Petkov, D., Edgar-Nevill, D., Madachy, R., & O'Connor, R. (2012). Towards a wider application of the systems approach in information systems and software engineering. In *Computer engineering: Concepts, methodologies, tools and applications* (pp. 1627–1645). Hershey, PA: IGI Global. doi:10.4018/978-1-61350-456-7.ch701

Petre, L., Sere, K., & Troubitsyna, E. (2012). *Dependability and computer engineering: Concepts for software-intensive systems*. Hershey, PA: IGI Global. doi:10.4018/978-1-60960-747-0

Poels, G., Decreus, K., Roelens, B., & Snoeck, M. (2013). Investigating goal-oriented requirements engineering for business processes. *Journal of Database Management, 24*(2), 35–71. doi:10.4018/jdm.2013040103

Polgar, J., & Adamson, G. (2011). *New generation of portal software and engineering: Emerging technologies*. Hershey, PA: IGI Global. doi:10.4018/978-1-60960-571-1

Power, C., Freire, A. P., & Petrie, H. (2011). Integrating accessibility evaluation into web engineering processes. In G. Alkhatib (Ed.), *Web engineered applications for evolving organizations: Emerging knowledge* (pp. 315–339). Hershey, PA: IGI Global. doi:10.4018/978-1-60960-523-0.ch018

Praeg, C. (2011). Framework for IT service value engineering: Managing value and IT service quality. In C. Praeg & D. Spath (Eds.), *Quality management for IT services: Perspectives on business and process performance* (pp. 274–297). Hershey, PA: IGI Global. doi:10.4018/978-1-61692-889-6.ch016

Prescott, J., & Bogg, J. (2013). *Gendered occupational differences in science, engineering, and technology careers*. Hershey, PA: IGI Global. doi:10.4018/978-1-4666-2107-7

Prpic, J. K., & Moore, G. (2012). E-portfolios as a quantitative and qualitative means of demonstrating learning outcomes and competencies in engineering. In K. Yusof, N. Azli, A. Kosnin, S. Yusof, & Y. Yusof (Eds.), *Outcome-based science, technology, engineering, and mathematics education: Innovative practices* (pp. 124–154). Hershey, PA: IGI Global. doi:10.4018/978-1-4666-1809-1.ch007

Puteh, M. M., & Ismail, K. (2012). Quality assurance through innovation policy: The pedagogical implications on engineering education. In *Human resources management: Concepts, methodologies, tools, and applications* (pp. 40–49). Hershey, PA: IGI Global. doi:10.4018/978-1-4666-1601-1.ch004

Puthanpurayil, A. M., Dhakal, R. P., & Carr, A. J. (2013). Optimal passive damper positioning techniques: State-of-the-art. In N. Lagaros, V. Plevris, & C. Mitropoulou (Eds.), *Design optimization of active and passive structural control systems* (pp. 85–111). Hershey, PA: IGI Global. doi:10.4018/978-1-4666-2029-2.ch004

Ramachandran, M. (2010). Knowledge engineering support for software requirements, architectures and components. In F. Meziane & S. Vadera (Eds.), *Artificial intelligence applications for improved software engineering development: New prospects* (pp. 129–145). Hershey, PA: IGI Global. doi:10.4018/978-1-60566-758-4.ch007

Ramachandran, M. (2011). *Knowledge engineering for software development life cycles: Support technologies and applications*. Hershey, PA: IGI Global. doi:10.4018/978-1-60960-509-4

Ramachandran, M., & de Carvalho, R. (2010). *Handbook of research on software engineering and productivity technologies: Implications of globalization*. Hershey, PA: IGI Global. doi:10.4018/978-1-60566-731-7

Rasul, M., Nouwens, F., Swift, R., Martin, F., & Greensill, C. V. (2012). Assessment of final year engineering projects: A pilot investigation on issues and best practice. In M. Rasul (Ed.), *Developments in engineering education standards: Advanced curriculum innovations* (pp. 80–104). Hershey, PA: IGI Global. doi:10.4018/978-1-4666-0951-8.ch005

Razali, Z. B., & Trevelyan, J. (2012). An evaluation of students' practical intelligence and ability to diagnose equipment faults. In K. Yusof, N. Azli, A. Kosnin, S. Yusof, & Y. Yusof (Eds.), *Outcome-based science, technology, engineering, and mathematics education: Innovative practices* (pp. 328–349). Hershey, PA: IGI Global. doi:10.4018/978-1-4666-1809-1.ch016

Reimann, D. (2011). Shaping interactive media with the sewing machine: Smart textile as an artistic context to engage girls in technology and engineering education. *International Journal of Art, Culture and Design Technologies, 1*(1), 12–21. doi:10.4018/ijacdt.2011010102

Rikkilä, J. (2014). Agile, lean, and service-oriented development, continuum, or chasm. In *Software design and development: concepts, methodologies, tools, and applications* (pp. 132–163). Hershey, PA: IGI Global. doi:10.4018/978-1-4666-4301-7.ch008

Rocci, L. (2010). Engineering and environmental technoethics. In R. Luppicini (Ed.), *Technoethics and the evolving knowledge society: Ethical issues in technological design, research, development, and innovation* (pp. 146–162). Hershey, PA: IGI Global. doi:10.4018/978-1-60566-952-6.ch008

Rodrigues, D., Estrella, J. C., Monaco, F. J., Branco, K. R., Antunes, N., & Vieira, M. (2012). Engineering secure web services. In V. Cardellini, E. Casalicchio, K. Castelo Branco, J. Estrella, & F. Monaco (Eds.), *Performance and dependability in service computing: Concepts, techniques and research directions* (pp. 360–380). Hershey, PA: IGI Global. doi:10.4018/978-1-60960-794-4.ch016

Rodríguez, J., Fernández-Medina, E., Piattini, M., & Mellado, D. (2011). A security requirements engineering tool for domain engineering in software product lines. In N. Milanovic (Ed.), *Non-functional properties in service oriented architecture: Requirements, models and methods* (pp. 73–92). Hershey, PA: IGI Global. doi:10.4018/978-1-60566-794-2.ch004

Rojko, A., Zürcher, T., Hercog, D., & Stebler, R. (2012). Implementation of remote laboratories for industrial education. In A. Azad, M. Auer, & V. Harward (Eds.), *Internet accessible remote laboratories: Scalable e-learning tools for engineering and science disciplines* (pp. 84–107). Hershey, PA: IGI Global. doi:10.4018/978-1-61350-186-3.ch005

Rosado, D. G., Mellado, D., Fernandez-Medina, E., & Piattini, M. G. (2013). *Security engineering for cloud computing: Approaches and tools*. Hershey, PA: IGI Global. doi:10.4018/978-1-4666-2125-1

Rose, S., Lauder, M., Schlereth, M., & Schürr, A. (2011). A multidimensional approach for concurrent model-driven automation engineering. In J. Osis & E. Asnina (Eds.), *Model-driven domain analysis and software development: Architectures and functions* (pp. 90–113). Hershey, PA: IGI Global. doi:10.4018/978-1-61692-874-2.ch005

Russell, C. (2012). Conceptual mapping, visualisation, and systems thinking in engineering. In A. Patil, H. Eijkman, & E. Bhattacharyya (Eds.), *New media communication skills for engineers and IT professionals: Trans-national and trans-cultural demands* (pp. 72–93). Hershey, PA: IGI Global. doi:10.4018/978-1-4666-0243-4.ch006

Russell, D., & Haghi, A. (2010). *Web-based engineering education: Critical design and effective tools*. Hershey, PA: IGI Global. doi:10.4018/978-1-61520-659-9

Russo, B., Scotto, M., Sillitti, A., & Succi, G. (2010). Improving agile methods. In B. Russo, M. Scotto, A. Sillitti, & G. Succi (Eds.), *Agile technologies in open source development* (pp. 189–231). Hershey, PA: IGI Global. doi:10.4018/978-1-59904-681-5.ch012

Russo, B., Scotto, M., Sillitti, A., & Succi, G. (2010). Requirements management. In B. Russo, M. Scotto, A. Sillitti, & G. Succi (Eds.), *Agile technologies in open source development* (pp. 268–286). Hershey, PA: IGI Global. doi:10.4018/978-1-59904-681-5.ch015

Russo, M. T. (2012). Humanities in engineering education. In M. Rasul (Ed.), *Developments in engineering education standards: Advanced curriculum innovations* (pp. 285–300). Hershey, PA: IGI Global. doi:10.4018/978-1-4666-0951-8.ch016

Saidane, A., & Guelfi, N. (2014). Towards test-driven and architecture model-based security and resilience engineering. In *Software design and development: Concepts, methodologies, tools, and applications* (pp. 2072–2098). Hershey, PA: IGI Global. doi:10.4018/978-1-4666-4301-7.ch096

Sala, N. (2012). Virtual reality in architecture, in engineering and beyond. In E. Abu-Taieh, A. El Sheikh, & M. Jafari (Eds.), *Technology engineering and management in aviation: Advancements and discoveries* (pp. 336–345). Hershey, PA: IGI Global. doi:10.4018/978-1-60960-887-3.ch020

Salzmann, C., Gillet, D., Esquembre, F., Vargas, H., Sánchez, J., & Dormido, S. (2012). Web 2.0 open remote and virtual laboratories in engineering education. In A. Okada, T. Connolly, & P. Scott (Eds.), *Collaborative learning 2.0: Open educational resources* (pp. 369–390). Hershey, PA: IGI Global. doi:10.4018/978-1-4666-0300-4.ch020

Samaras, G. (2012). Human-centered systems engineering: Managing stakeholder dissonance in healthcare delivery. In A. Kolker & P. Story (Eds.), *Management engineering for effective healthcare delivery: Principles and applications* (pp. 148–171). Hershey, PA: IGI Global. doi:10.4018/978-1-60960-872-9.ch007

Sampaio, A. Z., Henriques, P. G., Cruz, C. O., & Martins, O. P. (2011). Interactive models based on virtual reality technology used in civil engineering education. In G. Vincenti & J. Braman (Eds.), *Teaching through multi-user virtual environments: Applying dynamic elements to the modern classroom* (pp. 387–413). Hershey, PA: IGI Global. doi:10.4018/978-1-61692-822-3.ch021

Sanabria, S. J., Furrer, R., Neuenschwander, J., Niemz, P., & Sennhauser, U. (2014). Bonding defect imaging in glulam with novel air-coupled ultrasound testing. In A. Aguilera & J. Davim (Eds.), *Research developments in wood engineering and technology* (pp. 221–246). Hershey, PA: IGI Global. doi:10.4018/978-1-4666-4554-7.ch006

Sánchez de la Rosa, J. L., Miranda, S. A., & González, C. S. (2013). Evaluation of transversal competences of the engineering students and their relation to the enterprise requirements. In K. Patel & S. Vij (Eds.), *Enterprise resource planning models for the education sector: Applications and methodologies* (pp. 1–17). Hershey, PA: IGI Global. doi:10.4018/978-1-4666-2193-0.ch001

Sarirete, A., & Chikh, A. (2012). A knowledge management process in communities of practice of engineering based on the SECI model for knowledge. In E. Ng, N. Karacapilidis, & M. Raisinghani (Eds.), *Evaluating the impact of technology on learning, teaching, and designing curriculum: Emerging trends* (pp. 134–149). Hershey, PA: IGI Global. doi:10.4018/978-1-4666-0032-4.ch009

Sass, L. (2013). Direct building manufacturing of homes with digital fabrication. In *Industrial engineering: Concepts, methodologies, tools, and applications* (pp. 1231–1242). Hershey, PA: IGI Global. doi:10.4018/978-1-4666-1945-6.ch066

Sauvet, B., Boukhicha, M., Balan, A., Hwang, G., Taverna, D., Shukla, A., & Régnier, S. (2012). Selective pick-and-place of thin film by robotic micromanipulation. *International Journal of Intelligent Mechatronics and Robotics*, *2*(3), 24–37. doi:10.4018/ijimr.2012070103

Schnabel, M. A. (2013). Learning parametric designing. In *Industrial engineering: Concepts, methodologies, tools, and applications* (pp. 197–210). Hershey, PA: IGI Global. doi:10.4018/978-1-4666-1945-6.ch013

Schrödl, H., & Wind, S. (2013). Requirements engineering for cloud application development. In A. Bento & A. Aggarwal (Eds.), *Cloud computing service and deployment models: Layers and management* (pp. 137–150). Hershey, PA: IGI Global. doi:10.4018/978-1-4666-2187-9.ch007

Schulz, T., Radlinski, L., Gorges, T., & Rosenstiel, W. (2011). Software process model using dynamic bayesian networks. In M. Ramachandran (Ed.), *Knowledge engineering for software development life cycles: Support technologies and applications* (pp. 289–310). Hershey, PA: IGI Global. doi:10.4018/978-1-60960-509-4.ch016

Sevillian, D. B. (2013). Aircraft development and design: Enhancing product safety through effective human factors engineering design solutions. In *Industrial engineering: Concepts, methodologies, tools, and applications* (pp. 858–886). Hershey, PA: IGI Global. doi:10.4018/978-1-4666-1945-6.ch048

Shan, T. C., & Hua, W. W. (2010). Strategic technology engineering planning. In M. Hunter (Ed.), *Strategic information systems: Concepts, methodologies, tools, and applications* (pp. 414–434). Hershey, PA: IGI Global. doi:10.4018/978-1-60566-677-8.ch029

Sharma, A., & Maurer, F. (2014). A roadmap for software engineering for the cloud: Results of a systematic review. In *Software design and development: Concepts, methodologies, tools, and applications* (pp. 1–16). Hershey, PA: IGI Global. doi:10.4018/978-1-4666-4301-7.ch001

Sharma, N., Singh, K., & Goyal, D. (2014). Software engineering, process improvement, and experience management: Is the nexus productive? Clues from the Indian giants. In *Software design and development: Concepts, methodologies, tools, and applications* (pp. 1401–1414). Hershey, PA: IGI Global. doi:10.4018/978-1-4666-4301-7.ch068

Shirk, S., Arreola, V., Wobig, C., & Russell, K. (2012). Girls' e-mentoring in science, engineering, and technology based at the University of Illinois at Chicago women in science and engineering (WISE) program. In Computer engineering: Concepts, Methodologies, tools and applications (pp. 1144-1163). Hershey, PA: IGI Global. doi:10.4018/978-1-61350-456-7.ch505

Shukla, A., & Tiwari, R. (2011). *Biomedical engineering and information systems: Technologies, tools and applications*. Hershey, PA: IGI Global. doi:10.4018/978-1-61692-004-3

Sidhu, M. S. (2010). Challenges and trends of TAPS packages in enhancing engineering education. In M. Sidhu (Ed.), *Technology-assisted problem solving for engineering education: Interactive multimedia applications* (pp. 158–166). Hershey, PA: IGI Global. doi:10.4018/978-1-60566-764-5.ch011

Sidhu, M. S. (2010). Evaluation of TAPS packages. In M. Sidhu (Ed.), *Technology-assisted problem solving for engineering education: Interactive multimedia applications* (pp. 128–147). Hershey, PA: IGI Global. doi:10.4018/978-1-60566-764-5.ch009

Sidhu, M. S. (2010). Technology assisted problem solving packages: A new approach to learning, visualizing, and problem solving in engineering. In M. Sidhu (Ed.), *Technology-assisted problem solving for engineering education: Interactive multimedia applications* (pp. 69–90). Hershey, PA: IGI Global. doi:10.4018/978-1-60566-764-5.ch006

Sidhu, M. S. (2010). *Technology-assisted problem solving for engineering education: Interactive multimedia applications*. Hershey, PA: IGI Global. doi:10.4018/978-1-60566-764-5

Sidhu, M. S., & Kang, L. C. (2012). Emerging trends and technologies for enhancing engineering education: An overview. In L. Tomei (Ed.), *Advancing education with information communication technologies: Facilitating new trends* (pp. 320–330). Hershey, PA: IGI Global. doi:10.4018/978-1-61350-468-0.ch026

Sidhu, M. S., & Kang, L. C. (2012). New trends and futuristic information communication technologies for engineering education. In S. Chhabra (Ed.), *ICTs for advancing rural communities and human development: Addressing the digital divide* (pp. 251–262). Hershey, PA: IGI Global. doi:10.4018/978-1-4666-0047-8.ch017

Simonette, M. J., & Spina, E. (2014). Enabling IT innovation through soft systems engineering. In M. Pańkowska (Ed.), *Frameworks of IT prosumption for business development* (pp. 64–72). Hershey, PA: IGI Global. doi:10.4018/978-1-4666-4313-0.ch005

Singh, T., Verma, M. K., & Singh, R. (2014). Role of emotional intelligence in academic achievement: An empirical study on engineering students. In P. Ordóñez de Pablos & R. Tennyson (Eds.), *Strategic approaches for human capital management and development in a turbulent economy* (pp. 255–263). Hershey, PA: IGI Global. doi:10.4018/978-1-4666-4530-1.ch016

Smolnik, S., Teuteberg, F., & Thomas, O. (2012). *Semantic technologies for business and information systems engineering: Concepts and applications*. Hershey, PA: IGI Global. doi:10.4018/978-1-60960-126-3

Solemon, B., Sahibuddin, S., & Ghani, A. A. (2014). Requirements engineering process improvement and related models. In *Software design and development: Concepts, methodologies, tools, and applications* (pp. 203–218). Hershey, PA: IGI Global. doi:10.4018/978-1-4666-4301-7.ch011

Sorial, G. A., & Noroozi, B. (2012). Improvement of engineering students education by e-learning. In *Virtual learning environments: Concepts, methodologies, tools and applications* (pp. 870–883). Hershey, PA: IGI Global. doi:10.4018/978-1-4666-0011-9.ch414

Srinivasan, A., López-Ribot, J. L., & Ramasubramanian, A. K. (2011). Microfluidic applications in vascular bioengineering. In A. Shukla & R. Tiwari (Eds.), *Biomedical engineering and information systems: Technologies, tools and applications* (pp. 1–30). Hershey, PA: IGI Global. doi:10.4018/978-1-61692-004-3.ch001

Steinbuch, R. (2013). Evolutionary optimization of passive compensators to improve earthquake resistance. In N. Lagaros, V. Plevris, & C. Mitropoulou (Eds.), *Design optimization of active and passive structural control systems* (pp. 250–273). Hershey, PA: IGI Global. doi:10.4018/978-1-4666-2029-2.ch011

Stephenson, S. V., & Sage, A. P. (2010). Information and knowledge perspectives in systems engineering and management for innovation and productivity through enterprise resource planning. In M. Hunter (Ed.), *Strategic information systems: Concepts, methodologies, tools, and applications* (pp. 338–368). Hershey, PA: IGI Global. doi:10.4018/978-1-60566-677-8.ch025

Strasser, T., Zoitl, A., & Rooker, M. (2013). Zero-downtime reconfiguration of distributed control logic in industrial automation and control. In *Industrial engineering: Concepts, methodologies, tools, and applications* (pp. 2024–2051). Hershey, PA: IGI Global. doi:10.4018/978-1-4666-1945-6.ch109

Sudeikat, J., & Renz, W. (2010). Building complex adaptive systems: On engineering self-organizing multi-agent systems. In M. Hunter (Ed.), *Strategic information systems: Concepts, methodologies, tools, and applications* (pp. 767–787). Hershey, PA: IGI Global. doi:10.4018/978-1-60566-677-8.ch050

Sugiyama, S., & Burgess, L. (2012). Principle for engineering service based system by swirl computing. In X. Liu & Y. Li (Eds.), *Advanced design approaches to emerging software systems: Principles, methodologies and tools* (pp. 48–60). Hershey, PA: IGI Global. doi:10.4018/978-1-60960-735-7.ch003

Sun, Z., Han, J., Dong, D., & Zhao, S. (2010). Engineering of experience based trust for e-commerce. In M. Wang & Z. Sun (Eds.), *Handbook of research on complex dynamic process management: Techniques for adaptability in turbulent environments* (pp. 342–367). Hershey, PA: IGI Global. doi:10.4018/978-1-60566-669-3.ch014

Sunindyo, W. D., Moser, T., Winkler, D., Mordinyi, R., & Biffl, S. (2013). Workflow validation framework in collaborative engineering environments. In A. Loo (Ed.), *Distributed computing innovations for business, engineering, and science* (pp. 285–299). Hershey, PA: IGI Global. doi:10.4018/978-1-4666-2533-4.ch015

Szeto, A. Y. (2014). Assistive technology and rehabilitation engineering. In *Assistive technologies: Concepts, methodologies, tools, and applications* (pp. 277–331). Hershey, PA: IGI Global. doi:10.4018/978-1-4666-4422-9.ch015

Takruri-Rizk, H., Sappleton, N., & Dhar-Bhattacharjee, S. (2010). Progression of UK women engineers: Aids and hurdles. In A. Cater-Steel & E. Cater (Eds.), *Women in engineering, science and technology: Education and career challenges* (pp. 280–300). Hershey, PA: IGI Global. doi:10.4018/978-1-61520-657-5.ch013

Tekinerdogan, B., & Aksit, M. (2011). A comparative analysis of software engineering with mature engineering disciplines using a problem-solving perspective. In A. Dogru & V. Biçer (Eds.), *Modern software engineering concepts and practices: Advanced approaches* (pp. 1–18). Hershey, PA: IGI Global. doi:10.4018/978-1-60960-215-4.ch001

Tellioglu, H. (2012). About representational artifacts and their role in engineering. In G. Viscusi, G. Campagnolo, & Y. Curzi (Eds.), *Phenomenology, organizational politics, and IT design: The social study of information systems* (pp. 111–130). Hershey, PA: IGI Global. doi:10.4018/978-1-4666-0303-5.ch007

Terawaki, Y., Takahashi, Y., Kodama, Y., & Yana, K. (2013). The development of educational environment suited to the Japan-specific educational service using requirements engineering techniques: Case study of running Sakai with PostgreSQL. In V. Kumar & F. Lin (Eds.), *System and technology advancements in distance learning* (pp. 261–270). Hershey, PA: IGI Global. doi:10.4018/978-1-4666-2032-2.ch019

Thomas, K. D., & Muga, H. E. (2012). Sustainability: The new 21st century general education requirement for engineers. In M. Rasul (Ed.), *Developments in engineering education standards: Advanced curriculum innovations* (pp. 263–284). Hershey, PA: IGI Global. doi:10.4018/978-1-4666-0951-8.ch015

Trigo, B., Olguin, G., & Matai, P. (2010). The use of applets in an engineering chemistry course: Advantages and new ideas. In D. Russell & A. Haghi (Eds.), *Web-based engineering education: Critical design and effective tools* (pp. 108–118). Hershey, PA: IGI Global. doi:10.4018/978-1-61520-659-9.ch009

Tsadimas, A., Nikolaidou, M., & Anagnostopoulos, D. (2014). Model-based system design using SysML: The role of the evaluation diagram. In P. Fonseca i Casas (Ed.), *Formal languages for computer simulation: Transdisciplinary models and applications* (pp. 236–266). Hershey, PA: IGI Global. doi:10.4018/978-1-4666-4369-7.ch008

Unhelkar, B., Ghanbary, A., & Younessi, H. (2010). *Collaborative business process engineering and global organizations: Frameworks for service integration*. Hershey, PA: IGI Global. doi:10.4018/978-1-60566-689-1

Uziak, J., Oladiran, M. T., & Kommula, V. P. (2012). Integrating general education courses into engineering curriculum: Students' perspective. In M. Rasul (Ed.), *Developments in engineering education standards: Advanced curriculum innovations* (pp. 247–262). Hershey, PA: IGI Global. doi:10.4018/978-1-4666-0951-8.ch014

Valizadeh, M., Anzelotti, G., & Salehi, A. S. (2010). Web-based approaches in engineering education. In R. Luppicini & A. Haghi (Eds.), *Cases on digital technologies in higher education: Issues and challenges* (pp. 241–256). Hershey, PA: IGI Global. doi:10.4018/978-1-61520-869-2.ch017

Valizadeh, M., Anzelotti, G., & Salehi, S. (2010). Web-based training: An applicable tool for engineering education. In D. Russell & A. Haghi (Eds.), *Web-based engineering education: Critical design and effective tools* (pp. 186–198). Hershey, PA: IGI Global. doi:10.4018/978-1-61520-659-9.ch014

Valverde, R., Toleman, M., & Cater-Steel, A. (2010). Design science: A case study in information systems re-engineering. In M. Hunter (Ed.), *Strategic information systems: Concepts, methodologies, tools, and applications* (pp. 490–503). Hershey, PA: IGI Global. doi:10.4018/978-1-60566-677-8.ch033

Vaquero, L. M., Rodero-Merino, L., Cáceres, J., Chapman, C., Lindner, M., & Galán, F. (2012). Principles, methodology and tools for engineering cloud computing systems. In X. Liu & Y. Li (Eds.), *Advanced design approaches to emerging software systems: Principles, methodologies and tools* (pp. 250–273). Hershey, PA: IGI Global. doi:10.4018/978-1-60960-735-7.ch012

Vargas, E. P. (2014). Quality, improvement and measurements in high risk software. In *Software design and development: Concepts, methodologies, tools, and applications* (pp. 733–748). Hershey, PA: IGI Global. doi:10.4018/978-1-4666-4301-7.ch035

Vasant, P. (2013). *Meta-heuristics optimization algorithms in engineering, business, economics, and finance*. Hershey, PA: IGI Global. doi:10.4018/978-1-4666-2086-5

Venkatraman, S. (2013). Software engineering research gaps in the cloud. *Journal of Information Technology Research, 6*(1), 1–19. doi:10.4018/jitr.2013010101

Villazón-Terrazas, B. C., Suárez-Figueroa, M., & Gómez-Pérez, A. (2012). A pattern-based method for re-engineering non-ontological resources into ontologies. In A. Sheth (Ed.), *Semantic-enabled advancements on the web: Applications across industries* (pp. 17–54). Hershey, PA: IGI Global. doi:10.4018/978-1-4666-0185-7.ch002

Vinod, J. S. (2010). Dem simulations in geotechnical earthquake engineering education. *International Journal of Geotechnical Earthquake Engineering, 1*(1), 61–69. doi:10.4018/jgee.2010090804

Virdi, J. (2014). Business risk analysis: Obsolescence management in requirements engineering. In *Software design and development: Concepts, methodologies, tools, and applications* (pp. 1736–1763). Hershey, PA: IGI Global. doi:10.4018/978-1-4666-4301-7.ch083

Walk, S., Pöschko, J., Strohmaier, M., Andrews, K., Tudorache, T., Noy, N. F., ... Musen, M. A. (2013). PragmatiX: An interactive tool for visualizing the creation process behind collaboratively engineered ontologies. *International Journal on Semantic Web and Information Systems*, *9*(1), 45–78. doi:10.4018/jswis.2013010103 PMID:24465189

Wang, J. (2013). Architects and engineers. In *Challenging ICT applications in architecture, engineering, and industrial design education* (pp. 48–64). Hershey, PA: IGI Global. doi:10.4018/978-1-4666-1999-9.ch003

Wang, J. (2013). *Challenging ICT applications in architecture, engineering, and industrial design education*. Hershey, PA: IGI Global. doi:10.4018/978-1-4666-1999-9

Wang, J. (2013). Pedagogy and curriculum in architecture and engineering. In *Challenging ICT applications in architecture, engineering, and industrial design education* (pp. 65–92). Hershey, PA: IGI Global. doi:10.4018/978-1-4666-1999-9.ch004

Wang, J. (2013). Professionalism in architecture and engineering. In *Challenging ICT applications in architecture, engineering, and industrial design education* (pp. 137–156). Hershey, PA: IGI Global. doi:10.4018/978-1-4666-1999-9.ch007

Wang, J. (2013). Reviewing engineers and introducing industrial designers. In *Challenging ICT applications in architecture, engineering, and industrial design education* (pp. 111–136). Hershey, PA: IGI Global. doi:10.4018/978-1-4666-1999-9.ch006

Wang, Y. (2010). On cognitive properties of human factors and error models in engineering and socialization. In Y. Wang (Ed.), *Discoveries and breakthroughs in cognitive informatics and natural intelligence* (pp. 93–109). Hershey, PA: IGI Global. doi:10.4018/978-1-60566-902-1.ch006

Wang, Y. (2012). Convergence of software science and computational intelligence: A new transdisciplinary research field. In Y. Wang (Ed.), *Software and intelligent sciences: New transdisciplinary findings* (pp. 1–13). Hershey, PA: IGI Global. doi:10.4018/978-1-4666-0261-8.ch001

Wang, Y., & Patel, S. (2012). Exploring the cognitive foundations of software engineering. In Y. Wang (Ed.), *Software and intelligent sciences: New transdisciplinary findings* (pp. 232–251). Hershey, PA: IGI Global. doi:10.4018/978-1-4666-0261-8.ch014

Wei, J., Chen, J., & Zhu, Q. (2010). Service science, management and engineering education: A unified model for university. *International Journal of Service Science, Management, Engineering, and Technology*, *1*(2), 51–69. doi:10.4018/jssmet.2010040104

Weigang, L., Barros, A. D., & Romani de Oliveira, I. (2010). *Computational models, software engineering, and advanced technologies in air transportation: Next generation applications*. Hershey, PA: IGI Global. doi:10.4018/978-1-60566-800-0

Westh Nicolajsen, H. (2010). Limitations and perspectives on use of e-services in engineering consulting. In *Electronic services: Concepts, methodologies, tools and applications* (pp. 1280–1295). Hershey, PA: IGI Global. doi:10.4018/978-1-61520-967-5.ch078

Wilson, R., & Younis, H. (2013). *Business strategies for electrical infrastructure engineering: Capital project implementation*. Hershey, PA: IGI Global. doi:10.4018/978-1-4666-2839-7

Winter, A. (2010). The smart women – smart state strategy: A policy on women's participation in science, engineering and technology in Queensland, Australia. In A. Cater-Steel & E. Cater (Eds.), *Women in engineering, science and technology: Education and career challenges* (pp. 1–20). Hershey, PA: IGI Global. doi:10.4018/978-1-61520-657-5.ch001

Winter, R., & van Beijnum, I. (2014). Inter-domain traffic engineering using the origin preference attribute. In M. Boucadair & D. Binet (Eds.), *Solutions for sustaining scalability in internet growth* (pp. 18–38). Hershey, PA: IGI Global. doi:10.4018/978-1-4666-4305-5.ch002

Worth, D., Greenough, C., & Chin, S. (2014). Pragmatic software engineering for computational science. In *Software design and development: Concepts, methodologies, tools, and applications* (pp. 663–694). Hershey, PA: IGI Global. doi:10.4018/978-1-4666-4301-7.ch032

Wu, J. (2013). *Biomedical engineering and cognitive neuroscience for healthcare: Interdisciplinary applications*. Hershey, PA: IGI Global. doi:10.4018/978-1-4666-2113-8

Wu, Y., & Koszalka, T. A. (2013). Instructional design of an advanced interactive discovery environment: Exploring team communication and technology use in virtual collaborative engineering problem solving. In *Industrial engineering: Concepts, methodologies, tools, and applications* (pp. 117–136). Hershey, PA: IGI Global. doi:10.4018/978-1-4666-1945-6.ch009

Yadav, V. (2011). Research review: Globally distributed requirements engineering and agility. *International Journal of Innovation in the Digital Economy*, *2*(1), 1–11. doi:10.4018/jide.2011010101

Yang, J. J., Liu, J. F., Kurokawa, T., Kitamura, N., Yasuda, K., & Gong, J. P. (2013). Tough double-network hydrogels as scaffolds for tissue engineering: Cell behavior in vitro and in vivo test. In J. Wu (Ed.), *Technological advancements in biomedicine for healthcare applications* (pp. 213–222). Hershey, PA: IGI Global. doi:10.4018/978-1-4666-2196-1.ch023

Yeoh, W., Gao, J., & Koronios, A. (2010). Empirical investigation of critical success factors for implementing business intelligence systems in multiple engineering asset management organisations. In M. Hunter (Ed.), *Strategic information systems: Concepts, methodologies, tools, and applications* (pp. 2039–2063). Hershey, PA: IGI Global. doi:10.4018/978-1-60566-677-8.ch129

Yerrick, R., Lund, C., & Lee, Y. (2013). Online simulator use in the preparing chemical engineers. *International Journal of Online Pedagogy and Course Design*, *3*(2), 1–24. doi:10.4018/ijopcd.2013040101

Yilmaz, A. E., & Yilmaz, I. B. (2012). Natural language processing techniques in requirements engineering. In *Computer engineering: Concepts, methodologies, tools and applications* (pp. 533–545). Hershey, PA: IGI Global. doi:10.4018/978-1-61350-456-7.ch303

Yong, E. (2012). Literature review skills for undergraduate engineering students in large classes. In K. Yusof, N. Azli, A. Kosnin, S. Yusof, & Y. Yusof (Eds.), *Outcome-based science, technology, engineering, and mathematics education: Innovative practices* (pp. 240–261). Hershey, PA: IGI Global. doi:10.4018/978-1-4666-1809-1.ch012

Yun, H., Xu, J., Xiong, J., & Wei, M. (2013). A knowledge engineering approach to develop domain ontology. In V. Kumar & F. Lin (Eds.), *System and technology advancements in distance learning* (pp. 55–70). Hershey, PA: IGI Global. doi:10.4018/978-1-4666-2032-2.ch004

Zardari, S., Faniyi, F., & Bahsoon, R. (2013). Using obstacles for systematically modeling, analysing, and mitigating risks in cloud adoption. In I. Mistrik, A. Tang, R. Bahsoon, & J. Stafford (Eds.), *Aligning enterprise, system, and software architectures* (pp. 275–296). Hershey, PA: IGI Global. doi:10.4018/978-1-4666-2199-2.ch014

Zašcerinska, J., Ahrens, A., & Bassus, O. (2012). Enterprise 2.0 and 3.0 in education: Engineering and business students' view. In M. Cruz-Cunha, P. Gonçalves, N. Lopes, E. Miranda, & G. Putnik (Eds.), *Handbook of research on business social networking: Organizational, managerial, and technological dimensions* (pp. 472–494). Hershey, PA: IGI Global. doi:10.4018/978-1-61350-168-9.ch025

Zašcerinska, J., Ahrens, A., & Bassus, O. (2012). Enterprise 2.0 and 3.0 in education: Engineering and business students' view. In M. Cruz-Cunha, P. Gonçalves, N. Lopes, E. Miranda, & G. Putnik (Eds.), *Handbook of research on business social networking: Organizational, managerial, and technological dimensions* (pp. 472–494). Hershey, PA: IGI Global. doi:10.4018/978-1-61350-168-9.ch025

Zhang, D. (2012). Machine learning and value-based software engineering. In Y. Wang (Ed.), *Software and intelligent sciences: New transdisciplinary findings* (pp. 287–301). Hershey, PA: IGI Global. doi:10.4018/978-1-4666-0261-8.ch017

Zhang, H., Xu, C., Su, W., & Luo, H. (2014). Routing optimization for inter-domain traffic engineering under identifier network. In M. Boucadair & D. Binet (Eds.), *Solutions for sustaining scalability in internet growth* (pp. 127–147). Hershey, PA: IGI Global. doi:10.4018/978-1-4666-4305-5.ch007

Zhong, Y. (2013). Processing of 3D unstructured measurement data for reverse engineering. In S. Sirouspour (Ed.), *Advanced engineering and computational methodologies for intelligent mechatronics and robotics* (pp. 118–127). Hershey, PA: IGI Global. doi:10.4018/978-1-4666-3634-7.ch008

Zhou, Z., Wang, H., & Lou, P. (2010). *Manufacturing intelligence for industrial engineering: Methods for system self-organization, learning, and adaptation*. Hershey, PA: IGI Global. doi:10.4018/978-1-60566-864-2

Zhu, L., Jayaram, U., & Kim, O. (2011). Online semantic knowledge management for product design based on product engineering ontologies. *International Journal on Semantic Web and Information Systems*, *7*(4), 36–61. doi:10.4018/jswis.2011100102

Compilation of References

Abdelaziz, A. Y., Mekhamer, S. F., Badr, M. A. L., & Kamh, M. Z. (2008). Economic dispatch using an enhanced Hopfield neural network. *Electric Power Components and Systems, 36*(7), 719–732. doi:10.1080/15325000701881969

Ahuja, V., Farrens, M., & Ghosal, D. (2012). Cache-aware Affinitization on Commodity Multicores for High-speed Network Flows. In *Proceedings of the ACM/IEEE Symposium on Architectures for Networking and Communications Systems* (pp. 39-48). Austin, TX: IEEE. 10.1145/2396556.2396564

Alferov, G. V., & Malafeyev, O. A. (1996) The robot control strategy in a domain with dynamical obstacles. Lecture Notes in Computer Science, 1093, 211-217.

Alferov, G. V., Malafeyev, O. A., & Maltseva, A. S. (2015). Programming the robot in tasks of inspection and interception. 2015 Internetional Confference on Mechanics - Deventh Polyakhov's Reading, 7106713. doi:10.1109/POLYAKHOV.2015.7106713

Alferov, G., Ivanov, G., Efimova, P., & Sharlay, A. (2017). Study on the Structure of Limit Invariant Sets of Stationary Control Systems with Nonlinearity of Hysteresis Type. ICNAAM-16, Greece, AIP Conference Proceeding 2017. doi:10.1063/1.4992264

Alferov, G.V., & Gorizontov, A.M. (1987). Resource management in flexible automated production. *Kibernetika i Vychislitel'naya Tekhnika, 1987*(73), 76-81.

Allcock, W., Bresnahan, J., Kettimuthu, R., Link, M., Dumitrescu, C., Raicu, I., & Foster, I. (2005). The Globus Striped GridFTP Framework and Server. In *Proceedings of the ACM/IEEE Conference on Supercomputing* (pp. 54-65). Seattle, WA: IEEE. 10.1109/SC.2005.72

Al-Sumait, J.S., AL-Othman, A.K., & Sykulski, J.K. (2007). Application of pattern search method to power system valve-point economic load dispatch. *International Journal of Electrical Power & Energy Systems, 29*(10), 720–730.

Alsumait, J. S., Sykulski, J. K., & Al-Othman, A. K. (2010). A hybrid GA–PS–SQP method to solve power system valve-point economic dispatch problems. *Applied Energy, 87*(5), 1773–1781. doi:10.1016/j.apenergy.2009.10.007

Andersson, L., & Swallow, G. (2003). *The Multiprotocol Label Switching (MPLS) Working Group decision on MPLS signaling protocols*. IETF RFC 3468.

Ayyub, B. M., & Klir, G. J. (2006). *Uncertainty modelling and analysis in engineering and the sciences*. Chapman & Hall/CRC. doi:10.1201/9781420011456

Baklizi, A. (2003). Acceptance sampling based on truncated life tests in the Pareto distribution of the second kind. *Advances and Applications in Statistics*, *3*, 33–48.

Balakrishnan, N., Leiva, V., & Lopez, J. (2007). Acceptance sampling plans from truncated life tests based on the generalized Birnbaum-Saunders distribution. *Communications in Statistics. Simulation and Computation*, *36*(3), 643–656. doi:10.1080/03610910701207819

Banks, D. (2006). *Microengineering, MEMS, and Interfacing A Practical Guide*. Birmingham, UK: Taylor & Francis Group. doi:10.1201/9781420015416

Barisal, A. K., & Prusty, R. C. (2015). Large scale economic dispatch of power systems using oppositional invasive weed optimization. *Applied Soft Computing*, *29*, 122–137. doi:10.1016/j.asoc.2014.12.014

Barney, J. (1991). Firm resources and sustained competitive advantage. *Journal of Management*, *17*(1), 99–120. doi:10.1177/014920639101700108

Baron, R. M., & Kenny, D. A. (1986). The moderator-mediator variable distinction in social psychological research: Conceptual, strategic, and statistical considerations. *Journal of Personality and Social Psychology*, *51*(6), 1173–1182. doi:10.1037/0022-3514.51.6.1173 PMID:3806354

Bartram, D. (2000, December). Internet recruitment and selection: Kissing frogs to find princes. *International Journal of Selection and Assessment*, *8*(4), 261–274. doi:10.1111/1468-2389.00155

Becker, B., & Huselid, M. A. (1998). High performance work systems and firm performance: A synthesis of research and managerial implications. In G. R. Ferris (Ed.), *Research in personnel and human resources management* (Vol. 16, pp. 53–101). Greenwich, CT: JAI Press.

Bernard, M. (2013). *People Management: Data and Analytics Will Change It Forever, November 2013*. Retrieved from https://www.linkedin.com/pulse/20131111071432-64875646-2-keys-to-a-very-different-people-management

Bertsch, V., Treitz, M., Geldermann, J., & Rentz, O. (2007). Sensitivity analyses in multi-attribute decision support for off-site nuclear emergency and recovery management. *International Journal of Energy Sector Management*, *1*(4), 342–365. doi:10.1108/17506220710836075

Bhattacharya, A., & Chattopadhyay, P. (2010). Biogeography-based optimization for different economic load dispatch problems. *IEEE Transactions on Power Systems*, *25*(2), 1064–1077. doi:10.1109/TPWRS.2009.2034525

Bhattacharya, A., & Chattopadhyay, P. K. (2010). Hybrid differential evolution with biogeography-based optimization for solution of economic load dispatch. *IEEE Transactions on Power Systems*, *25*(4), 1955–1964. doi:10.1109/TPWRS.2010.2043270

Bliese, P. D. (1998). Team size, ICC values, team-level correlations: A simulation. *Organizational Research Methods*, *1*(4), 355–373. doi:10.1177/109442819814001

Bolus, R., & Pitts, J. (1999, April). Patient satisfaction: The indispensable outcome. *Managed Care (Langhorne, Pa.)*, *8*(4), 24–28. PMID:10387373

Borghans, I., Kleefstra, S. M., Kool, R. B., & Westert, G. P. (2012). Is the length of stay in hospital correlated with patient satisfaction? *International Journal for Quality in Health Care*, *24*(5), 443–451. doi:10.1093/intqhc/mzs037 PMID:22789666

Boulding, W., Glickman, S. W., Manary, M. P., Schulman, K. A., & Staelin, R. (2011). Relationship between patient satisfaction with inpatient care and hospital readmission within 30 days. *The American Journal of Managed Care*, *17*(1), 41–48. PMID:21348567

Burt, R. S. (1982). *Toward a Structural Theory of Action: Network Models of Social Structure, Perception and Action (Quantitative studies in social relations)*. New York: Academic Press. doi:10.1016/B978-0-12-147150-7.50014-4

Carr-Hill, R., & McIvor, S. (1989). *The NHS and its Customers*. Centre for Health Economics: University of York.

Chakrabarty, K., & Zeng, J. (2014). *Design automation methods and tools for microfluidics-based biochips*. Springer Netherlands.

Chakraborty, S., Senjyu, T., Yona, A., Saber, A. Y., & Funabashi, T. (2011). Solving economic load dispatch problem with valve-point effects using a hybrid quantum mechanics inspired particle swarm optimization. *IET Generation, Transmission & Distribution*, *5*(10), 1042–1052. doi:10.1049/iet-gtd.2011.0038

Chase, J., Gallatin, A., & Yocum, K. (2001). End System Optimizations for High-speed TCP. *IEEE Communications Magazine*, *39*(4), 68–74. doi:10.1109/35.917506

Chiang, C. L. (2005). Improved genetic algorithm for power economic dispatch of units with valve-point effects and multiple fuels. *IEEE Transactions on Power Systems*, *20*(4), 1690–1699. doi:10.1109/TPWRS.2005.857924

Ciornei, I., & Kyriakides, E. (2012). A GA-API solution for the economic dispatch of generation in power system operation. *IEEE Transactions on Power Systems*, *27*(1), 233–242. doi:10.1109/TPWRS.2011.2168833

Cleary, P. D., & McNeil, B. J. (1998). *Patient satisfaction as an indicator of quality care inquiry*. Academic Press.

Codreanu, I., Martowicz, A., Gallina, A., Pieczonka, L., & Uhl, T. (2009). Study of the effect of process induced uncertainties on the performance of a micro-comb resonator. *Solid State Phenomena, 147-149*, 716-725.

Coelho, L. S., & Mariani, V. C. (2006). Combining of chaotic differential evolution and quadratic programming for economic dispatch optimization with valve-point effect. *IEEE Transactions on Power Systems*, *21*(2), 989–996. doi:10.1109/TPWRS.2006.873410

Cohen, D., & Prusak, L. (2001). *In good company: How social capital makes organizations work*. Boston: Harvard Business School Press.

Cooren, Y. (2008). *Perfectionnement d'un algorithme adaptatif d'Optimisation par Essaim Particulaire. Applications en génie médical et en électronique* (Ph.D. Thesis). 12 Val de Marne University, Paris, France.

Der Kiureghian, A., & Ke, J.-B. (1988). The stochastic finite element method in structural reliability. *Probabilistic Engineering Mechanics*, *3*(2), 83–91. doi:10.1016/0266-8920(88)90019-7

Dinh, L. L., Ngoc, D. V., Nguyen, T. H., & Anh, D. L. (2013).A hybrid differential evolution and harmony search for nonconvex economic dispatch problems. *Proceedings of the Power Engineering and Optimization Conference (PEOCO)*, 238-243.

Dodson, B. (1994). *Weibull Analysis*. Milwaukee, WI: ASQ Quality Press.

Doherty, R. D. (2010). Getting social with recruitment, Strategic HR Review. 9(6), 11-15. *International Journal of Selection and Assessment*, *8*(4), 261–274.

Duman, S., Güvenç, U., & Yörükeren, N. (2010). Gravitational search algorithm for economic dispatch with valve-point effects. *International Review of Electrical Engineering*, *5*(6), 2890–2895.

Duvvuru, N., & Swarup, K. S. (2011). A hybrid interior point assisted differential evolution algorithm for economic dispatch. *IEEE Transactions on Power Systems*, *26*(2), 541–549. doi:10.1109/TPWRS.2010.2053224

Epstein, B. (1954). Truncated life tests in the exponential case. *Annals of Mathematical Statistics*, *68*(3), 555–564. doi:10.1214/aoms/1177728723

Esnet. (2017). *Energy Sciences Network*. Retrieved August 28, 2017, from http://www.es.net/

Eurocode 3: Design of steel structures. (1992). Brussels: European Committee for Standardization.

Fair, R. B. (2007). Digital microfluidics: Is a true lab-on-a-chip possible? *Microfluidics and Nanofluidics*, *3*(3), 245–281. doi:10.100710404-007-0161-8

Farrell, R. (1962). Bounded Length Confidence Intervals for the Zero of a Regression Function. *Annals of Mathematical Statistics*, *33*(1), 237–247. doi:10.1214/aoms/1177704727

Federal Aviation Administration. (2008). Aeronautical Decision Making. In Pilot's Handbook of Aeronautical Knowledge. FAA-H-8083-25. 17.1–17.32. FAA.

Feldman, D. C., & Klaas, B. S. (2002). Internet job hunting: A field study of applicant experiences with on-line recruiting. *Human Resource Management*, *41*(2), 175–192. doi:10.1002/hrm.10030

Fertig, F. W., & Mann, N. R. (1980). Life-test sampling plans for two-parameter Weibull populations. *Technometrics*, *22*(2), 165–177. doi:10.1080/00401706.1980.10486131

Flap, H. D., & Volker, B. (2004). Social networks and performance at work. In H. D. Flap & B. Volker (Eds.), *Creation and returns of social capital* (pp. 172–196). London, UK: Routledge.

Fletcher, R. (2000). *Practical Methods of Optimization* (2nd ed.). John Wiley and Sons. doi:10.1002/9781118723203

Florian, W. (2009). E-Recruitment Vs. Traditional Recruitment: A Descriptive Analysis. GRIN Verlag.

Flueler, T. (2006). Decision Making for Complex Socio-Technical Systems: Robustness from Lessons Learned in Long-Term Radioactive Waste Governance. In *Environment & Policy*. Springer.

Frangi, A. (2008). *Advances in multiphysics simulation and experimental testing of MEMS*. Imperial College Press. doi:10.1142/p522

Frenkel, D. (2004). *Introduction to Monte Carlo Methods. In Computational Soft Matter: From Synthetic Polymers to Proteins, Lecture Notes, John von Neumann Institute for Computing* (Vol. 23, pp. 29–60). Julich: NIC Series.

Friedken, N. E., & Johnson, E. C. (1998). *A structural theory of social influence*. Cambridge, MA: Cambridge University Press. doi:10.1017/CBO9780511527524

Gaing, Z. L. (2003). Particle swarm optimization to solving the economic dispatch considering the generator constraints. *IEEE Transactions on Power Systems*, *18*(3), 1187–1195. doi:10.1109/TPWRS.2003.814889

Gianchandani, Y. B., Tabata, O., & Zappe, H. P. (2008). *Comprehensive Microsystems*. Elsevier.

Giniger, S., Dispenzieri, A., & Eisenberg, J. (1983). Age, experience and performance on speed and skill jobs in an applied setting. *The Journal of Applied Psychology*, *68*(3), 469–475. doi:10.1037/0021-9010.68.3.469

Globoforce. (2012). SHRM/Globforce Employee Recognition survey FALL 2012 Report. In *The business impact of employee recognition, 2012*. Author.

Glover, F. (1986). Future Paths for Integer Programming and Links to Artificial Intelligence. *Computers & Operations Research*, *13*(5), 533–549. doi:10.1016/0305-0548(86)90048-1

Goode, H. P., & Kao, J. H. K. (1961). Sampling plans based on the Weibull distribution. In *Proceedings of Seventh National Symposium on Reliability and Quality Control* (pp. 24–40). Philadelphia: Academic Press.

Granovetter, M. S. (1985). Economic action and social structure: The problem of embeddedness. *American Journal of Sociology*, *91*(3), 481–510. doi:10.1086/228311

Grigsby, L. L. (2007). *Power System Stability and Control* (2nd ed.). CRC Press. doi:10.1201/9781420009248

Gu, Y., Hong, X., & Grossman, R. (2004). Experiences in Design and Implementation of a High Performance Transport Protocol. In *Proceedings of the ACM/IEEE Supercomputing Conference* (pp. 22-35). Pittsburgh, PA: IEEE. doi: 10.1109/SC.2004.24

Gupta, S. S. (1962). Life test sampling plans for normal and lognormal distribution. *Technometrics*, *4*(2), 151–175. doi:10.1080/00401706.1962.10490002

Gupta, S. S., & Groll, P. A. (1961). Gamma distribution in acceptance sampling based on life tests. *Journal of the American Statistical Association*, *56*(296), 942–970. doi:10.1080/01621459.1961.10482137

Gu, Y., & Grossman, R. (2003). SABUL: A Transport Protocol for Grid Computing. *Journal of Grid Computing*, *1*(4), 377–386. doi:10.1023/B:GRID.0000037553.18581.3b

Gu, Y., & Grossman, R. (2007). UDT: UDP-based Data Transfer for High-speed Wide Area Networks. *Computer Networks*, *51*(7), 1777–1799. doi:10.1016/j.comnet.2006.11.009

Haimeecode. (2012). *The State of Employer Branding, October 2012*. Retrieved from https://www.slideshare.net/haimeecode/state-of-employer-brand

Hair, J. F., Black, W. C., Babin, B. J., Anderson, R. E., & Tatham, R. L. (2006). *Multivariate data analysis* (6th ed.). Upper Saddle River, NJ: Prentice-Hall.

Hanford, N., Ahuja, V., Farrens, M., Ghosal, D., Balman, M., Pouyoul, E., & Tierney, B. (2014). Analysis of the Effect of Core Affinity on High-Throughput Flows. In *Proceedings of the International Workshop on Network-Aware Data Management* (pp. 9-15). New Orleans, LA: Academic Press. 10.1109/NDM.2014.10

Hasofer, A. M., & Lind, N. C. (1974). Exact and invariant second-moment code format. *ASCE Journal of the Engineering Mechanics Division*, *100*, 111–121.

Hazra, J., & Sinha, A. K. (2008). Environmental constrained economic dispatch using bacteria foraging optimization. *Proceedings of Power System Technology and IEEE Power India Conference, Joint International Conference*, 1-6. 10.1109/ICPST.2008.4745330

Hazra, S., Roy, P. K., & Sinha, A. (2015). An efficient evolutionary algorithm applied to economic load dispatch problem. *Proceedings of the CCC and IT*, 1-6.

Heike, H. D., Tarcolea, C., Schüte, M., & Manescu, B. (1999). *Grundlagen der Statistik und Wahrscheinlichkeitsrechnung*. Oldenbourg.

Helton, J. C., & Davis, F. J. (2003). Latin hypercube sampling and the propagation of uncertainty in analyses of complex systems. *Reliability Engineering & System Safety*, *81*(1), 23–69. doi:10.1016/S0951-8320(03)00058-9

Hojat, M., Louis, D. Z., Maxwell, K., Markham, F. W., Wender, R. C., & Gonnella, J. S. (2011, June). A Brief Instrument to Measure Patients' Overall Satisfaction With Primary Care Physicians. *Family Medicine*, *43*(6), 412–417. PMID:21656396

Hong, C. C., Choi, J. W., & Ahn, C. H. (2003). A novel in-plane passive microfluidic mixer with modified tesla structures. *Lab on a Chip*, *4*(2), 109–113. doi:10.1039/b305892a PMID:15052349

International Civil Aviation Organization. (2002). Human Factors Guidelines for Safety Audits Manual. Doc. ICAO 9806-AN/763. Canada, Montreal: Author.

International Civil Aviation Organization. (2004). *Cross-Cultural Factors in Aviation Safety: Human Factors Digest Nº 16*. Circ. ICAO 302-AN/175. Montreal, Canada: Author. doi:10.1007/1-4020-3529-2

International Civil Aviation Organization. (2009). *Manual on Global Performance of the Air Navigation System. Doc. 9883*. Montreal, Canada: Author.

International Civil Aviation Organization. (2013). Safety Management Manual (SMM) (3rd ed.). Doc. ICAO 9859-AN 474. Montreal: Author.

International Civil Aviation Organization. (2013). *State of Global Aviation Safety*. Montreal, Canada: Author.

Internet2. (2017). *Internet2 Official Website*. Retrieved August 28, 2017, from http://www.internet2.edu

Iperf3. (2017). *ESnet iperf3*. Retrieved August 28, 2017, from https://github.com/esnet/iperf

ISO 31000 (2009) / ISO Guide 73:2002

Ivanov, G., Alferov, G., Sharlay, A., & Efimova, P. (2017). Conditions of Asymptotic Stability for Linear Homogeneous Switched Systems. ICNAAM-16, Greece, AIP Conference Proceeding 2017. doi:10.1063/1.4992263

Ivanov, G. G., & Sharlay, A. S. (2015). On Stability of Linear Homogeneous Switched Systems. *2015 International Conference "Stability and Control Processes" in Memory of V. I. Zubov, SCP 2015*, 13-15. 10.1109/SCP.2015.7342039

Jain, S., Kumar, A., Mandal, S., Ong, J., Poutievski, L., Singh, A., ... Vahdat, A. (2013). B4: Experience with a Globally-Deployed Software Defined WAN. In *Proceedings of the ACM SIGCOMM Conference* (pp. 3-14). Hong Kong, China: ACM. 10.1145/2486001.2486019

Jason, B. (2013a). *20 Mobile Recruiting Insights #mrec13, September 2013*. Retrieved from http://www.talenthq.com/2013/09/20-mobile-recruiting-insights

Jason, B. (2013b). *45 Mobile Recruiting Tips and Trends From #mrec, September 2013*. Retrieved from http://www.talenthq.com/2013/09/45-mobile-recruiting-tips-and-trends-from-mrec

Jayabarathi, T., Jayaprakash, K., Jeyakumar, D. N., & Raghunathan, T. (2005). Evolutionary programming techniques for different kinds of economic dispatch problems. *Electric Power Systems Research*, *73*, 2169–2176.

Jeanne, M. (2013). *Gamification In Leadership Development: How Companies Use Gaming To Build Their Leader Pipeline, September 2013*. Retrieved from https://www.forbes.com/sites/jeannemeister/2013/09/30/gamification-in-leadership-development-how-companies-use-gaming-to-build-their-leader-pipeline/#7bfa74325a5

Jen, D. (2012). *How to use Social Media for Employer Branding, July 2012*. Retrieved from https://www.lever.co/blog/how-to-use-social-media-for-employer-branding

Jeon, N. L., Dertinger, S. K. W., Chiu, D. T., Choi, I. S., Stroock, A. D., & Whitesides, G. M. (2000). Generation of solution and surface gradients using microfluidic systems. *Langmuir*, *16*(22), 8311–8316. doi:10.1021/la000600b

Jobvite. (2012). *2012 Social Job Seeker Survey*. London: Forum Jobs, Inc.

Juilo, R. (2013). *HMV tweets their own sacking, January 2013*. Retrieved from https://www.marketing-technews.net/news/2013/feb/01/hmv-tweets-their-own-sacking

Kadry. (2006). A Proposed Technique to evaluate the Stochastic Mechanical Response based on Transformation with Finite Element Method. *International Journal of Applied Mathematics and Mechanics*.

Kadry, S. (2006). A Proposed Technique to evaluate the Stochastic Mechanical Response based on Transformation with Finite Element Method. *International. Journal of Applied Mathematics and Mechanics*, *2*, 94–108.

Kantam, R. R. L., & Rosaiah, K. (1998). Half logistic distribution in acceptance sampling based on life tests. *IAPQR Transactions*, *23*, 117–125.

Kantam, R. R. L., Rosaiah, K., & Srinivasa Rao, G. (2001). Acceptance sampling based on life tests: Log-logistic model. *Journal of Applied Statistics*, *28*(1), 121–128. doi:10.1080/02664760120011644

Keating, C., Fernandez, A. A., Jacobs, D. A., & Kauffmann, P. (2001). A methodology for analysis of complex sociotechnical processes. *Business Process Management Journal*, *7*(1), 33–50. doi:10.1108/14637150110383926

Keegan, D. P., Eiler, R. G., & Jones, C. R. (1995). Are your performance measures obsolete? *Management Accounting*, *70*(12), 45–50.

Kharchenko, V., Shmelova, T., & Sikirda, Y. (2012). *Decision-Making of Operator in Air Navigation System*. Kirovograd: KFA of NAU.

Kharchenko, V., Shmelova, T., & Sikirda, Y. (2012). Methodology for analysis of flight situation development using GERT's and Markov's networks: In *Proceedings of the 5 World Congress Aviation in the XXIst century.*

Kharchenko, V., Shmelova, T., & Zubrytskyi, V. (2017). Decision making in air navigation system as a socio-technical system. *Proceedings of the International Conference "AVIA 2017"*.

Kharchenko, V., Shmelova, T., & Sikirda, Y. (2011). Graphanalytical Models of Decision-Making by Air Navigation System's Human-Operator. *Proceedings of the National Aviation University*, *1*, 5–17.

Kharchenko, V., Shmelova, T., & Sikirda, Y. (2011). Methodology for Analysis of Decision Making in Air Navigation System. *Proceedings of the National Aviation University*, *3*, 85–94.

Kharchenko, V., Shmelova, T., & Sikirda, Y. (2012). Modeling of Behavioral Activity of Air Navigation System's Human-Operator in Flight. *Proceedings of the National Aviation University*, *2*, 5–17.

Kharchenko, V., Shmelova, T., & Sikirda, Y. (2016). *Decision-Making in Socio-Technical Systems*. Kyiv: NAU.

Kiefer, J., & Wolfowitz, J. (1952). Stochastic Estimation of the Maximum of A Regression Function. *Annals of Mathematical Statistics*, *23*(3), 462–466. doi:10.1214/aoms/1177729392

Kroese, D. P., Taimre, T., & Botev, Z. I. (2011). *Handbook of Monte Carlo Methods*. Wiley. doi:10.1002/9781118014967

Kulakov, F., Alferov, G. V., Efimova, P., Chernakova, S., & Shymanchuk, D. (2015) Modeling and control of robot manipulators with the constraints at the moving objects. *2015 International Conference on "Stability and Control Processes" in Memory of V.I. Zubov,* 102-105.

Kulakov, F., Alferov, G., & Efimova, P. (2015). Method of remote control over space robots. 2015 Internetional Confference on Mechanics - Deventh Polyakhov's Reading, 7106742. doi:10.1109/POLYAKHOV.2015.7106742

Kulakov, F., Sokolov, B., Shalyto, A., & Alferov, G. (2016). Robot Master Slave and Supervisory Control with Large Time Delays of Control Signals and Feedback. *Applied Mathematical Sciences, 10*(33-36), 1783–1796. doi:10.12988/ams.2016.6380

Lam, A. Y. S., & Li, V. O. K. (2010). Chemical-reaction-inspired metaheuristic for optimization. *IEEE Transactions on Evolutionary Computation, 14*(3), 381–399. doi:10.1109/TEVC.2009.2033580

Lars, S. (2012). *Three Days, Three Key Themes: My Review of Talent Connect 2012, October 2012*. Retrieved from https://business.linkedin.com/talent-solutions/blog/2012/10/talent-connect-2012-3-themes

Lawler, G., & Limic, V. (2010). *Random Walk: A Modern Introduction*. New York, NY: Cambridge University Press. doi:10.1017/CBO9780511750854

Lefebvre, B. (2008, January-June). Intentional-Reflective Model of the Agent. *Reflexion Processes and Control, 1,* 69–78.

Lefebvre, B., & Adams-Webber, J. (2002, January-June). Functions of Fast Reflexion in Bipolar Choice. *Reflexion Processes and Control, 1,* 29–40.

Lefebvre, V. A. (2001). Algebra of Conscience (2nd ed.). Dordrecht, The Netherlands: Kluwer Publishers.

Leitao, B. (2009). Tuning 10Gb Network Cards on Linux. In *Proceedings of the Linux Symposium* (pp. 169-184). Ottawa, Canada: Academic Press.

Lemaire, Mohamed, & Flores-Macias. (1997). The use of finite element codes for the reliability of structural systems. *Frangopol.*

Leychenko, S., Malishevskiy, A., & Mikhalic, N. (2006). *Human Factors in Aviation: monograph in two books*. Kirovograd: YMEKS.

Lin, R. M., & Wang, W. J. (2006). Structural dynamics of microsystems – current state of research and future direction. *Mechanical Systems and Signal Processing, 20*(5), 1015–1043. doi:10.1016/j.ymssp.2005.08.013

Llewellyn-Thomas, H. A., Arshinoff, R., Bell, M., Williams, J. I., & Naylor, C. D. (1998). In the queue for total joint replacement: Patients' perspectives on waiting times. *Journal of Evaluation in Clinical Practice, 4*(1), 63–74. doi:10.1046/j.1365-2753.1998.t01-1-00006.x PMID:9524913

Luis, S. (2013). *2013: The Year of Social HR, January 2013*. Retrieved from http://www.elsua.net/2013/01/09/2013-the-year-of-social-hr-2

Madsen, H. O., Kenk, S., & Lind, N. C. (1986). *Methods of structural safety*. Prentice-Hall Inc.

Malafeyev, O., Alferov, G., & Andreyeva, M. (2015) Group strategy of robots in game - theoretic model of interception with incomplete information. *2015 Internetional Conference on Mechanics - Seventh Polyakhov's reading*, 7106751. 10.1109/POLYAKHOV.2015.7106751

Mawardi, A., & Pitchumani, R. (2005). Design of Microresonators under Uncertainty. *Journal of Microelectromechanical Systems*, *14*(1), 63–69. doi:10.1109/JMEMS.2004.839021

McKeown, N., Anderson, T., Balakrishnan, H., Parulkar, G., Peterson, L., Rexford, J., ... Turner, J. (2008). OpenFlow: Enabling Innovation in Campus Networks. *Computer Communication Review*, *38*(2), 69–74. doi:10.1145/1355734.1355746

Mehrabian, A. R., & Lucas, C. (2006). A novel numerical optimization algorithm inspired from weed colonization. *Ecological Informatics*, *1*(4), 355–366. doi:10.1016/j.ecoinf.2006.07.003

Meilakhs, A. M. (1975). O stabilizatsii upravlayemih system v usloviyah neopredelennosti. *Avtomatika I Telemechanika, 2*, 182-189.

Mike, W. (2012). *Making the Quantum Leap: UPS Social Media Recruitment ROI 2012, January 2012*. Retrieved from https://www.slideshare.net/MikeVangel/real-social-media-recruitment-roi-from-ups-11278020

Moens, D., & Vandepitte, D. (2005). A survey of non-probabilistic uncertainty treatment in finite element analysis. *Computer Methods in Applied Mechanics and Engineering*, *194*(12-16), 1527–1555. doi:10.1016/j.cma.2004.03.019

Mohamed, A., & Lemaire, M. (1995). Linearzed mechanical model to evaluate reliability of offshore structures. *Journal of Structural Safety*, *17*(3), 167–193. doi:10.1016/0167-4730(95)00009-S

Mohammadi, A., & Varahram, M. H. (2006). Using neural network for solving of on-line economic dispatch problem. *Proceedings of International Conference on Intelligent Agents for Modelling, Control and Automation*, 81 – 87. 10.1109/CIMCA.2006.228

Molchanov, A. P. & Pyatnitskii, E. S. (1986). Funkcii Lyapunova, opredelyaushie neobhodimiye I dostatochniye usloviya absolutnoi ustoichivosty nelineinih nestacionarnih system upravleniya. *Avtomatika I Telemechanika, 4*, 5-15.

Muscolino, G., Ricciardi, G., & Impollonia, N. (2000). Improved dynamic analysis of structures with mechanical uncertainties under deterministic input. *Structural Safety*, *15*, 199–212.

Nechval, N. A., & Vasermanis, E. K. (2004). *Improved Decisions in Statistics*. Riga: Izglitibas soli.

Nechval, N. A., Berzins, G., Purgailis, M., & Nechval, K. N. (2008). Improved estimation of state of stochastic systems via invariant embedding technique. *WSEAS Transactions on Mathematics*, *7*, 141–159.

Nechval, N. A., Purgailis, M., Berzins, G., Cikste, K., Krasts, J., & Nechval, K. N. (2010). Invariant embedding technique and its applications for improvement or optimization of statistical decisions. In K. Al-Begain, D. Fiems, & W. Knottenbelt (Eds.), *Analytical and Stochastic Modeling Techniques and Applications, (LNCS)* (Vol. 6148, pp. 306–320). Berlin: Springer-Verlag. doi:10.1007/978-3-642-13568-2_22

Neverova, E. G., Malafeyev, O. A., Alferov, G. V., & Smirnova, T. E. (2015) Model of interaction between anticorruption authorities and corruption groups. *2015 International Conference on "Stability and Control Processes" in Memory of V.I. Zubov*, 488-490.

Niknam, T., Mojarrad, H. D., & Meymand, H. Z. (2011). Non-smooth economic dispatch computation by fuzzy and self-adaptive particle swarm optimization. *Applied Soft Computing*, *11*(2), 2805–2817. doi:10.1016/j.asoc.2010.11.010

Nomana, N., & Iba, H. (2008). Differential evolution for economic load dispatch problems. *Electric Power Systems Research*, *78*(8), 1322–1331. doi:10.1016/j.epsr.2007.11.007

OSCARS. (2017). *On-Demand Secure Circuits and Advance Reservation System*. Retrieved August 28, 2017, from http://www.es.net/oscars

Ouhimmou, El Hami, Ellaia, & Tkiouat. (2010). Approximate solution of the structural problems using probabilistic transformation. *International Journal Key Engineering Materials, 446*, 91-99.

Ouhimmou,, S., El Hami, A., Ellaia, R., & Tkiouat, M. (2012). Contribution to development of Reliability and Optimization methods applied to Mechanical Structures. *Journal of Applied Mathematics*.

Ouhimmou, S., El Hami, A., Ellaia, R., & Tkiouat, M. (2010). Reliability method coupled with Finite Element Analysis for structural problems with implicit response functions. *International Journal of Advances and Applications in Mechanical Engineering and Technology*, *1*, 109–131.

Ouhimmou, S., El Hami, A., Ellaia, R., & Tkiouat, M. (2011). Reliability Problems with Implicit Response Functions. *Recent Patents on Mechanical Engineering*, *4*, 37–46.

Oxford Economics. (2012). *The Digital Transformation of People Management", March 2012*. Retrieved from https://www.oxfordeconomics.com/my-oxford/projects/129033

Park, J. B., Lee, K. S., Shin, J. R., & Lee, K. Y. (2005). A particle swarm optimization for economic dispatch with nonsmooth cost functions. *IEEE Transactions on Power Systems*, *20*(1), 34–42. doi:10.1109/TPWRS.2004.831275

Philipov, A. F. (1985). *Differencialnye uravneniya s nerazrivnoy pravoy chastyu*. Moscow: Nauka.

Pichugin, Yu., Alferov, G., & Malafeyev, O. (2016). Parameters estimation in mechanism design. *Contemporary Engineering Sciences*, *9*(1-4), 175–185. doi:10.12988/ces.2016.512319

Ployhard. (2006). Social media in the workplace: issues and strategic questions. SHRM Foundation: shaping the future of HR. In *Human resources and social media*. KMPG International Cooperative.

Pothiya, S., Ngamroo, I., & Kongprawechnon, W. (2010). Ant colony optimization for economic dispatch problem with non-smooth cost functions. *International Journal of Electrical Power & Energy Systems, 32*(5), 478–487. doi:10.1016/j.ijepes.2009.09.016

Prasad, R., Murray, M., Dovrolis, C., & Claffy, K. (2003). Bandwidth estimation: Metrics, measurement techniques, and tools. *IEEE Network, 17*(6), 27–35. doi:10.1109/MNET.2003.1248658

Qin, Y. (2010). *Micromanufacturing Engineering and Technology*. William Andrew.

Quintana, J. M., Gonzalez, N., Bilbao, A., Aizpuru, F., Escobar, A., & Esteban, C. (2006). Predictors of patient satisfaction with hospital care. *BMC Health Services Research, 6*(102). PMID:16914046

Rackwitz, R. (2001). Reliability analysis: A review and some perspectives. *Structural Safety, 23*(4), 365–395. doi:10.1016/S0167-4730(02)00009-7

Randles, R. H., & Douglas, A. W. (1979). *Introduction to the theory of nonparametric statistics* (Vol. 1). New York: Wiley.

Reddy, A. S., & Vaisakh, K. (2013). Shuffled differential evolution for large scale economic dispatch. *Electric Power Systems Research, 96*, 237–245. doi:10.1016/j.epsr.2012.11.010

Rosaiah, K., & Kantam, R. R. L. (2005). Acceptance sampling based on the inverse Rayleigh distribution. *Economic Quality Control, 20*(2), 277–286. doi:10.1515/EQC.2005.277

Rosaiah, K., Kantam, R. R. L., & Santosh Kumar, Ch. (2006). Reliability test plans for exponentiated log-logistic distribution. *Economic Quality Control, 21*(2), 165–175. doi:10.1515/EQC.2006.279

Roy, P. K., Bhui, S., & Paul, C. (2014). Solution of economic load dispatch using hybrid chemical reaction optimization approach. *Applied Soft Computing, 24*, 109–125. doi:10.1016/j.asoc.2014.07.013

Rubinstein, R. Y., & Kroese, D. P. (2008). *Simulation and the Monte Carlo Method*. Wiley-Interscience.

Ryan, H. (2013). *Inside HMV's Twitter Disaster: What went wrong (and how to keep your company safe), February 2013*. Retrieved from https://www.linkedin.com/pulse/20130214190038-2967511-inside-hmv-s-twitter-disaster-what-went-wrong-and-how-to-keep-your-company-safe

Santra, D., Mondal, A., & Mukherjee, A. (2014). Study of economic load dispatch by various hybrid optimization techniques. In Hybrid Soft Computing: Research and Applications (pp. 37-74). Springer India.

Sarah, P. (2012). *With 40% of Traffic from Outside the U.S., Glassdoor Raises $20M Series D to Expand Internationally, October 2012*. Retrieved from https://techcrunch.com/2012/10/31/with-40-of-traffic-from-outside-u-s-glassdoor-raises-20-million-series-d-to-expand-internationally

Sayah, S., & Hamouda, A. (2013). A hybrid differential evolution algorithm based on particle swarm optimization for nonconvex economic dispatch problems. *Applied Soft Computing, 13*(4), 1608–1619.

Shesser, R., Kirsch, T., Smith, J., & Kirsch, R. (1991). An analysis of emergency department use by patients with minor illness. *Annals of Emergency Medicine, 20*(7), 743-747.

Shmelova, T., & Konovalova, A. (2017). Decision making of ATCO in emergency situation "aircraft decompression". *Proceedings of the International conference "AVIA -2017".*

Shmelova, T., & Sikirda, Y. (2016). Calculation the scenarios of the flight situation development using GERT's and Markov's networks. *Proceedings of the Conference "Management of high-speed moving objects training operators of complex systems".*

Shmelova, T., Bondarev, D., & Znakovska, Y. (2016). Modelling of the Decision Making by UAV's Operator in Emergency Situations. In *Proceedings of the IEEE 4th International Conference on Methods and Systems of Navigation and Motion Control (MSNMC-2016)*. Kyev: IEEE.

Shmelova, T., Dolhov, D., & Muliar, P. (2015). Mathematical models of operators activities in aeronavigation sociotechnical system. In *Proceedings of the Conference "Techniques and technology. Science yesterday, today, tomorrow".* Warsaw.

Shmelova, T., Mirskova, M., & Krepak, D. (2017). Decision making in emergency situation of air navigation system's operator. *Proceedings of the International conference "AVIA -2017".*

Shmelova, T., Sikirda, Y., & Sunduchkov, K. (2013). Socio-technical Analysis of Air Navigation System. *Science and technology of the Air Forces of the Armed Forces of Ukraine*, *4*(13), 34–39.

Shmelova, T., & Shulimov, O. (2017). Method of determination of optimal place of landing for remotely piloted aircraft. *International Conference "AVIA 2017".*

Shmelova, T., & Sikirda, Y. (2012). Analysis of Human-Operator's Decision-Making in Air Navigation System. *Radioelectronic and Computer Systems*, *7*(59), 319–324.

Shmelova, T., Sikirda, Y., & Jafarzade, T. (2013). Models of Flight Situations Development while Decision-Making by Air Navigation System's Human-Operator. *Information Processing Systems*, *8*(115), 136–142.

Siarry, P., Dréo, J., Pétrowski, A., & Taillard, E. (2003). *Métaheuristiques pour l'optimisation difficile*. Eyrolles.

Sinha, N., Chakrabarti, R., & Chattopadhyay, P. K. (2003). Evolutionary programming techniques for economic load dispatch. *IEEE Transactions on Evolutionary Computation*, *7*(1), 83–94. doi:10.1109/TEVC.2002.806788

SKYbrary Aviation Safety. (n.d.). Retrieved from http://www.skybrary.aero/index.php/Main_Page

Snell, S. A., Youndt, M. A., & Wright, P. M. (1996). Establishing a framework for research in strategic human resource management: Merging resource theory and organizational learning. In G. R. Ferris (Ed.), *Research in personnel and human resource management* (Vol. 14, pp. 61–90). Greenwich, CT: JAI Press.

Sobel, M., & Tischendrof, J. A. (1959). Acceptance sampling with life test objective. In *Proceedings of Fifth National Symposium on Reliability and Quality Control* (pp. 108–118). Philadelphia: Academic Press.

Spall, J. (1992). Multivariate Stochastic Approximation Using a Simultaneous Perturbation Gradient Approximation. *IEEE Transactions on Automatic Control*, *37*(3), 332–341. doi:10.1109/9.119632

Spall, J. (1998). Implementation of the Simultaneous Perturbation Algorithm for Stochastic Optimization. *IEEE Transactions on Aerospace and Electronic Systems*, *34*(3), 817–823. doi:10.1109/7.705889

Spall, J. (2003). *Introduction to Stochastic Search and Optimization: Estimation, Simulation, and Control*. Hoboken, NJ: John Wiley & Sons, Inc. doi:10.1002/0471722138

Squeglia, N. L. (2008). *Zero Acceptance Number Sampling Plans* (5th ed.). Milwaukee, WI: ASQ Quality Press.

Srinivasa, R. G., & Kantam, R. R. L. (2010). Acceptance sampling plans from truncated life tests based on the log-logistic distribution for percentiles. *Economic Quality Control*, *25*, 153–167.

Stanciu, I. (2012). Dispersion of SU-8 Photolithography Process for Fabrication of Microfluidic Devices. *Scientific Bulletin, Series C: Electrical Engineering and Computer Science, University Politehnica Bucharest*, *74*(2), 127–142.

Stanciu, I. (2015). Uncertainty Analysis of Mixing Efficiency Variation in Passive Micromixers due to Geometric Tolerances. *Modelling and Simulation in Engineering*, *2015*, 343087. doi:10.1155/2015/343087

Stanciu, I. (2017). *Uncertainty Analysis for Improved Design of Microfluidic Devices*. LAP LAMBERT Academic Publishing.

Stephens, K. S. (2001). *The Handbook of Applied Acceptance Sampling Plans, Procedures and Principles*. Milwaukee, WI: ASQ Quality Press.

Subathra, M. S. P., Selvan, S. E., Victoire, T. A. A., Christinal, A. H., & Amato, U. (2015). A hybrid with cross- entropy method and sequential quadratic programming to solve economic load dispatch problem. *IEEE Systems Journal*, *9*(3), 1031–1044. doi:10.1109/JSYST.2013.2297471

Sudreta, B., & Der Kiureghian, A. (2002). Comparison of finite element reliability methods. *Probabilistic Engineering Mechanics*, *17*(4), 337–348. doi:10.1016/S0266-8920(02)00031-0

Swarup, K. S. (2005). Ant colony optimization for economic generator scheduling and load dispatch. *Proceedings of the 6th WSEAS International Conference on Evolutionary Computing*, 167-175.

Sykes, T. A., Venkatesh, V., & Gosain, S. (2009). Model of acceptance with peer support: A social network perspective to understand employees' system use. *Management Information Systems Quarterly*, *33*(2), 371–393.

Thompson, J. D. (1967). Organizations in Action: Social Science Bases of Administrative Theory Classics in Organization and Management Series. New York: McGraw-Hill.

Toma, G., Triner, W., & McNutt, L.-A. (2009). Patient satisfaction as a function of emergency department previsit expectations. *Annals of Emergency Medicine*, *54*(3), 360–367. doi:10.1016/j.annemergmed.2009.01.024 PMID:19282060

Torcson, P. J. (2005). Patient Satisfaction: The Hospitalist's Role. *The Hospitalist*, 27–30.

Tsai, T. R., & Wu, S. J. (2006). Acceptance sampling based on truncated life tests for generalized Rayleigh distribution. *Journal of Applied Statistics, 33*(6), 595–600. doi:10.1080/02664760600679700

Tsai, W. (2002). Social structure of "coopetition" within a multiunit organization: Coordination, Competition, and intra organizational knowledge sharing. *Organization Science, 13*(2), 179–190. doi:10.1287/orsc.13.2.179.536

Tsai, W., & Ghosal, S. (1998). Social capital and value creation: The role of intrafirm networks. *Academy of Management Journal, 41*(4), 464–476. doi:10.2307/257085

UDT. (2017a). *A List of Projects/Software/Companies that use UDT*. Retrieved September 1, 2017, from http://udt.sourceforge.net/poweredby.html

UDT. (2017b). *UDT: Breaking the Data Transfer Bottleneck*. Retrieved August 28, 2017, from http://udt.sourceforge.net/

Vivian, G. (2013). *Gamification' Techniques Increase Your Employees' Ability To Learn By 40%, September 2013*. Retrieved from http://www.businessinsider.com/gamification-techniques-increase-your-employees-ability-to-learn-by-40-2013-9

Wada, T., & Fujisaki, Y. (2013). A Stopping Rule for Simultaneous Perturbation Stochastic Approximation. In *Proceedings of the European Control Conference* (pp. 644-649). Zurich, Switzerland: Academic Press.

Waldman, D. A., & Avolio, B. J. (1986). A meta-analysis of age differences and job performance. *The Journal of Applied Psychology, 71*(1), 33–38. doi:10.1037/0021-9010.71.1.33

Walker, G., Kogut, B., & Shan, W. (1997). Social capital, structural holes and the formation of an industry network. *Organization Science, 8*(2), 109–125. doi:10.1287/orsc.8.2.109

Wang, Q., & Spall, J. (2011). Discrete Simultaneous Perturbation Stochastic Approximation on Loss Function with Noisy Measurements. In *Proceedings of the American Control Conference* (pp. 4520-4525). San Francisco, CA: Academic Press. 10.1109/ACC.2011.5991407

Wang, Q., & Spall, J. (2013). Rate of Convergence Analysis of Discrete Simultaneous Perturbation Stochastic Approximation Algorithm. In *Proceedings of the American Control Conference* (pp. 4771-4776). Washington, DC: Academic Press. 10.1109/ACC.2013.6580576

Wang, H., Iovenitti, P., Harvey, E., & Masood, S. (2003). Passive mixing in microchannels by applying geometric variations. In *Micromachining and Microfabrication* (pp. 282–289). International Society for Optics and Photonics. doi:10.1117/12.472888

Wang, S. K., Chiou, J. P., & Liu, C. W. (2007). Non-smooth/non-convex economic dispatch by a novel hybrid differential evolution algorithm. *IET Generation, Transmission & Distribution, 1*(5), 793–803. doi:10.1049/iet-gtd:20070183

Wasko, M. M., & Faraj, S. (2005). Why should I share? Examining social capital and knowledge contribution in electronic networks of practice. *Management Information Systems Quarterly, 29*(1), 35–57.

Welbourne, T. M., & Andrews, A. O. (1996). Predicting the performance of initial public offerings: Should human resource management be in the equation? *Academy of Management Journal*, *39*(4), 891–919. doi:10.2307/256716

Won, J. R., & Park, Y. M. (2003). Economic dispatch solutions with piecewise quadratic cost functions using improved genetic algorithm. *International Journal of Electrical Power & Energy Systems*, *25*(5), 355–361. doi:10.1016/S0142-0615(02)00098-4

Wright, P. M., Dunford, B. B., & Snell, S. A. (2001). Human resources and the resource-based view of the firm. *Journal of Management*, *27*(6), 701–721. doi:10.1177/014920630102700607

Wu, C. J., & Tsai, T. R. (2005). Acceptance sampling plans for Birnbaum-Saunders distribution under truncated life tests. *International Journal of Reliability Quality and Safety Engineering*, *12*(06), 507–519. doi:10.1142/S0218539305001987

XDD. (2017). *XDD: The eXtreme dd toolset*. Retrieved August 28, 2017, from https://github.com/bws/xdd

Xinchao, Z. (2010). A perturbed particle swarm algorithm for numerical optimization. *Applied Soft Computing*, *10*(1), 119–124. doi:10.1016/j.asoc.2009.06.010

Yang, S., Lee, H., & Kurnia, S. (2009). Social capital in information and communications technology research: Past, present, and future. *Communications of the AIS*, *25*, 184–220.

Yang, X. S., Hosseini, S. S. S., & Gandomi, A. H. (2012). Firefly algorithm for solving non-convex economic dispatch problems with valve loading effect. *Applied Soft Computing*, *12*(3), 1180–1186. doi:10.1016/j.asoc.2011.09.017

Yun, D., Wu, C., Rao, N., Liu, Q., Kettimuthu, R., & Jung, E.-S. (2017). Data Transfer Advisor with Transport Profiling Optimization. In *Proceedings of the IEEE Conference on Local Computer Networks* (pp. 269-277). Singapore: IEEE. doi: 10.1109/LCN.2017.23

Yun, D., Wu, C., Rao, N., Settlemyer, B., Lothian, J., Kettimuthu, R., & Vishwanath, V. (2015). Profiling Transport Performance for Big Data Transfer over Dedicated Channels. In *Proceedings of the International Conference on Computing, Networking and Communications* (pp. 858-862). Anaheim, CA: Academic Press. 10.1109/ICCNC.2015.7069458

Yun, D., & Wu, C. (2015). An Integrated Transport Solution to Big Data Movement in High-performance Networks. In *Proceedings of the IEEE International Conference on Network Protocols* (pp. 460-462). San Francisco, CA: IEEE. 10.1109/ICNP.2015.55

About the Contributors

Seifedine Kadry, PhD, has been an Associate Professor with Beirut Arab University in Lebanon since 2017. He serves as Editor-in-Chief of the Research Journal of Mathematics and Statistics, the ARPN Journal of Systems and Software, and the International Journal of Mathematical Sciences and Computing (IJMSC). Currently, he is the department chair of Math and Statistics in the American University of the Middle East. In addition, he has published several authored and edited books in Elsevier, Springer, IGI, and Taylor and Francis publishers. He is the author of more than 100 papers on applied math, computer science, stochastic systems, and education in peer-reviewed journals. Since 2008, he is the symposium chair in ICNAAM international conference. He received PhD in computational and applied mathematics in 2007 from the Blaise Pascal University (Clermont-II) - Clermont-Ferrand in France, MS in applied mathematics in 2001 from École Polytechnique Fédérale de Lausanne (EPFL) and BS in applied mathematics in 1999 from the Lebanese University.

* * *

Maryam Haghshenas was born in 1985 and has earned a B.Sc. in Computer Hardware Engineering from the Islamic Azad University, South Tehran branch, Tehran, Iran in 2007. She graduated with a M.Sc. in IT Management at Islamic Azad University, Science & Research branch, Tehran, Iran in 2011. In 2017, she completed a PhD in Media Management at the University of Tehran, Iran. Her thesis is titled: The presentation of a model to integrate social media in LMS (case study: University of Tehran E-learning Center). She achieved top student status during her PhD course. She worked for Magfa IT Development Center Co. which is an affiliate of the Industrial Development and Renovation Organization (IDRO) in Iran as a senior expert in implementing IT projects. Dr. Haghshenas is currently the project manager of e-learning courses at the prestigious IRI Broadcasting Organization. She has over 100 papers presented at national and international conferences and has been published in prestigious journals. She is passionate about teaching in the e-learning, social media, and learning management systems fields. Dr. Haghshenas is currently active on the editorial boards of over 20 international journals and conferences. She obtained the "Top Researcher" ranking at the University of Tehran's 22nd annual Research Festival in 2013 and the Silver Statue Award from the Iranian Researchers Network in 2017.

Subrata Mondal is working as an Assistant Professor of Power Engineering Department, Jadavpur University. He completed his BE in Mechanical Engineering from Jadavpur University in 2004. He completed his M.Tech. in Mechanical Engineering from IIT Kanpur in 2007 and completed his Ph.D.(Engg.) from School of Laser Science & Engineering from Jadavpur University in 2013. He is having 5 years teaching as well as research experience. He is having 1 year industrial experience. He published 12 international Journal paper and paper presented in 7 international conferences.

Anirban Mukherjee did his Bachelors in Civil Engineering in 1994 from Jadavpur University, Kolkata. He obtained his PhD on 'Automatic Diagram Drawing based on Natural Language Text Understanding' from Indian Institute of Engineering, Science and Technology (IIEST), Shibpur in 2014. He has over 20 years of experience in academics and is currently an Associate Professor of Engineering Science & Management at RCCIIT. His research interest includes computer graphics, computational intelligence, optimization and assistive technology. He has co-authored two UG engineering textbooks, co-edited three edited reference books and co-authored more than 18 books on Computer Graphics/ Multimedia for distance learning courses of different Universities of India. He has published papers in few international journals and conferences and also few book chapters. He is in the editorial board of International Journal of Ambient Computing and Intelligence (IJACI).

Konstantin Nechval (Ph.D. in Mechanical Engineering) is an Associate Professor in the Department of Applied Mathematics at the Transport and Telecommunication Institute, Riga, Latvia. He has authored and coauthored more than 79 papers. His research interests include stochastic processes, pattern recognition, operations research, statistical decision theory, adaptive control, and mechanical engineering. Dr. K. Nechval was awarded the "CASYS'07 Best Paper Award" for his paper: "Dual Control of Education Process" presented at the Eight International Conference on Computing Anticipatory Systems (Liege, Belgium, August 6-11, 2007) and the "MM2009 Best Paper Award" for his paper: "Optimal Statistical Decisions in a New Product Lifetime Testing" presented at the Fourth International Conference on Maintenance and Facility Management (Rome, Italy, April 22-24, 2009).

Nicholas Nechval (Ph.D. in Automatic Control and Systems Engineering, and Dr.Habil.Sc. in Radio Engineering) is a Professor in the Department of Mathematics at the Baltic International Academy, Riga, Latvia, and Principal Investigator in the BVEF Research Institute at the University of Latvia, Riga, Latvia. Before his current position, from 1993 to 1999, Dr. Nechval was a Professor of Mathematics and Computer Science and the Head of the Research Laboratory at the Riga Aviation University. From 1999 to 2014, Dr. Nechval was a Professor and Head (Statistics Department and EVF Research Institute) at the University of Latvia, Riga, Latvia. He was also a Principal Investigator in the Institute of Mathematics and Computer Science at the University of Latvia. Dr. Nechval, in 1991, was awarded a Silver Medal of the Exhibition Committee (Moscow, Russia) for his research on the problem of Prevention of Collisions between Aircraft and Birds. His research interests include mathematics, econometrics, stochastic processes, pattern recognition, digital radar signal processing, operations research, statistical decision theory, adaptive control, and mechanical engineering. Professor Nechval has authored and coauthored more than 455 papers and 9 books, including the book: Aircraft Protection from Birds (Moscow: Russian Academy of Science, 2007) coauthored with V. D. Illyichev (Academician of the Russian Academy of Science), and the book: Improved Decisions in Statistics (Riga: Izglitibas Soli, 2004) coauthored with Prof. E. K. Vasermanis (University of Latvia). This book was awarded the "2004 Best Publication Award" by the Baltic Operations Research Society. Dr. Nechval is also an Editorial Board Member of several international journals.

Siham Ouhimmou received his Ph.D. degree in mechanical engineering from Mohammedia School of Engineers Rabat - Morocco in 2011, some training research are made during the Ph.D in laboratory LOFIMS of National Institute of Applied Sciences in Rouen - France . He is currently an assistant professor of mechanical science at Hassan II University of Science in Casablanca – Moroco. His research interests include Reliability methods and optimization methods for resolution of mechanical problems, Reliability Based Design Optimsation (RBDO), Structural Nonlinear Dynamics and Diagnosis, and multi-objective optimization.

Dipankar Santra received the Bachelor degree in electrical engineering from Jadavpur University, Kolkata, India, in 2002 and the M.Tech degree in Electrical Engineering (Electrical Power) from University of Calcutta, India, in 2009. He is currently pursuing the PhD. degree at Jadavpur University, Kolkata, India; he has more than 16 years' experience in academics and is currently an Assistant Professor in the department of electrical engineering at RCC Institute of Information Technology, Kolkata, India. His areas of interest include power system economic load dispatch, soft computing applications to different power system problems and electrical machines.

Jayanti Sarker received the ME degree in Power Engineering from Jadavpur University, Kolkata, India, in 2009.Currently, she is an Assistant Professor of Electrical Engineering at Techno India. Her research interest includes power system and power electronics (FACTS).

Krishna Sarker received the M. Tech degree in Electrical Power from University of Calcutta, India in 2009. Currently, he is an Assistant Professor of Electrical Engineering at Techno India Group. His research interest includes power system, power electronics (FACTS), DFIG and renewable energy.

Irina Stanciu is a lecturer at the Faculty of Engineering in Foreign Languages, University "Politehnica" of Bucharest. She has a PhD in electronic engineering. Her main fields of scientific interests are finite element modelling and simulation and microfabrication technologies for MEMS devices.

Chase Q. Wu completed his Ph.D. dissertation at Oak Ridge National Laboratory and received his Ph.D. degree in computer science from Louisiana State University in 2003. He was a research fellow at Oak Ridge National Laboratory during 2003-2006 and an associate professor at University of Memphis during 2006-2015. He is currently an associate professor at New Jersey Institute of Technology. His research interests include big data, parallel and distributed computing, high-performance networking, sensor networks, and cyber security.

Daqing Yun received his Ph.D. degree in computer science from New Jersey Institute of Technology in 2016. He is currently an assistant professor of computer science at Harrisburg University of Science and Technology. His research interests include high-performance networking, parallel and distributed computing, green networking, and big data.

Index

www.ingramcontent.com/pod-product-compliance
Lightning Source LLC
LaVergne TN
LVHW080528300926
849162LV00007B/1558